Proteins: Computational Approaches and Bioinformatics Analysis

Proteins: Computational Approaches and Bioinformatics Analysis

Editor: Taylor Watts

www.callistoreference.com

Callisto Reference,
118-35 Queens Blvd., Suite 400,
Forest Hills, NY 11375, USA

Visit us on the World Wide Web at:
www.callistoreference.com

ISBN: 978-1-64116-800-7 (Hardback)

Cataloging-in-Publication Data

Proteins : computational approaches and bioinformatics analysis / Taylor Watts.
     p. cm.
Includes bibliographical references and index.
ISBN 978-1-64116-800-7
1. Proteins. 2. Proteins--Data processing. 3. Computational biology. 4. Bioinformatics. I. Watts, Taylor.
QP551 .P76 2023
572.6--dc23

Table of Contents

Preface

Proteins are large and intricate molecules that perform a variety of vital functions in the human body. They are crucial for the structure, control and function of the body's organs and tissues, while carrying out the majority of the work inside the cells. They are composed of a large number of smaller components known as amino acids that are linked with each other in lengthy chains. The search for novel protein molecules can be time consuming in case of certain proteins, such as enzymes, anticancer peptides, cytokines, G protein-coupled receptors, cell-penetrating peptides and cancerlectins. Understanding the function of proteins requires the computational identification of special protein molecules. Bioinformatics is a vital part of protein analysis, including structure analysis, sequence analysis and evolution analysis. This book contains some path-breaking studies on proteins. It aims to shed light on the computational approaches and bioinformatics analysis of proteins. As this field is emerging at a rapid pace, the contents of this book will help the readers understand the modern concepts and applications of the subject.

This book has been the outcome of endless efforts put in by authors and researchers on various issues and topics within the field. The book is a comprehensive collection of significant researches that are addressed in a variety of chapters. It will surely enhance the knowledge of the field among readers across the globe.

It gives us an immense pleasure to thank our researchers and authors for their efforts to submit their piece of writing before the deadlines. Finally in the end, I would like to thank my family and colleagues who have been a great source of inspiration and support.

Editor

Precursor Intensity-Based Label-Free Quantification Software Tools for Proteomic and Multi-Omic Analysis within the Galaxy Platform

Subina Mehta [1],*, Caleb W. Easterly [1], Ray Sajulga [1], Robert J. Millikin [2], Andrea Argentini [3], Ignacio Eguinoa [3], Lennart Martens [3], Michael R. Shortreed [2], Lloyd M. Smith [2], Thomas McGowan [4], Praveen Kumar [1], James E. Johnson [4], Timothy J. Griffin [1] and Pratik D. Jagtap [1],*

[1] Department of Biochemistry, Molecular Biology and Biophysics, University of Minnesota, Minneapolis, MN 55455, USA; caleb_easterly@med.unc.edu (C.W.E.); rsajulga@umn.edu (R.S.); kumar207@umn.edu (P.K.); tgriffin@umn.edu (T.J.G.)
[2] Department of Chemistry, University of Wisconsin, Madison, WI 53706, USA; rmillikin@wisc.edu (R.J.M.); mshort@chem.wisc.edu (M.R.S.); smith@chem.wisc.edu (L.M.S.)
[3] VIB-UGent Center for Medical Biotechnology, VIB, Ghent University, 9000 Ghent, Belgium; a.argentini@gmail.com (A.A.); ignacio.eguinoa@psb.vib-ugent.be (I.E.); lennart.martens@vib-ugent.be (L.M.)
[4] Minnesota Supercomputing Institute, University of Minnesota, Minneapolis, MN 55455, USA; jj@umn.edu (J.E.J.); mcgo0092@umn.edu (T.M.)
* Correspondence: smehta@umn.edu (S.M.); pjagtap@umn.edu (P.D.J.)

Abstract: For mass spectrometry-based peptide and protein quantification, label-free quantification (LFQ) based on precursor mass peak (MS1) intensities is considered reliable due to its dynamic range, reproducibility, and accuracy. LFQ enables peptide-level quantitation, which is useful in proteomics (analyzing peptides carrying post-translational modifications) and multi-omics studies such as metaproteomics (analyzing taxon-specific microbial peptides) and proteogenomics (analyzing non-canonical sequences). Bioinformatics workflows accessible via the Galaxy platform have proven useful for analysis of such complex multi-omic studies. However, workflows within the Galaxy platform have lacked well-tested LFQ tools. In this study, we have evaluated moFF and FlashLFQ, two open-source LFQ tools, and implemented them within the Galaxy platform to offer access and use via established workflows. Through rigorous testing and communication with the tool developers, we have optimized the performance of each tool. Software features evaluated include: (a) match-between-runs (MBR); (b) using multiple file-formats as input for improved quantification; (c) use of containers and/or conda packages; (d) parameters needed for analyzing large datasets; and (e) optimization and validation of software performance. This work establishes a process for software implementation, optimization, and validation, and offers access to two robust software tools for LFQ-based analysis within the Galaxy platform.

Keywords: proteomics; label-free quantification; galaxy framework; workflows

1. Introduction

Peptide- and protein-level quantification (either labeled or label-free) is routinely used in mass spectrometry (MS)-based shotgun proteomics data analysis workflows to determine the relative abundance of peptides or proteins in a given sample [1], including post-translationally modified peptides [2] and amino acid sequence variants identified by proteogenomics [3,4]. In the field of metaproteomics, where protein samples obtained from environmental microbiomes are studied, the quantification of microbial peptides or "metapeptides" (peptides obtained from shotgun sequencing of microbial communities) is essential to perform taxonomic and functional quantification of proteins expressed from the microbiome [5].

In the case of the label-free quantification (LFQ) methods, the peak intensity or area under the curve of a detected peptide ion allows the relative quantification of peptides across different samples. LFQ [6,7] is a useful method for quantification when the introduction of stable isotopes is impractical (for example, in human or animal model studies) or for applications such as proteogenomics or metaproteomics, which rely on peptide-level quantification. Currently, there are several software packages available for LFQ analysis [8]. LFQ analysis can be performed by public domain software suites such as MaxQuant [9] and Skyline [10], or by commercial software such as PEAKS [11] and Progenesis [12]. Although commercial and actively-supported software offers reliability and ease of use, its usage comes with a cost and usually includes canned features that are used for most standard datasets. Open-source software, on the other hand, has the benefit of being amenable to testing and optimization for emerging disciplines to offer economical options for data analysis.

In this study, through a rigorous testing and evaluation process, we have incorporated and optimized two established, open-source tools, moFF [13] and FlashLFQ [14] in the Galaxy platform. In order to achieve this, we worked with the software developers of these tools and tested features using two benchmark datasets, the ABRF Proteomics Research Group (PRG) 2015 dataset [15] and a Universal Proteomics Standard (UPS) dataset [1], and compared the outputs with results from MaxQuant, a highly used standalone software platform capable of LFQ analysis. Based on feedback, the tool developers of moFF and FlashLFQ made changes to the software's capabilities, which included (a) using match-between-runs (MBR); (b) ability to process and analyze large input datasets; (c) compatibility with a variety of input file formats. After this rigorous evaluation and optimization, these tools were implemented in the accessible and reproducible Galaxy [16] platform. Galaxy tools are maintained and developed by an international community of developers (https://galaxyproject.org/iuc/) so as to facilitate ease of usage and maintain its contemporary status for any emerging software tools or applications. An additional advantage of having these tools available via the Galaxy platform is the ability to process the data in workflows, wherein multiple tools can be used in a sequential manner to generate processed outputs from the input data. The Galaxy for proteomics (Galaxy-P) team has developed workflows related to MS-based multi-omic studies such as, proteogenomics [17,18] and metaproteomics [19,20]. The addition of these precursor intensity-based LFQ tools to the existing workflows will facilitate peptide level quantification for multi-omics research studies, as well as more standard proteomics applications.

As a result of this study, we made two quantitative software tools available to researchers via the Galaxy platform. These software are available via the Galaxy Tool Shed [21,22], GitHub, and on Galaxy public instances.

2. Methods

We used two datasets, (a) an ABRF dataset [15] and (b) a spiked-in benchmark UPS dataset [1] to determine the accuracy of each tool with regards to their calculated protein fold-changes. We obtained the MS (raw) data from publicly available repositories and converted them to MGF (Mascot generic format) files using MSConvert (vendor support) (Galaxy Version 3.0.19052.0) [23] to make it compatible with search algorithms within the Galaxy Platform. moFF and FlashLFQ processing were performed within the Galaxy platform (Version 19.09).

2.1. (A) ABRF Dataset

The spiked-in dataset from the ABRF PRG 2015 study was used to determine the accuracy of each software tool. This dataset, generated through the collaborative work of the ABRF Proteomics Research Group (https://abrf.org/research-group/proteomics-research-group-prg) contains four proteins added to human cell lysate samples: ABRF-1 (beta galactosidase from *Escherichia coli*), ABRF-2 (lysozyme from

Gallus gallus), ABRF-3 (amylase from *Aspergillus niger*) and ABRF-4 (protein G from *Streptococcus*) [15]. Each sample contained the four proteins at the same concentration, while the concentrations varied across the four samples: 0 (blank/negative control), 20, 100, and 500 fmol. The peptide raw data for the ABRF dataset was acquired on the LTQ orbitrap Velos with EASY-nLC.

2.2. (B) Spiked-In UPS Benchmark Dataset

To evaluate these tools, we downloaded publicly available data [1] (PRIDE #5412; ProteomeXchange repository PXD000279), wherein UPS1 and UPS2 standards (Sigma-Aldrich, St. Louis, MO, USA) were spiked into *E. coli* K12 strain samples. Based on the dynamic benchmark dataset protocol, the UPS and *E. coli* peptides we quantified using nanodrop spectrophotometer at 280 nm, and 2 µg of *E. coli* peptides were spiked with 0.15 µg of UPS1 or UPS2 peptides. About 1.6 µg of the mix was analyzed on the Q Exactive (Thermo Fisher, Waltham, MA, USA) mass spectrometer [1]. The UPS1 and UPS2 standards contain 48 human proteins at either the same (5000 fmol, UPS1) or varying concentration (50,000 fmol to 0.5 fmol, UPS2), respectively.

2.3. Peptide Identification

For both datasets, we used SearchGUI (SG) [24] (version 3.3.3.0) and Peptide Shaker (PS) [25] (version 1.16.26) to search the MS/MS spectra against respective protein FASTA databases along with contaminants from cRAP database (https://www.thegpm.org/crap/). Although SearchGUI has the option to use as many as eight search algorithms, we used only four search algorithms (X!tandem, OMSSA, MSGF+, and Comet) for this evaluation study.

For the spiked-in ABRF PRG dataset, a protein FASTA file was generated by merging the UniProt human reference database with spiked-in proteins and contaminant proteins (73,737 protein sequences database generated on 6 February 2019). Search parameters used were trypsin enzyme for digestion, where two missed cleavages were allowed. Carbamidomethylation of cysteine was selected as a fixed modification and methionine oxidation was selected as a variable modification. The precursor mass tolerance was set to 10 ppm and the fragment mass tolerance to 0.5 Da, with minimum charge as 2 and maximum charge of 6. For Peptide Shaker, the false discovery rate (FDR) was set at 1% at the PSM, peptide, and protein level, along with filtering the peptide length ranging from 6–65 peptides.

For the spiked-in UPS dataset, the mass spectra were searched against a protein FASTA database provided by Cox. et al., 2014 [1], (4494 protein sequences database generated on 25 July 2019). The parameters for SearchGUI-Peptide Shaker analysis were as follows: precursor mass tolerance was set to 10 ppm and the fragment mass tolerance to 20 ppm with minimum and maximum charge as 2 and 6, respectively.

For MaxQuant analysis (version 1.6.7.0, Cox lab, Max Planck institute of Biochemistry, Martinsried, Germany), the built-in Andromeda search engine [26] was used. The parameters for MaxQuant were matched with the SearchGUI-PeptideShaker search. The fixed modification was set for carbamidomethylation of cysteine and oxidation of methionine as a variable modification. The FDR was set at 1% and the MS/MS tolerance was set at 10 ppm. The tabular output data from Peptide Shaker (PSM.tab) and Andromeda (msms.txt) were used for protein quantification.

2.4. Quantification Tools

moFF and FlashLFQ, were initially tested outside of the Galaxy platform. We tested various releases for moFF (versions 1.2.0 to 2.0.2) and FlashLFQ (versions 0.1.99 to 1.0.3) and provided developers with feedback to improve software stability and data quality. We then implemented these updated tools within Galaxy. The results from moFF (version 2.0.2) and FlashLFQ (version 1.0.3) were then compared with MaxQuant (version 1.6.0.16), a widely used LFQ quantification software suite. For testing, all the quantification tools were set at monoisotopic tolerance of 10 ppm and run with or without MBR, where indicated.

2.5. Normalization and Protein Quantification

After peptide-level precursor intensity values were generated, normalization was performed using limma [27], and peptides were summarized into protein-level abundances with protein expression control analysis (PECA) [28]. Specifically, the "normalizeBetweenArrays" limma function was used for most normalization methods (i.e., scale, cyclic loess, and quantile). For VSN (variance stabilizing normalization), the "normalizeVSN" limma function was used [27,29]. After normalization, PECA was used to combine the peptide-level measurements to protein-level values for the detection of differentially expressed proteins. These two tools were run via custom R scripts (version 1.3), which can be accessed via the Supplementary Document 2 (https://github.com/galaxyproteomics/quant-tools-analysis).

3. Results

Both moFF and FlashLFQ are established software tools and contain useful features such as amenability to Galaxy implementation, compatibility with existing Galaxy upstream and downstream tools, ability to read mzML and Thermo raw file formats, open-source code, MBR functionality, and results that can be easily evaluated with performance metrics.

moFF is an extensible quantification tool amenable to any operating system. The input for moFF is peptide search engine output and Thermo raw files and/or mzML files; it performs both MS/MS as well as MBR quantification. moFF tool also has a novel filtering option for MBR peak intensities [30]. moFF has been wrapped in Galaxy (Figure 1A) using a Bioconda package [31]. The Galaxy version of moFF is available via Galaxy toolshed [21], GitHub [32] and Galaxy public instances (proteomics.usegalaxy.eu, usegalaxy.be and z.umn.edu/metaproteomicsgateway).

FlashLFQ is a peptide and protein LFQ algorithm developed for proteomics data analysis. It was developed to quantify peptides and proteins from any search tool, including MetaMorpheus, which also performs PTM identification from MS/MS data. It uses Bayesian statistics to estimate the difference in the abundance of inferred proteins between samples, though this feature was not evaluated here. FlashLFQ can normalize fractionated datasets by using a bounded Nelder–Mead optimizer [33] to find a normalization coefficient for each fraction, similar to MaxLFQ. FlashLFQ was implemented in Galaxy (Figure 1B) within a Singularity container [34] as FlashLFQ is a Windows application requiring the NET core framework for deployment in the Unix-based Galaxy environment. Singularity provides a secure means of running such tools in Galaxy. The Galaxy version of FlashLFQ is available via Galaxy toolshed [22] GitHub [35] and via Galaxy public instances (proteomics.usegalaxy.eu and z.umn.edu/metaproteomicsgateway).

Relevant features of moFF and FlashLFQ, as well as the design of the evaluation study, are summarized in Figure 2. An essential aspect of this study was both of these tools being in active development by groups amenable to collaboration, which greatly helped optimization-based tests and feedback from the Galaxy-P team members.

To generate peptide identification inputs for moFF and FlashLFQ, datasets were searched against appropriate protein databases using SearchGUI/PeptideShaker. The speed and accuracy of FlashLFQ and moFF were evaluated in comparison to MaxQuant, a popular software tool used for LFQ. For MaxQuant, searches were performed by MaxQuant's built-in Andromeda search algorithm. All three software programs have an MBR feature, where unidentified peaks are "matched" to identify peaks in other runs based on similar *m/z* and retention time. MaxLFQ, an algorithm within MaxQuant, normalizes raw intensities, and also aggregates them into protein groups [1]. For moFF and FlashLFQ, limma was used to normalize peptide intensities, and PECA was used to determine protein fold-changes and associated *p*-values. The limma tool within Galaxy implements different normalization techniques such as quantile, VSN, cyclic LOESS, and scale normalization. Users can choose between these normalization methods. FlashLFQ also has built-in normalization and protein quantification functions, which we have used in this study.

(A)

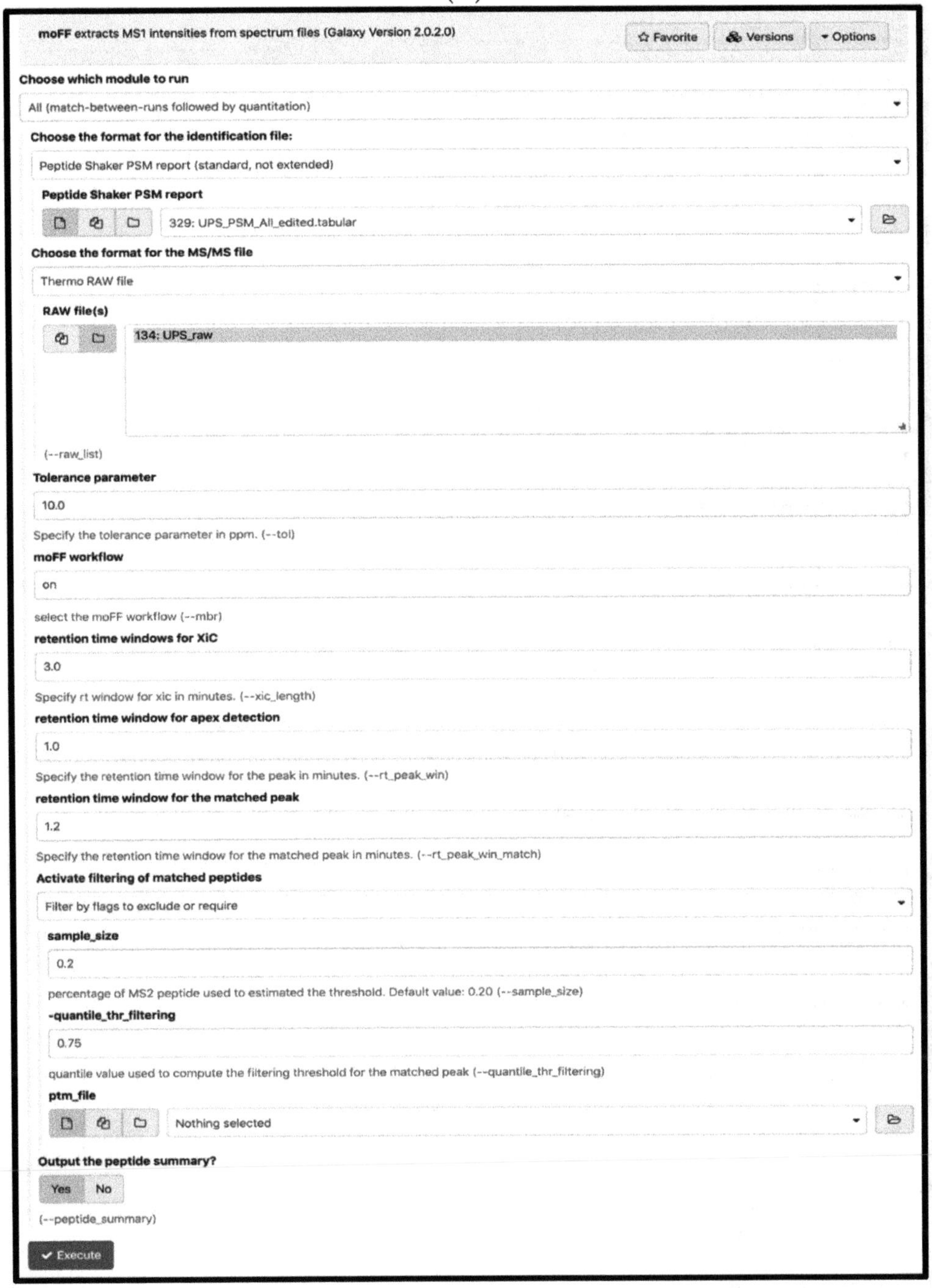

Figure 1. *Cont.*

(B)

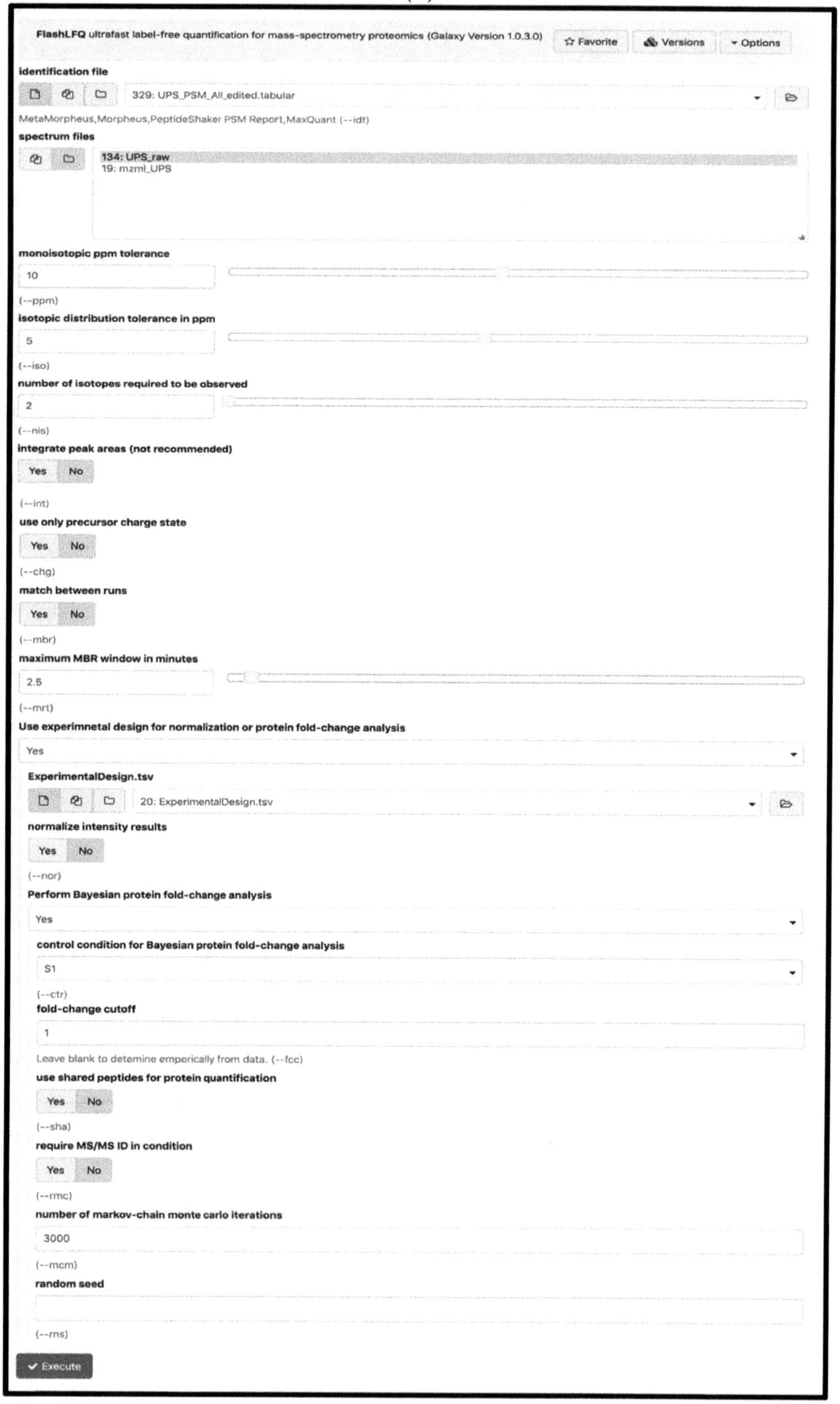

Figure 1. Galaxy interface of moFF and FlashLFQ: (**A**) Bioconductor package of moFF is wrapped within Galaxy and available via Galaxy toolshed [21] and Galaxy public instances (proteomics.usegalaxy.eu). (**B**) A docker/singularity container of FlashLFQ is wrapped within Galaxy and available via Galaxy toolshed [22] and Galaxy public instances (proteomics.usegalaxy.eu).

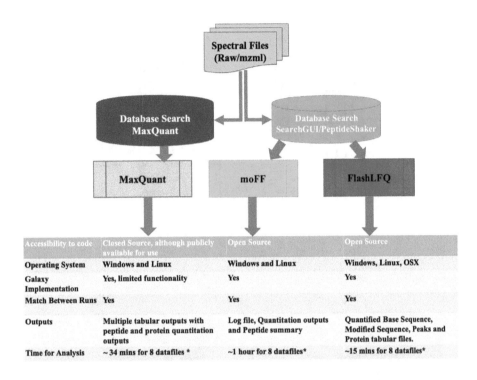

Figure 2. Experimental design of the evaluation study: spectra files are converted to MGF before mass spectra are matched with peptides using respective search engines. Each of the quantification tools use RAW files and the peptide identification tabular output as inputs. The figure also shows the features of each tool. The outputs from all of the tools were then compared against each other. The asterisk symbol (*) denotes that the files were run on same computing device.

After ascertaining that moFF and FlashLFQ results correlate well with MaxQuant results (Supplementary Figure S1), we set out to evaluate the MBR feature of the software tools. For this, we used the ABRF PRG datasets, with four spiked-in proteins at three different concentrations (20 fmol, 100 fmol, and 500 fmol) and a negative control (see methods). The spiked-in proteins should not be detected in the negative control, either with or without MBR. We observed that moFF and FlashLFQ outputs showed non-zero intensity values for the spiked-in proteins in blank control samples if MBR was enabled (Figure 3A, left). The match-between runs (MBR) module for the earlier version of both the software, moFF and FlashLFQ, simply predicted the retention time for the matched peptide in the target run by looking for intensity in the target m/z and RT window. If the target peptide was not present in that run, overlapping eluting peptides or noise would result in the assignment of spurious background signals. As a result, ABRF spiked in proteins were detected in the blank control sample whole using the MBR mode in the earlier versions of the software tools. We worked with the developers to improve their MBR algorithms so that intensity values for these proteins in the blank control samples were correctly reported as zero (Figure 3B, right). moFF's new version removes spurious matches by filtering scans for the similarity between the theoretical isotopic envelope of target peptide and corresponding envelope found in the target run. This method has been published in the moFF 2.0 version [30]. The FlashLFQ developers implemented an optional setting that requires a protein to have at least one peptide assigned to an MS/MS spectrum in a sample group so that a non-zero value can be assigned to the peptides in the sample group. Due to the above-mentioned changes in the algorithms and subsequent filtering steps, the newer versions of these tools do not detect spurious signals for target peptides in blank control samples.

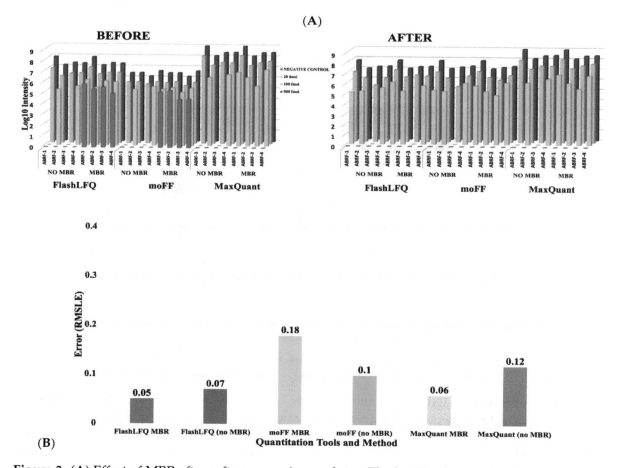

Figure 3. (**A**) Effect of MBR after software version updates: The log10 values of the intensities (blue bars) from each of the four ABRF spiked-in proteins (ABRF-1: beta Galactosidase from *E. coli*, ABRF-2: Lysozyme from *Gallus gallus*, ABRF-3: amylase from *Aspergillus*, ABRF-4: protein G *Streptococcus*) were plotted. The results from prior versions of moFF (v1.2.1) and FlashLFQ (v0.1.99) (before) shows that MBR detects ABRF proteins (shown in red) in the negative control sample in both software. The results from the current versions of moFF (v2.0.2) and FlashLFQ (v1.0.3.0) implemented in Galaxy (after), shows that the MBR feature does not detect ABRF proteins in the negative control. (**B**) Accuracy of fold-change estimation: for evaluating the accuracy of quantified results, we estimated the fold change of the spiked-in proteins in the 500 fmol sample as compared to 100 fmol sample. The root mean squared log error (RMSLE) was calculated for fold change estimation. For this dataset, moFF with MBR displayed significantly higher RMSLE value, whereas FlashLFQ's MBR performed similarly to MaxQuant's MBR.

The 500 fmol and 100 fmol datasets from the ABRF dataset were used to determine the fold-change accuracy (Figure 3B). In order to determine the accuracy of the fold-change, root mean squared log error (RMSLE) [36] was calculated,

$$RMSLE = \sqrt{\frac{\sum_{i=1}^{N}(log_{10} \, r_i - log_{10}\hat{r}_i)^2}{N}} \tag{1}$$

where, r_i is the true ratio, $\hat{r}_i$ is the estimated ratio and N is the number of proteins identified via sequence database searching of the sample.

Root mean squared log error is a metric to evaluate the difference between predicted and observed values. In this case, the values being compared are the predicted (known) fold-changes and the observed fold-changes. RMSLE being an error based metric provides the true picture of prediction quality, however, deciding a suitable threshold value is challenging. The objective was to obtain an

RMSLE value closer to zero for all the tools. The RMSLE values for the three tools are shown in Figure 3B, with the MBR feature enabled. We observed that moFF with MBR had slightly higher error compared to the other tools. MaxQuant's MBR and FlashLFQ's MBR perform quite similarly, though all three tools show a low error when the MBR is enabled.

Although MaxLFQ and FlashLFQ have their own in-built methods for normalizing peptide abundance, for a more direct comparison of moFF and FlashLFQ performance, we normalized the peptide intensity levels using limma and obtained differentially expressed proteins through the PECA bioconductor package. Normalized peptide intensity values from moFF and FlashLFQ were input into PECA, wherein, the tool calculates the p-values of the peptide level data and then groups the values into protein level data. The PECA output was then compared with MaxQuant values using the UPS benchmark dataset. For this, quantitative information for the 48 proteins from the UPS dataset was extracted using an R-script to generate a tabular output with fold-change values.

The fold-change accuracy of all quantified UPS proteins after normalization was calculated by comparing the estimated protein fold-change with the true fold-change using the RMSLE (Figure 4A).

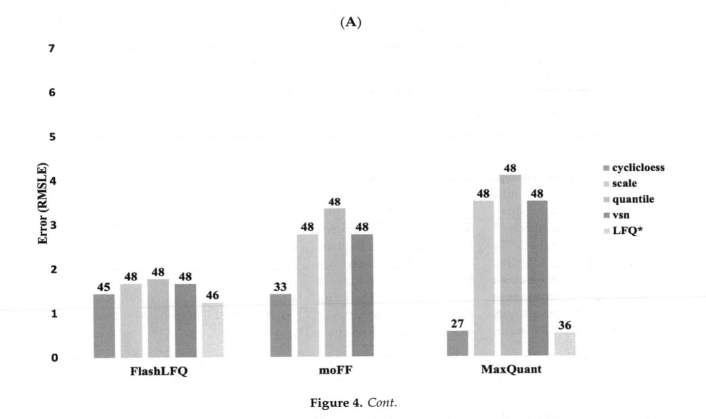

(A)

Figure 4. *Cont.*

(B)

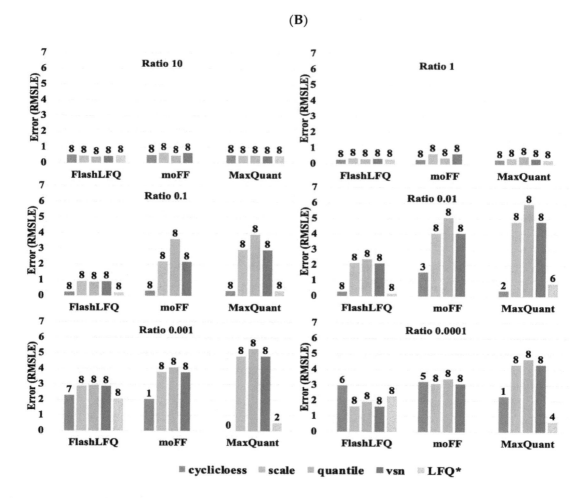

Figure 4. (**A**) Fold-change accuracy (MBR) of all proteins: after normalization, the estimated protein abundance ratios for all the identified UPS proteins were compared to the true abundance ratios, using the root mean squared log error (RMSLE). The plot represents the RMSLE values using different normalization methods. *LFQ denotes the LFQ values represent FlashLFQ's and MaxQuant's inbuilt normalization value. The value on the top of the bars denotes the number of proteins that were quantified. (**B**) Fold change accuracy (MBR) of proteins with similar estimated ratios: In total there are 48 UPS proteins, we classified the UPS proteins into different groups based on the UPS2/UPS1 ratio estimation, the true ratios run from 10 to 10^{-4}. The value on the top of the bars denotes the number of proteins that were quantified using each normalization method. The RMSLE of the intensity ratio was used to measure the accuracy of the estimated fold changes.

Figure 4A shows the comparison of different normalizations using the MBR values. Although the bar graph shows that MaxQuant's MaxLFQ performed the best compared to all, it did so at the cost of the number of proteins quantified. However, we noticed that MBR from FlashLFQ, with its in-built normalization (light blue bar in Figure 4A), performed better overall in terms of quantification and the number of proteins identified. Meanwhile, moFF and FlashLFQ provided higher numbers of quantified proteins while still maintaining low RMSLE values. We also performed comparison studies of MBR vs. no MBR, the results of which are shown in Supplementary Figure S3A. We also found that moFF and FlashLFQ quantified similar numbers of peptides across the ABRF and UPS datasets (Supplementary Figure S2).

After evaluating the RMSLE for all proteins, we estimated the accuracy of similarly abundant UPS proteins (Figure 4B). We categorized the UPS standards by their concentration ratio (UPS2/UPS1), which resulted in 6 different categories (i.e., ratios of 1 to 0.0001). The results showed that MaxQuant quantification works optimally for high and medium abundant proteins. However, for low abundance proteins, the fold-accuracy was lower, presumably because of missing intensity values. Another important observation was that MaxLFQ denotes a smaller error compared to the other tools in the low abundance samples, but quantifies fewer proteins. An evaluation study for MBR vs. noMBR was also performed and showed a similar trend represented in Supplementary Figure S3B.

After evaluation of the moFF and FlashLFQ tools, we worked with the developers of these tools to implement their optimized software in Galaxy, enabling integration into diverse MS-based proteomics workflows and promoting their usage by the Galaxy community. Our implementation will allow the users to choose their choice of tool and normalization (Supplementary Figure S4), which will benefit their research.

4. Discussion

Protein and peptide-level quantification has been used by proteomics researchers to determine how the proteome responds to biological perturbation [37]. In particular, precursor-intensity based LFQ has enabled researchers to perform quantitative proteogenomics analyses [38]. Quantitative changes in the proteome can also be correlated with transcript abundance changes [39] to get a more complete picture of how an organism responds to a stimulus. For example, in cancer proteogenomics studies, these abundance measurements help identify differential expression patterns of variant peptides that may have functional significance in cancer [40,41].

Peptide-level quantification also aids in functional studies of microbial communities and microbiomes using metaproteomics. For example, in metaproteomics studies, metapeptides or metaproteins detected from environmental [42] or host-derived samples [43] can be quantified to shed light on the dynamics of the taxa, biological function, and their abundance [44]. Our group has developed and optimized Galaxy-based tools and workflows for proteogenomics [45] and metaproteomics analyses [19,20]. Tools implemented through this study will extend these workflows to enable quantification of metapeptides and/or metaproteins.

In our analyses, the three LFQ tools, moFF, FlashLFQ, and MaxQuant, correlate well in their results according to our evaluation. In this study, we have added moFF and FlashLFQ to the Galaxy framework, which not only facilitates the dissemination of these tools but also enables automated data analysis by using them within workflows [16]. We also highlight the importance of the process of careful user evaluation, feedback to developers, and optimization of the tools and workflows. Preliminary testing was performed on the command line or GUI versions of the tool. These tools were then packaged into the Galaxy platform, where results were compared to the command-line/GUI versions and also optimized more usage in automated workflows.

Open-source software usage has faced challenges due to dependencies such as operating system (Windows, Linux, OSX), language or platforms (Python, C++, Java), lack of adhering to HUPO standards [46], and installation or usability issues [47]. To overcome these issues, the Galaxy-P project, as well as others in the Galaxy community, have sought collaborations with many research groups that have developed these tools, following a protocol which includes defining key input and output data types, establishing key operating parameters for the Galaxy tool, overcoming operating system compatibility issues (e.g., Singularity containers for Windows tools), along with rigorous testing and optimization. This collaborative and iterative process of development and optimization ensures the software performs accurately and efficiently within the Galaxy platform.

Ideally, software tools that are UNIX based, such as moFF, are easier for deployment within Galaxy. We also demonstrated here that tools such as FlashLFQ could be packaged within a singularity container to enable easy and secure implementation within Galaxy. MaxQuant, which is a popular, public-domain proteomics software package, is available in both Windows and Linux-compatible versions [9]. Although in the early development and testing phase, the LFQ module that uses MS1 precursor intensity data within MaxQuant (MaxLFQ) was made available within Galaxy toolshed [48]. Once fully tested and evaluated, accessibility to this software via the Galaxy platform will offer even more choices for precursor-intensity based quantification. Offering users a choice of multiple validated software tools also highlights the benefits of a workflow engine such as Galaxy, where users can easily develop parallel workflows using different combinations of tools to determine methods that provide optimal results based on user requirements.

In our tests, we observed that FlashLFQ has a faster runtime as compared to the other two tools. MaxQuant processing time is longer, presumably since it performs peptide identification and quantitation simultaneously. For example, on the same computing device, the UPS dataset was processed by FlashLFQ in approximately 15 min for quantification only, whereas MaxQuant and moFF took 34 min and 3 h, respectively. Our evaluation and availability of these tools within a unified platform such as Galaxy offers users a choice for their workflows where the speed of analysis can also be considered.

5. Conclusions

This study demonstrates a successful collaborative effort in software tool development and dissemination, which is a hallmark of the Galaxy community and the Galaxy-P project [19]. This community-driven approach brings together users and software developers who work together to validate and make the tool accessible and usable for other researchers across the world. The study described here provides a model of success for the process used to ultimately provide optimized, well-validated tools for community use. We did not seek a goal to determine the single best tool for LFQ use, but rather focused on offering users a choice of validated quantification tools amenable to customizable analytical workflows. In addition to our work here, others from the Galaxy community are also working on integrating tools within the MaxQuant suite [49], which will extend the choices for LFQ quantification available. As a result of this study, Galaxy users can now confidently use two rigorously validated LFQ software tools (moFF and FlashLFQ) for their quantitative proteomic studies. We are currently working on incorporating the quantitative capabilities of moFF and FlashLFQ within existing metaproteomics and proteogenomics workflows, so that they can be used by the research community in their quantitative multi-omics studies.

Supplementary Materials:
Figure S1: Peptide Correlation: The raw intensities of the ABRF peptides were correlated using Pearson Correlation coefficient. Output from FlashLFQ and moFF correlated well with MaxQuant. Figure S2: Peptide Overlap across FlashLFQ and moFF: A Venn diagram of the quantified ABRF (a) and UPS (b) unique peptides are shown to display the coverage of the peptides across FlashLFQ and moFF. The input for both the tools was the PSM report from the Peptide shaker containing 18172(ABRF) and 33497 (UPS) unique peptides. Figure S3. (A) Fold-change accuracy (MBR vs no-MBR) of all proteins: After normalization, the estimated protein ratios for all the identified UPS proteins were compared to the true ratios, using the Root mean squared log error (RMSLE). The plot represents the comparison between MBR and no-MBR using normalization methods. Note that LFQ values represent MaxQuant's and FlashLFQ's normalized value. The value on the top of the bars denotes the number of proteins that were quantified. Although the bar graph shows that MaxQuant's MaxLFQ performs the best compared to all, we notice that the number of UPS proteins identified and used for the calculation were less compared to other tools. (B) Fold change accuracy (MBR vs no-MBR) of proteins with similar estimated ratios: In total there are 48 UPS proteins, we classified the UPS proteins into different groups based on the UPS2/UPS1 ratio estimation, the true ratios run from 10 to 10^{-4}. Value on the top of the bars denote the number of proteins that were quantified using each normalization method. This is a comparative study between MBR and no MBR. The RMSLE of the intensity ratio was used to measure the accuracy of the estimated fold change. The figure shows that moFF and FlashLFQ work well in comparison to MaxQuant for ratio estimation of low abundance proteins (<0.1). Figure S4: Fold Accuracy of Tools with inbuilt and external normalization and protein quantitation: This figure displays the errors in Fold change accuracy while using inbuilt and external normalization and protein level quantitation features of the evaluated tools. Supplementary document 2 (https://github.com/galaxyproteomics/quant-tools-analysis) is the GitHub repository of the Rscripts.

Author Contributions: S.M. compared the tools, communicated with software tool developers, performed data analysis, and wrote the manuscript. C.W.E. and R.S. developed the R-scripts and ran statistical data analysis. R.J.M., A.A., and I.E. helped update their quantitation tools according to the feedbacks provided. T.M. and J.E.J. helped package these tools within the Galaxy platform. P.K. provided inputs during manuscript writing. L.M., M.R.S., and L.M.S. provided scientific inputs and helped in manuscript writing. T.J.G. provided scientific directions and helped in manuscript writing. P.D.J. conceived of the project, led the project and helped in manuscript writing. All authors have read and agreed to the published version of the manuscript.

Funding: This research was funded by National Cancer Institute-Informatics Technology for Cancer Research (NCI-ITCR) grant 1U24CA199347 and National Science Foundation (U.S.) grant 1458524 to T.G. We would also like to acknowledge the Extreme Science and Engineering Discovery Environment (XSEDE) research allocation BIO170096 to P.D.J. and use of the Jetstream cloud-based computing resource for scientific computing (https://jetstream-cloud.org/) maintained at Indiana University. The European Galaxy server that was used for some calculations is in part funded by Collaborative Research Centre 992 Medical Epigenetics (DFG grant SFB 992/1 2012) and German Federal Ministry of Education and Research (BMBF grants 031 A538A/A538C RBC, 031L0101B/031L0101C de.NBI-epi, 031L0106 de.STAIR (de.NBI)). Part of the work was performed by the Belgian ELIXIR node, also hosting the tools at the Belgian Galaxy instance, which is funded by the Research Foundation, Flanders (FWO) grant I002919N.

Acknowledgments: We would like to thank the European Galaxy team for the help in the support during Galaxy implementation. We would also like to thank Carlo Horro (from Barnes Group, University of Bergen, Norway) and Björn A. Grüning (University of Freiburg, Germany) for helping us during the quantification tools analysis. We thank Emma Leith for proofreading the manuscript. We acknowledge funding for this work from the grant We also acknowledge the support from the Minnesota Supercomputing Institute for the maintenance and update of the Galaxy instances.

References

1. Cox, J.; Hein, M.Y.; Luber, C.A.; Paron, I.; Nagaraj, N.; Mann, M. Accurate proteome-wide label-free quantification by delayed normalization and maximal peptide ratio extraction, termed MaxLFQ. *Mol. Cell. Proteom.* **2014**, *13*, 2513–2526. [CrossRef] [PubMed]
2. Zhang, C.; Liu, Y. *Retrieving Quantitative Information of Histone PTMs by Mass Spectrometry, In Methods Enzymology*; Academic Press Inc.: Cambridge, MA, USA, 2017; pp. 165–191. [CrossRef]
3. Alfaro, J.A.; Ignatchenko, A.; Ignatchenko, V.; Sinha, A.; Boutros, P.C.; Kislinger, T. Detecting protein variants by mass spectrometry: A comprehensive study in cancer cell-lines. *Genome Med.* **2017**, *9*, 62. [CrossRef] [PubMed]
4. Yeom, J.; Kabir, M.H.; Lim, B.; Ahn, H.S.; Kim, S.Y.; Lee, C. A proteogenomic approach for protein-level evidence of genomic variants in cancer cells. *Sci. Rep.* **2016**, *6*. [CrossRef] [PubMed]
5. Easterly, C.W.; Sajulga, R.; Mehta, S.; Johnson, J.; Kumar, P.; Hubler, S.; Mesuere, B.; Rudney, J.; Griffin, T.J.; Jagtap, P.D. metaQuantome: An Integrated, Quantitative Metaproteomics Approach Reveals Connections Between Taxonomy and Protein Function in Complex Microbiomes. *Mol. Cell. Proteom.* **2019**, *18*, S82–S91. [CrossRef] [PubMed]
6. Krey, J.F.; Wilmarth, P.A.; Shin, J.B.; Klimek, J.; Sherman, N.E.; Jeffery, E.D.; Choi, D.; David, L.L.; Barr-Gillespie, P.G. Accurate label-free protein quantitation with high- and low-resolution mass spectrometers. *J. Proteome Res.* **2014**, *13*, 1034–1044. [CrossRef] [PubMed]
7. Välikangas, T.; Suomi, T.; Elo, L.L. A comprehensive evaluation of popular proteomics software workflows for label-free proteome quantification and imputation. *Brief. Bioinform.* **2017**, *19*, 1344–1355. [CrossRef]
8. Chawade, A.; Sandin, M.; Teleman, J.; Malmström, J.; Levander, F. Data processing has major impact on the outcome of quantitative label-free LC-MS analysis. *J. Proteome Res.* **2015**, *14*, 676–687. [CrossRef]
9. Tyanova, S.; Temu, T.; Cox, J. The MaxQuant computational platform for mass spectrometry-based shotgun proteomics. *Nat. Protoc.* **2016**, *11*, 2301–2319. [CrossRef] [PubMed]
10. Schilling, B.; Rardin, M.J.; MacLean, B.X.; Zawadzka, A.M.; Frewen, B.E.; Cusack, M.P.; Sorensen, D.J.; Bereman, M.S.; Jing, E.; Wu, C.C.; et al. Platform-independent and label-free quantitation of proteomic data using MS1 extracted ion chromatograms in skyline: Application to protein acetylation and phosphorylation. *Mol. Cell. Proteom.* **2012**. [CrossRef]
11. Ma, B.; Zhang, K.; Hendrie, C.; Liang, C.; Li, M.; Doherty-Kirby, A.; Lajoie, G. PEAKS: Powerful software for peptide de novo sequencing by tandem mass spectrometry. *Rapid Commun. Mass Spectrom.* **2003**, *17*, 2337–2342. [CrossRef]
12. Progenesis QI for Proteomics: Waters. Available online: https://www.waters.com/waters/en_US/Progenesis-QI-for-Proteomics/nav.htm?locale=en_US&cid=134790665 (accessed on 8 October 2019).

13. Argentini, A.; Goeminne, L.J.E.; Verheggen, K.; Hulstaert, N.; Staes, A.; Clement, L.; Martens, L. MoFF: A robust and automated approach to extract peptide ion intensities. *Nat. Methods* **2016**, *13*, 964–966. [CrossRef]

14. Millikin, R.J.; Solntsev, S.K.; Shortreed, M.R.; Smith, L.M. Ultrafast Peptide Label-Free Quantification with FlashLFQ. *J. Proteome Res.* **2018**, *17*, 386–391. [CrossRef] [PubMed]

15. Van Riper, S.; Chen, E.; Remmer, H.; Chien, A.; Stemmer, P.; Wang, Y.; Jagtap, P. Identification of Low Abundance Proteins in a Highly Complex Protein Mixture. 2016. Available online: https://zenodo.org/record/3563207#.XwRcuhJ5t2E (accessed on 7 July 2020). [CrossRef]

16. Afgan, E.; Baker, D.; van den Beek, M.; Blankenberg, D.; Bouvier, D.; Čech, M.; Chilton, J.; Clements, D.; Coraor, N.; Eberhard, C.; et al. The Galaxy platform for accessible, reproducible and collaborative biomedical analyses: 2016 update. *Nucleic Acids Res.* **2016**, *44*, W3–W10. [CrossRef] [PubMed]

17. Jagtap, P.D.; Johnson, J.E.; Onsongo, G.; Sadler, F.W.; Murray, K.; Wang, Y.; Shenykman, G.M.; Bandhakavi, S.; Smith, L.M.; Griffin, T.J. Flexible and accessible workflows for improved proteogenomic analysis using the galaxy framework. *J. Proteome Res.* **2014**, *13*, 5898–5908. [CrossRef]

18. Anderson, K.J.; Vermillion, K.L.; Jagtap, P.; Johnson, J.E.; Griffin, T.J.; Andrews, M.T. Proteogenomic Analysis of a Hibernating Mammal Indicates Contribution of Skeletal Muscle Physiology to the Hibernation Phenotype. *J. Proteome Res.* **2016**, *15*, 1253–1261. [CrossRef]

19. Blank, C.; Easterly, C.; Gruening, B.; Johnson, J.; Kolmeder, C.A.; Kumar, P.; May, D.; Mehta, S.; Mesuere, B.; Brown, Z.; et al. Disseminating metaproteomic informatics capabilities and knowledge using the galaxy-P framework. *Proteomes* **2018**, *6*, 7. [CrossRef] [PubMed]

20. Jagtap, P.D.; Blakely, A.; Murray, K.; Stewart, S.; Kooren, J.; Johnson, J.E.; Rhodus, N.L.; Rudney, J.; Griffin, T.J. Metaproteomic analysis using the Galaxy framework. *Proteomics* **2015**, *15*, 3553–3565. [CrossRef]

21. Proteomics moFF Galaxy Toolshed Repository. Available online: https://toolshed.g2.bx.psu.edu/view/galaxyp/proteomics_moff/7af419c90f5f (accessed on 7 July 2020).

22. Proteomics FlashLFQ Galaxy Toolshed Repository. Available online: https://toolshed.g2.bx.psu.edu/view/galaxyp/flashlfq/908ab13490dc (accessed on 7 July 2020).

23. Adusumilli, R.; Mallick, P. Data conversion with proteoWizard msConvert. In *Methods in Molecular Biology*; Humana Press Inc.: Totowa, NJ, USA, 2017; pp. 339–368. [CrossRef]

24. Barsnes, H.; Vaudel, M. SearchGUI: A Highly Adaptable Common Interface for Proteomics Search and de Novo Engines. *J. Proteome Res.* **2018**, *17*, 2552–2555. [CrossRef] [PubMed]

25. Vaudel, M.; Burkhart, J.M.; Zahedi, R.P.; Oveland, E.; Berven, F.S.; Sickmann, A.; Martens, L.; Barsnes, H. PeptideShaker enables reanalysis of MS-derived proteomics data sets: To the editor. *Nat. Biotechnol.* **2015**, *33*, 22–24. [CrossRef]

26. Cox, J.; Neuhauser, N.; Michalski, A.; Scheltema, R.A.; Olsen, J.V.; Mann, M. Andromeda: A peptide search engine integrated into the MaxQuant environment. *J. Proteome Res.* **2011**, *10*, 1794–1805. [CrossRef]

27. Ritchie, M.E.; Phipson, B.; Wu, D.; Hu, Y.; Law, C.W.; Shi, W.; Smyth, G.K. Linear Models for Microarray and RNA-Seq Data. *Nucleic Acids Res.* **2015**, *43*, e47. [CrossRef] [PubMed]

28. Suomi, T.; Corthals, G.L.; Nevalainen, O.S.; Elo, L.L. Using peptide-level proteomics data for detecting differentially expressed proteins. *J. Proteome Res.* **2015**, *14*, 4564–4570. [CrossRef] [PubMed]

29. Välikangas, T.; Suomi, T.; Elo, L.L. A systematic evaluation of normalization methods in quantitative label-free proteomics. *Brief. Bioinform.* **2018**, *19*, 1–11. [CrossRef]

30. Argentini, A.; Staes, A.; Grüning, B.; Mehta, S.; Easterly, C.; Griffin, T.J.; Jagtap, P.; Impens, F.; Martens, L. Update on the moFF Algorithm for Label-Free Quantitative Proteomics. *J. Proteome Res.* **2019**, *18*, 728–731. [CrossRef]

31. Dale, R.; Grüning, B.; Sjödin, A.; Rowe, J.; Chapman, B.A.; Tomkins-Tinch, C.H.; Valieris, R.; Batut, B.; Caprez, A.; Cokelaer, T.; et al. Bioconda: Sustainable and comprehensive software distribution for the life sciences. *Nat. Methods* **2018**, *15*, 475–476. [CrossRef]

32. Proteomics moFF GitHub Repository. Available online: https://github.com/compomics/moFF (accessed on 7 July 2020).

33. Luersen, M.A.; le Riche, R.; Guyon, F. A constrained, globalized, and bounded Nelder-Mead method for engineering optimization. *Struct. Multidiscip. Optim.* **2004**, *27*, 43–54. [CrossRef]

34. Kurtzer, G.M.; Sochat, V.; Bauer, M.W. Singularity: Scientific containers for mobility of compute. *PLoS ONE* **2017**, *12*, e0177459. [CrossRef] [PubMed]

35. Proteomics FlashLFQ GitHub Repository. Available online: https://github.com/smith-chem-wisc/FlashLFQ (accessed on 7 July 2020).

36. Walther, B.A.; Moore, J.L. The concepts of bias, precision and accuracy, and their use in testing the performance of species richness estimators, with a literature review of estimator performance. *Ecography* **2005**, *28*, 815–829. [CrossRef]

37. Bensimon, A.; Heck, A.J.R.; Aebersold, R. Mass Spectrometry–Based Proteomics and Network Biology. *Annu. Rev. Biochem.* **2012**, *81*, 379–405. [CrossRef]

38. Vermillion, K.L.; Jagtap, P.; Johnson, J.E.; Griffin, T.J.; Andrews, M.T. Characterizing cardiac molecular mechanisms of mammalian hibernation via quantitative proteogenomics. *J. Proteome Res.* **2015**, *14*, 4792–4804. [CrossRef]

39. Ori, A.; Toyama, B.H.; Harris, M.S.; Bock, T.; Iskar, M.; Bork, P.; Ingolia, N.T.; Hetzer, M.W.; Beck, M. Integrated Transcriptome and Proteome Analyses Reveal Organ-Specific Proteome Deterioration in Old Rats. *Cell Syst.* **2015**, *1*, 224–237. [CrossRef]

40. Roumeliotis, T.I.; Williams, S.P.; Gonçalves, E.; Alsinet, C.; Velasco-Herrera, M.D.; Aben, N.; Ghavidel, F.Z.; Michaut, M.; Schubert, M.; Price, S.; et al. Genomic Determinants of Protein Abundance Variation in Colorectal Cancer Cells. *Cell Rep.* **2017**, *20*, 2201–2214. [CrossRef]

41. Schlaffner, C.N.; Pirklbauer, G.J.; Bender, A.; Steen, J.A.J.; Choudhary, J.S. A fast and quantitative method for post-translational modification and variant enabled mapping of peptides to genomes. *J. Vis. Exp.* **2018**. [CrossRef]

42. Gans, J.; Wolinsky, M.; Dunbar, J. Microbiology: Computational improvements reveal great bacterial diversity and high toxicity in soil. *Science* **2005**, *309*, 1387–1390. [CrossRef] [PubMed]

43. Li, X.; LeBlanc, J.; Truong, A.; Vuthoori, R.; Chen, S.S.; Lustgarten, J.L.; Roth, B.; Allard, J.; Ippoliti, A.; Presley, L.L.; et al. A metaproteomic approach to study human-microbial ecosystems at the mucosal luminal interface. *PLoS ONE* **2011**, *6*. [CrossRef]

44. Cheng, K.; Ning, Z.; Zhang, X.; Li, L.; Liao, B.; Mayne, J.; Stintzi, A.; Figeys, D. MetaLab: An automated pipeline for metaproteomic data analysis. *Microbiome* **2017**, *5*, 157. [CrossRef] [PubMed]

45. Chambers, M.C.; Jagtap, P.D.; Johnson, J.E.; McGowan, T.; Kumar, P.; Onsongo, G.; Guerrero, C.R.; Barsnes, H.; Vaudel, M.; Martens, L.; et al. An Accessible Proteogenomics Informatics Resource for Cancer Researchers. *Cancer Res.* **2017**, *77*, e43–e46. [CrossRef]

46. Mayer, G.; Montecchi-Palazzi, L.; Ovelleiro, D.; Jones, A.R.; Binz, P.A.; Deutsch, E.W.; Chambers, M.; Kallhardt, M.; Levander, F.; Shofstahl, J.; et al. The HUPO proteomics standards initiativemass spectrometry controlled vocabulary. *Database* **2013**. [CrossRef] [PubMed]

47. Smith, R. Conversations with 100 Scientists in the Field Reveal a Bifurcated Perception of the State of Mass Spectrometry Software. *J. Proteome Res.* **2018**, *17*, 1335–1339. [CrossRef]

48. Proteomics MaxQuant Galaxy Toolshed Repository. Available online: https://toolshed.g2.bx.psu.edu/view/galaxyp/maxquant/175e062b6a17 (accessed on 7 July 2020).

49. Proteomics MaxQuant GitHub Repository. Available online: https://github.com/galaxyproteomics/tools-galaxyp/tree/master/tools/maxquant (accessed on 7 July 2020).

2

Disseminating Metaproteomic Informatics Capabilities and Knowledge using the Galaxy-P Framework

Clemens Blank [1] (iD), Caleb Easterly [2], Bjoern Gruening [1], James Johnson [3], Carolin A. Kolmeder [4], Praveen Kumar [2], Damon May [5], Subina Mehta [2], Bart Mesuere [6], Zachary Brown [2], Joshua E. Elias [7], W. Judson Hervey [8], Thomas McGowan [3], Thilo Muth [9], Brook L. Nunn [5] (iD), Joel Rudney [10], Alessandro Tanca [11] (iD), Timothy J. Griffin [2] and Pratik D. Jagtap [2,*] (iD)

[1] Bioinformatics Group, Department of Computer Science, University of Freiburg, 79110 Freiburg im Breisgau, Germany; blankclemens@gmail.com (C.B.); gruening@informatik.uni-freiburg.de (B.G.)

[2] Department of Biochemistry, Molecular Biology and Biophysics, University of Minnesota, Minneapolis, MN 55455, USA; easte080@umn.edu (C.E.); kumar207@umn.edu (P.K.); smehta@umn.edu (S.M.); brow4261@umn.edu (Z.B.); tgriffin@umn.edu (T.J.G.)

[3] Minnesota Supercomputing Institute, University of Minnesota, Minneapolis, MN 55455, USA; jj@umn.edu (J.J.); mcgo0092@umn.edu (T.M.)

[4] Institute of Biotechnology, University of Helsinki, 00014 Helsinki, Finland; carolin.kolmeder@helsinki.fi

[5] Department of Genome Sciences, University of Washington, Seattle, WA 98195, USA; damonmay@uw.edu (D.M.); brookh@uw.edu (B.L.N.)

[6] Computational Biology Group, Ghent University, Krijgslaan 281, B-9000 Ghent, Belgium; Bart.Mesuere@ugent.be

[7] Department of Chemical & Systems Biology, Stanford University, Stanford, CA 94305, USA; josh.elias@stanford.edu

[8] Center for Bio/Molecular Science & Engineering, Naval Research Laboratory, Washington, DC 20375, USA; Judson.Hervey@nrl.navy.mil

[9] Bioinformatics Unit (MF1), Department for Methods Development and Research Infrastructure, Robert Koch Institute, 13353 Berlin, Germany; MuthT@rki.de

[10] Department of Diagnostic and Biological Sciences, University of Minnesota, Minneapolis, MN 55455, USA; jrudney@umn.edu

[11] Porto Conte Ricerche Science and Technology Park of Sardinia, 07041 Alghero, Italy; tanca@portocontericerche.it

* Correspondence: pjagtap@umn.edu

Abstract: The impact of microbial communities, also known as the microbiome, on human health and the environment is receiving increased attention. Studying translated gene products (proteins) and comparing metaproteomic profiles may elucidate how microbiomes respond to specific environmental stimuli, and interact with host organisms. Characterizing proteins expressed by a complex microbiome and interpreting their functional signature requires sophisticated informatics tools and workflows tailored to metaproteomics. Additionally, there is a need to disseminate these informatics resources to researchers undertaking metaproteomic studies, who could use them to make new and important discoveries in microbiome research. The Galaxy for proteomics platform (Galaxy-P) offers an open source, web-based bioinformatics platform for disseminating metaproteomics software and workflows. Within this platform, we have developed easily-accessible and documented metaproteomic software tools and workflows aimed at training researchers in their operation and disseminating the tools for more widespread use. The modular workflows encompass the core requirements of metaproteomic informatics: (a) database generation; (b) peptide spectral matching; (c) taxonomic analysis and (d) functional analysis. Much of the software available via the Galaxy-P platform was selected, packaged and deployed through an online metaproteomics "Contribution Fest" undertaken by a unique consortium of expert software developers and users from the metaproteomics research community, who have co-authored this manuscript. These resources are documented on GitHub and freely available through the Galaxy Toolshed, as well as a publicly accessible metaproteomics gateway Galaxy instance. These documented workflows are well suited for the training of novice metaproteomics researchers, through online resources such as the Galaxy Training Network, as well as hands-on training workshops. Here, we describe the metaproteomics tools available within these Galaxy-based resources, as well as the process by which they were selected and implemented in our community-based work. We hope this description will increase access to and utilization of metaproteomics tools, as well as offer a framework for continued community-based development and dissemination of cutting edge metaproteomics software.

Keywords: metaproteomics; functional microbiome; bioinformatics; software workflow development; Galaxy platform; mass spectrometry; community development

---

1. Introduction

Microbiome research has offered promising insights into microbial contributions to human health [1] and environmental dynamics [2]. Microbiome responses can be studied by a variety of approaches, including genome and transcriptome sequencing (metagenomics and metatranscriptomics, respectively), protein expression profiling (metaproteomics), and metabolite characterization (metabolomics). Over the years, the metagenomics-based approach has been the major approach for most microbiome studies, mainly because of the advances in sequencing technology [3] and development of statistical and analytical tools [4].

Recent trends in microbiome research have shown the promise of other "omic" approaches, with metaproteomics receiving much attention as an approach with great promise as a complement to more mature metagenomics approaches [5,6]. Metaproteomic studies identify the proteins that are actively being expressed by a community of microbiota under specific conditions [7]. Researchers have been promoting the potential benefits of metaproteomics for a better understanding of microbiome dynamics—particularly since it can provide insights into the functional state of the microbial community, beyond what can just be predicted by metagenomics [6,8].

Although the metaproteomics approach has been used for more than a decade, it is still emerging and has not yet become an approach routinely utilized by the microbiome research community. This has been primarily due to the technical difficulties associated with the approach. However, with recent advances in sample preparation, improved sensitivity of protein detection by mass spectrometry (MS), and new informatics tools for data analysis and interpretation, more researchers are turning to metaproteomics and realizing its potential in microbiome research [9].

Metaproteomics research holds promise in its ability to offer mechanistic insights into microbiome activity by performing functional analysis on identified peptides and proteins [10]. For example, microbiome studies have shown that the suite of metabolic pathways within microbiota from different persons tends to remain relatively consistent, even though microbial taxa may display considerable variation between individuals [11].

One of the key areas of advancement in metaproteomics over the past decade lies within the branch of informatics. New approaches continue to emerge across all the core areas of metaproteomics informatics, which include: (a) protein sequence database generation methods for microbial communities [12–16]; (b) database search methods for matching tandem mass spectrometry (MS/MS) data to peptide sequences [17,18]; and (c) interpretation methods and tools for taxonomic and functional analysis (Figure 1) [19–21].

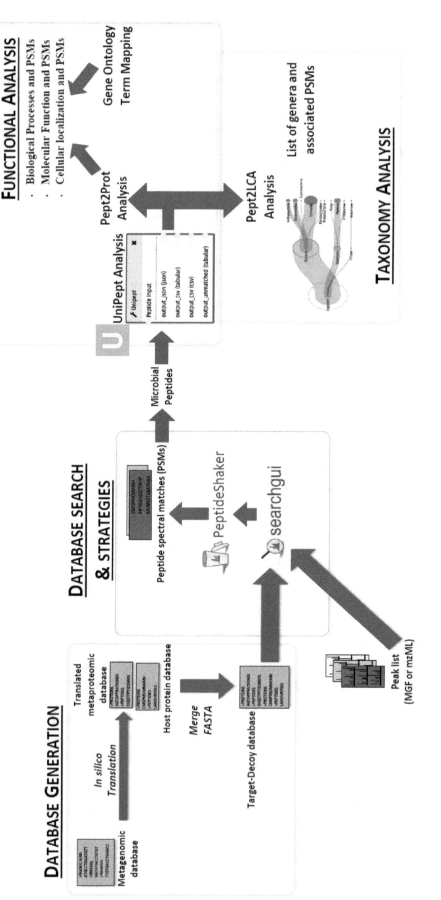

Figure 1. Generalized metaproteomics schema: Identification of metaproteome peptides is a complex workflow consisting of metaproteome sequence database generation (in FAST-ALL (FASTA) format) and peak processing of tandem mass spectrometry (MS/MS) data (in Mascot Generic Format (MGF) of mzML format). These two output files are used to match observed MS/MS spectra to predicted peptide sequences. This generates a list of bacterial peptide–spectral matches (PSMs). Later, the bacterial PSMs can be parsed out and subjected to functional analysis and taxonomic analysis for biological insight.

Despite these many advances, metaproteomic informatics remains very much a work in progress because of many unresolved challenges. Unlike single-organism proteomics, the protein sequence FAST-All (FASTA) databases for metaproteomics, which contain the predicted proteomes of multiple organisms, can be extremely large and complex [18]. It is not uncommon for the in silico translation of metagenome assemblies to a predicted metaproteome to contain hundreds of thousands to millions of predicted protein sequence entries. To reduce the possibility of mis-assigned spectra, it is common practice to include a FASTA-formatted host database and common laboratory contaminant proteins (e.g., skin keratins, proteases). For example, the study of the human oral microbiome would contain human epithelial cells proteins as part of the host database, in addition to microbial proteins from the consortia that form dental plaque. Algorithms and strategies for matching MS/MS to peptide sequences by database searching have been modified to address this challenge—in particular, addressing the decreased sensitivity of peptide matches due to increased false discovery rates in large databases [18], and challenges of protein identity inference due to sharing of proteins across multiple organisms in the database (e.g., the meta-protein concept in [22]).

Another significant challenge presented by metaproteomic informatics is that many disparate, specialized software tools must be used within each of the core areas required for successful data analysis and interpretation. For most researchers, these programs are difficult to access, master and operate. This reality offers a significant barrier for many researchers who could otherwise benefit from using metaproteomics approaches in their research.

Here we introduce new resources aimed at increasing access to advanced metaproteomic informatics tools and facilitating training in their use, thereby breaking down the barriers that hold back many researchers seeking to utilize metaproteomics in their work. The tools are housed in the Galaxy for proteomics (Galaxy-P) platform [23,24], which offers a user-friendly interface. Disparate software tools can be accessed and operated in an automated manner within a unified operating environment, which can be scaled to meet the demands of large-scale data analysis and informatics, as is often required in multi-omic approaches such as metaproteomics [24,25]. These resources were developed via a unique community-based effort, which leverages a consortium of leading experts from the metaproteomics research community, including a mixture of developers, data scientists and wet-bench researchers. These researchers participated in a contribution-fest (see z.umn.edu/mphack2016 for more information), wherein specific software was selected, deployed, tested and optimized within the Galaxy framework. In this manuscript, we describe not only the resources we have made available through this community-based effort, but also the process used to successfully achieve our goals. The accessible resources should help to increase wider adoption of metaproteomic informatics tools, as well as provide a framework for future collaborative efforts to make cutting-edge metaproteomic informatics tools available to the greater research community.

2. The Metaproteomics Gateway

2.1. Description of the Accessible Resources

Metaproteomics analysis of mass spectrometry data involves multiple core steps including database generation, MS/MS spectral matching to peptide sequences, taxonomic analysis and functional analysis. Below, we describe the general strategies and software currently available within these core areas, along with the process by which our consortium selected tools for deployment and dissemination via Galaxy-P. Since the main goal of this work, was to provide documentation to facilitate training and mastery of these software and workflows, we have provided step-by-step training instructions and related information in Supplement S (z.umn.edu/supps1). We have built a publicly accessible metaproteomics instance, or gateway (z.umn.edu/metaproteomicsgateway), for the purposes of providing access to documentation and other instructional materials, and an opportunity for hands-on training using example datasets and optimized metaproteomics workflows

(See Table 1). Full instructions are provided at this site for registering in this gateway and gaining access to all materials.

Table 1. Links to the resources for metaproteomics training.

Metaproteomics Gateway	*z.umn.edu/metaproteomicsgateway*
Galaxy Training Network	*http://galaxyproject.github.io/training-material/topics/ proteomics/tutorials/metaproteomics/tutorial.html*
Documentation	*Supplement S1*
Introductory video	*z.umn.edu/mpvideo2018*
Galaxy toolshed	*https://toolshed.g2.bx.psu.edu/*
GitHub	*https://github.com/galaxyproteomics*

2.2. The Playground: The Galaxy-P Platform

Galaxy-P is an extension of the open-source, Galaxy bioinformatics platform, which utilizes a web-based interface to access any instance, whether housed locally or remotely. The Galaxy interface includes a **Tool menu** (on the left of the screen—Figure 2), **Central main viewing pane** and the **History menu** (on the right side of the screen—Figure 2).

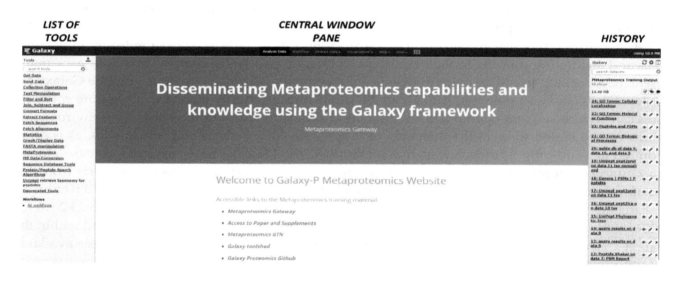

Figure 2. Galaxy interface and metaproteomics gateway. The Galaxy interface includes a tool menu, which consists of the list of available customized software within the instance in use. The central main viewing pane offers an area to view parameters for tools, edit workflows, and to visualize the results. The history menu maintains a real-time record of inputs and intermediate or final outputs from active software operations as the data is processed.

2.3. The First Step: Protein Sequence Database Generation Using a Galaxy-Based Tool

The composition of the protein sequence database used to match MS/MS spectra to sequences has a profound effect on the depth and reliability of identified peptides and inferred proteins in metaproteomics [14]. The source of the sample, sample preparation methods utilized, and the focus of the specific study all play a role in determining the composition of the protein sequence database. The results are only as good as the sequence database used—for example if a peptide sequence present in the sample is not present in the database, neither the peptide, nor the protein it is associated with can be identified. Conversely, if the protein sequence database includes many proteins that are not actually contained in the sample being analyzed (e.g., a database containing all known bacterial proteins), the database size can be so large that it decreases the sensitivity for identifying peptides that are truly in the sample. Thus, generating optimized databases for metaproteomics is not trivial. Ideally, the database would be constructed based on the known taxonomic makeup of the sample being analyzed—which can be achieved by metagenomic analysis of the sample or by selecting publicly available taxonomic metagenomics databases, if these exist for the sample in question.

During the contribution fest, several options for protein sequence database generation were considered. We first looked at options already available within the Galaxy-P platform. One option was the use of publicly available taxonomic repositories specific to certain sample types or environments [26–31]. A tool in Galaxy-P (Protein Database Downloader) was already in place for automated generation of databases based on information available from repositories including the Human Microbiome Project, the Human Oral Microbiome database, and the EBI metagenomics resource.

Another option already available within the Galaxy-P suite of tools is a tool for generating customized protein sequence databases from a list of genera thought to be in a sample. In some cases, a list of genera is available through previous published studies and can be useful in generating a protein sequence database [32,33]. In particular, 16S rRNA sequencing is used to assign operational taxonomic units (OTUs) in the form of species, genera or phyla. This can serve as a guide for generating a customized protein sequence database. Galaxy-P houses a tool to work through the UniProt Application Programming Interface (API) and extract protein sequences for all of the genera or phyla within a given list, generating a customized database for the metaproteomic analysis.

Given these already existing tools, we decided to direct our efforts to deploying more cutting-edge tools for database generation, which follows recent trends in using metagenomics information to generate more accurate protein sequence databases tailored to the taxonomic make-up of any given sample [33–37]. In particular, whole metagenome sequencing offers increased taxonomic resolution over 16S rRNA sequencing, thus enabling more accurate taxonomic and functional categorization of identified sequences [38].

Targeting tools that leveraged emerging methods in whole metagenome sequencing, we considered two approaches. One is the recently described Omega (overlap-graph metagenome assembler), a software tool for assembly of shotgun metagenome data that can be used along with the Sipros algorithm for database generation and matching to MS/MS data [39]. The second was a novel method and software (called Sixgill) described by May et al. that uses a 'metapeptide database' derived from shotgun metagenomics sequencing [15]. The database generated using this method is optimized for MS/MS data, thereby providing a more rapid and accurate peptide to spectrum matching. In the original publication, the method was used on two ocean samples that had undergone whole genome metagenomics sequencing, and was shown to offer a significant increase in the number of identifications (presumably due to a more accurate and compact database) as compared to a metaproteome sequence database assembled using standard methods, as well as using the comprehensive sequence database from the NCBI repository.

Given its demonstrated performance and optimized algorithm for utilizing large-scale, whole genome sequence data, we chose to implement the Sixgill software in Galaxy-P (Figure 3). We have provided step-by-step instructions for the use of Sixgill to create a metapeptide database, as well as the necessary input data, as described in Supplement S1. The deployed Sixgill tool provides a 'build' function, which generates a tab separated value (TSV) file containing the amino acid sequence of metapeptides along with other metrics. The Sixgill 'makefasta' function utilizes this information to generate a FASTA-formatted peptide database, which is compatible with database searching programs.

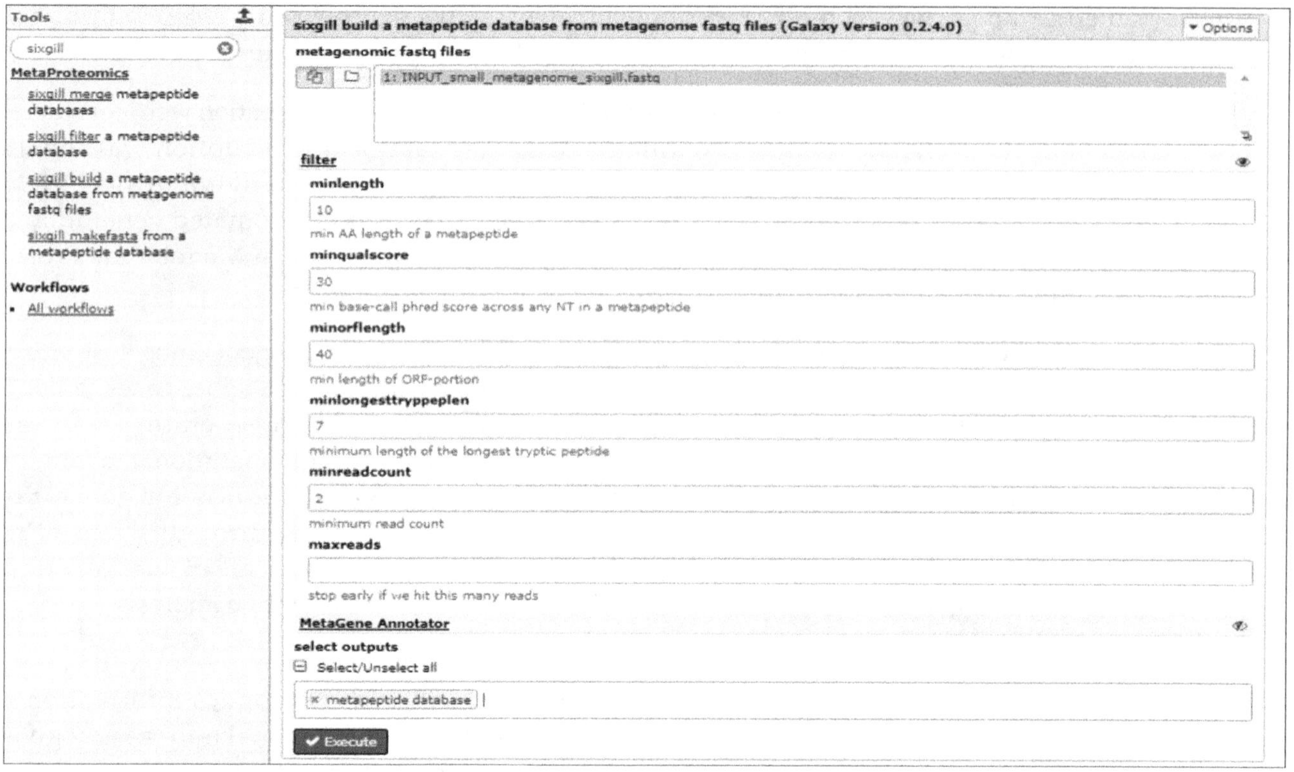

Figure 3. Sixgill tool within Galaxy. The Sixgill tool within Galaxy shows the build module, which uses a shotgun sequencing generated FASTQ file as an input, and generates a Tab-Separated Values (TSV) format file as an output. The filtering parameters aid in determining the quality and features of the output and are dependent on minimum length of the gene sequence, quality score, etc.

2.4. The Next Steps: Using a Galaxy Workflow

Galaxy also offers an option of generating a Galaxy 'workflow' which contains all the processing steps and software tool parameters for a particular analysis—except for the input or output data. Usually, workflows consist of multiple software tools, which are run in an automated, sequential manner, where outputs from one tool provide the input data for the next tool—ideally suited for multi-step analyses that are inherent to metaproteomic data analysis. Once built and optimized, workflows can be saved such that they become a main operational unit for analyzing different datasets in an efficient manner. Saved workflows can be also shared with other Galaxy users—thus promoting dissemination, reproducibility and collaboration.

The remaining three steps comprising our metaproteomics informatics resource (spectral matching, taxonomy analysis and functional analysis) are encapsulated in a single workflow (Figure 4). The starting data inputs to this workflow are MS/MS data files (in the form of mascot generic files, MGFs) and the FASTA-formatted metapeptide sequence database generated in step 1 above. The second step (spectral matching) yields identified metapeptides that act as inputs for the third step (taxonomy analysis) and fourth step (functional analysis). For functional analysis, an additional input file with Gene Ontology (GO) terms is also required.

In our specific workflow built for training purposes, MGF files (from Bering Strait ocean samples) are searched against the metapeptide database (generated using Sixgill software on metagenomics data) as inputs. In order to save time, we have trimmed the MGF datasets and the Bering Strait metapeptide database from those provided in the manuscript by May et al. [15]. Users are recommended to refer to Supplement S1 for detailed instructions on how to use the workflow on the example dataset.

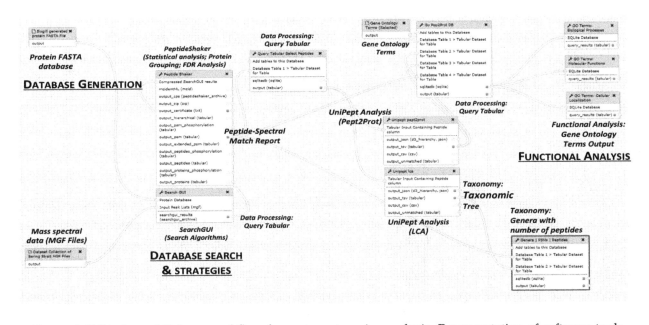

Figure 4. Edit view of Galaxy workflow for metaproteomics analysis. Representation of software tools used in a Galaxy metaproteomics workflow to identify bacterial peptides from the metaproteomic dataset. The first part of workflow includes database generation, followed by peak processing. The outputs from these sections are used for database search to generate a list of both bacterial peptide-spectral matches (PSMs). Later, bacterial PSMs were parsed out and subjected to Unipept analysis using Pept2Pro algorithm to generate outputs for functional analysis. Gene ontology categories such as biological processes, cellular localization and molecular function are generated. Additionally, bacterial PSMs were subjected to Unipept analysis using the lowest common ancestor algorithm to generate outputs for taxonomic analysis.

2.5. The Second Step: Spectral Matching

Sequence database searching algorithms that are able to match MS/MS spectra to peptide sequences contained in large databases (e.g., 10^6 or more sequences) have also been developed specifically for metaproteomics applications [40,41]. Selecting from the available software for metaproteomic sequence database searching must balance the following factors: (a) ability to effectively use large databases while still sensitively matching spectra to peptide sequences; (b) speed of the core algorithm, along with scalability for execution on parallel computing infrastructure, enabling the processing of large datasets using large sequence databases in a reasonable timeframe; and (c) the ability to generate outputs with robust false discovery rate (FDR) estimations, that are also compatible with downstream processing steps for taxonomic and functional analysis.

Multiple strategies have been suggested to increase the sensitivity of peptide identifications for the large sequence databases encountered in metaproteomics. This includes an iterative database searching workflow [42], a cascaded database search method [43] and a two-step method for searching large databases [44,45]. Muth et al. have recommended using a database sectioning approach, such that searches against subsets of a large database may increase the number of high confidence identifications [18]. The same group has proposed the use of de novo spectral matching in tandem with traditional sequence database-dependent methods [18], as well as the use of multiple database search algorithms, such as those offered by the SearchGUI tool [46], to increase the numbers of confident metapeptide identifications.

For the workflow deployed in our informatics resource, we chose a relatively straightforward approach for spectral matching. We used the SearchGUI tool already deployed in Galaxy, utilizing X!Tandem as the sequence database search algorithm of choice. Although the Galaxy-deployed SearchGUI tool offers the use of multiple database search algorithms (e.g., MS-GF+, Myrimatch,

OMSSA, Comet, Myrimatch, MS-Amanda and Novor), X!Tandem was determined to have a balance of speed and sensitivity that made it a good choice, especially for a training resource. The outputs from SearchGUI are further filtered and statistically analyzed using the companion PeptideShaker tool [47], which provides outputs compatible with downstream processing. Supplement S1 provides detailed instructions on the sequence database-searching step in this workflow, including a description of the small-scale input data we have provided for training purposes.

2.6. The Third Step: Taxonomic Classification

In metaproteomic studies, the identified microbial peptides can be used to determine the taxonomic composition of the sample. A number of options exist for taxonomic classification from the metapeptide data, some which were already deployed in Galaxy-P. The Unipept tool, deployed previously in Galaxy-P [24], maps sequences to annotated microbial organisms contained in the UniProt knowledgebase and subjects these to lowest common ancestor (LCA) analysis to provide a list of taxon identifications (at the level of kingdom, phylum, genus or species, if possible). The BLAST-P tool, also previously implemented in Galaxy-P [23], can match peptides to microbial proteins contained in the comprehensive NCBI non-redundant (nr) database, followed by taxonomic classification using MEGAN software [48] for metaproteomics data analysis [44].

During the metaproteomics contribution-fest, a number of new tools and extensions to new tools were considered for deployment in Galaxy. For example, taxonomy classification tools from the MetaProteomeAnalyzer [22] were considered, which process peptides identified via multiple database searching engines using information from the UniProt and National Center for Biotechnology Information (NCBI) repositories. Another tool under consideration was Prophane (https://mikrobiologie.uni-greifswald.de/en/resources/metaproteomics-data-analyses/prophane/), which uses the CLUSTAL W sequence alignment tool and other annotation tools to perform taxonomic classification.

Ultimately, the work stemming from the contribution fest focused on extending the functionality in Galaxy-P of the Unipept tool [19,49,50]. As mentioned above, Unipept was already deployed in Galaxy-P, providing textual outputs of taxonomic classes (Figure 5). We extended this function, adding the capability of visualizing taxonomic groups by packaging recently added visualization capabilities of Unipept into the Galaxy-based tool (Figure 5). With this functionality, the outputs from the metaproteomics workflow run in Galaxy-P now offers the user the option of launching a visualization window of the taxonomic results. (Figure 4). Details about this functionality within the workflow are provided in Supplement S1.

2.7. The Fourth Step: Functional Analysis

Metaproteomics has a distinct advantage in determining the functional signature associated with a microbial community under a specific condition based on identification of the proteins that are actually being expressed [44]. However, characterizing the functional state from a collection of expressed proteins is not trivial. Functional annotation based on a protein profile requires several components: a controlled vocabulary (or ontology) that represents protein function, databases containing annotations of known proteins or protein families with terms from these vocabularies, and alignment tools that map functional annotations within data repositories to the experimentally identified peptides or proteins. Many ontologies exist and often focus on different aspects of function: the Gene Ontology [51] and Kyoto Encyclopedia of Genes and Genomes (KEGG) pathways [52] are two of the most prominent. A number of databases use diverse methodologies to assign function to proteins and or its groups—these include the InterPro [53], and "evolutionary genealogy of genes: Non-supervised Orthologous Groups" (eggnog) [54] databases. Finally, tools to map functional annotations from these databases to experimentally identified proteins are often database-specific, such as the eggNOG-mapper [55] and InterProScan [56]. In addition, MEGAN6 can be used to carry out InterPro2GO, KEGG, SEED and EggNOG analysis to determine the distribution of functions amongst expressed proteins in the microbiome [48].

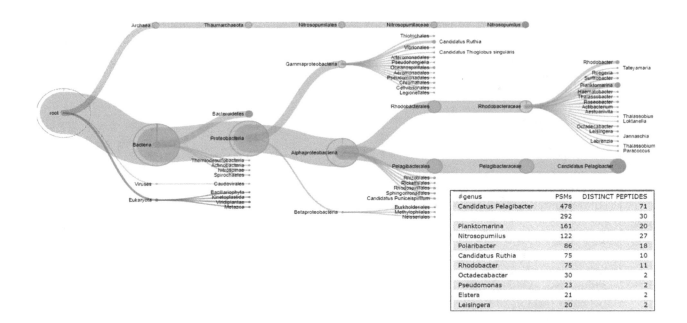

#genus	PSMs	DISTINCT PEPTIDES
Candidatus Pelagibacter	478	71
	292	30
Planktomarina	161	20
Nitrosopumilus	122	27
Polaribacter	86	18
Candidatus Ruthia	75	10
Rhodobacter	75	11
Octadecabacter	30	2
Pseudomonas	23	2
Elstera	21	2
Leisingera	20	2

Figure 5. Taxonomy analysis using Unipept. Bacterial PSMs were subjected to Unipept analysis against UniProt database using lowest common ancestor algorithm to generate outputs for taxonomic analysis. These outputs include a Unipept Viewer which is an interactive visualization plugin that can be used to visualize taxonomic distribution of the ocean metaproteomic dataset. Unipept also generates a Comma-Separated Values (CSV) format file that lists the peptide assignments to taxa. That file then can be parsed to generate a tabular output (lower right).

Beyond mapping functional annotations to identified proteins, visualization of the collective functional categories is the next desirable step. Here, various options are available—with potential for deployment in the Galaxy platform. For single GO terms of interest, the QuickGO browser [57] enables the user to view the full term definition, as well as to browse closely related terms. For large lists of GO terms, the 'reduce and visualize Gene Ontology' (REVIGO) tool [58] allows the reduction of GO terms to a representative subset and several visualizations of the resulting smaller list. Moreover, the Prophane suite of tools can also be used to determine the distribution of functions in a microbiome sample and visualize them. MetaProteomeAnalyzer provides enzyme and pathway display options where proteins grouped by UniProt ontologies (e.g., biological process or molecular function), EC (Enzyme Commission) numbers and KEGG pathways can be visualized [22].

Although all of these options have great potential for functional annotation and visualization, our community-based efforts focused on utilizing the Galaxy-deployed Unipept tool and its Pept2Prot option, which maps identified peptide sequences to proteins. The proteins are then mapped to GO terms for molecular function, biological processes and cellular localization, followed by using the GO term mapping information (Figure 6). The grouping into functional categories was performed using a Galaxy tool, query tabular tailored to automate extraction and grouping of tabular data results. These results are presented as a tabular output for further downstream analysis, such as visualization software. Details are provided in Supplement S1 about the tools involved in this functional annotation step, along with instructions.

A

#description	bering_peptides	bering_psms
structural constituent of ribosome	31	168
ATP binding	31	136
DNA binding	23	240
transporter activity	21	148
rRNA binding	19	60
metal ion binding	14	215
receptor activity	12	57
oxidoreductase activity	7	18
GTP binding	6	32
GTPase activity	6	32

B

#description	bering_peptides	bering_psms
cytoplasm	40	265
ribosome	24	152
outer membrane-bounded periplasmic space	18	234
membrane	12	53
integral component of membrane	10	35
plasma membrane	6	44
small ribosomal subunit	6	17
ATP-binding cassette (ABC) transporter complex	5	29
cell outer membrane	4	18
intracellular	3	12

C

#description	bering_peptides	bering_psms
translation	32	171
transport	24	262
protein refolding	16	63
transcription, DNA-templated	14	120
protein folding	12	53
regulation of transcription, DNA-templated	11	146
transmembrane transport	10	61
amino acid transport	8	36
carbohydrate transport	7	69
chromosome condensation	6	59

Figure 6. Functional analysis using Unipept and GO (Gene Ontology) terms. Bacterial PSMs were subjected to Unipept analysis against Pept2Pro algorithm to generate outputs for functional analysis. Using PSM report, gene ontology mapping files and Unipept outputs, the query tabular file generates tabular outputs for gene ontology categories. The generated tabular outputs for molecular function (**A**), cellular localization; (**B**) and biological processes; (**C**) and also enlist the number of associated peptides and PSMs with each gene ontology category.

2.8. Links to Accessible Resources for Training

The main goal of our contribution fest was to provide an instrument for researchers to access and learn the operation of cutting-edge metaproteomics tools. We have provided several means for researchers to access and train in the operation of these tools (See Table 1). We have established a Metaproteomics Gateway, composed of a publicly accessible Galaxy instance containing the tools, workflows and example data described in this manuscript. Supplement S1 provides a detailed description for the use of this gateway. We have also provided our documentation and training instructions within the Galaxy Training Network repository (http://galaxyproject.github.io/training-material/), a central resource for providing documentation on Galaxy-based tools and platforms.

Our tools and workflows have also been made openly available through the Galaxy Tool Shed and on GitHub. We hope that the available resources that also include an introductory video will encourage researchers to incorporate metaproteomics studies into their current expertise of research.

In conclusion, we have described accessible resources aimed at training researchers in the use of advanced metaproteomic informatics tools, with the intent of increasing the adoption of metaproteomics by the wider research community. These tools have been made available through a unique, community-based process, which has leveraged a community of metaproteomic informatics experts, as well as the powerful Galaxy platform. We would like to emphasize that the use of Galaxy was highly enabling for this work, as it provides a unified environment for operating many disparate tools required in metaproteomics, as well as a platform that can be used to promote training and usage by the larger community.

Several other points are worth noting from the work we have described here. It is evident that, for each of the core steps described in the metaproteomic data analysis pipeline, there are many valuable software tools that already exist. During our contribution fest, our consortium of researchers were only able to deploy, test and optimize a select few of these tools. Work is ongoing on implementing additional tools. In the future, we anticipate increased need for visualization, quantitation and statistical tools in metaproteomics research, which will aid in biological interpretation. It is our hope that this manuscript serves as an invitation to others to join our collaborative community and help to make additional high-value tools for metaproteomics available. Again, our usage of the open source Galaxy-P platform for deployment and dissemination provides a playground for developers to come 'play' in, and collaborate with other like-minded researchers from around the world. We also hope that the 'shareable' workflows developed will facilitate the undertaking of global scale research projects.

It is our hope that this manuscript will help establish a framework for continued, community-based efforts at making cutting-edge metaproteomics tools available to others, along with the necessary documentation and hands-on training resources to educate researchers in their use. Ultimately, we hope this approach will yield great dividends in increasing the adoption of metaproteomic approaches by more researchers, which will help catalyze a better understanding of the molecular characteristics of dynamic microbial communities and microbiome.

Acknowledgments: This work is supported by grants from the NSF award 1458524 and NIH award U24CA199347 to T.J. Griffin and the core Galaxy-P research team. BLN and DM were supported by NSF grant OCE-1633939. The authors would also like to thank Intergalactic Utilities Commission (https://galaxyproject.org/iuc/) for encouraging and supporting the contribution fest. We would also like to thank the organizers of Association of Biomolecular Research Facilities (ABRF), American Society of Mass Spectrometry (ASMS), Galaxy Community Conference (GCC), and International Metaproteomics Symposium (IMS) for providing us a platform for conducting workshops and facilitating interesting discussions during the conferences.

Author Contributions: C.B. generated the material for the Galaxy Training Site and provided edits for manuscript. C.E. worked on the functional analysis portion in the manuscript and helped in writing the manuscript. B.G. led the metaproteomics contribution-fest and oversaw the work on Galaxy Training Network. J.J. packaged most of the tools that were used in the workflow and optimized the workflow. C.A.K. provided scientific inputs and helped in manuscript draft work. P.K. generated trimmed versions of datasets and databases and helped in manuscript writing. D.M. developed the Sixgill software and along with B.L.N. provided the datasets and scientific inputs during workflow development. S.M. helped in supplemental data documentation and testing the tools and workflows. B.M., the developer of the Unipept tool, provided scientific inputs and provided inputs during manuscript writing. Z.B., J.E., W.J.H., Thomas McGowan provided scientific inputs during the contribution-fest or provided edits during manuscript writing. Thilo Muth helped in maintenance of tools and workflows on the metaproteomics gateway. B.L.N., J.R., A.T. provided scientific inputs and helped in manuscript writing. T.J.G. provided scientific directions and helped in manuscript writing. P.D.J. conceived of and led the project and wrote the manuscript.

References

1.   Knight, R.; Callewaert, C.; Marotz, C.; Hyde, E.R.; Debelius, J.W.; McDonald, D.; Sogin, M.L. The Microbiome and Human Biology. *Annu. Rev. Genom. Hum. Genet.* **2017**, *31*, 65–86. [CrossRef] [PubMed]

2.   Foo, J.L.; Ling, H.; Lee, Y.S.; Chang, M.W. Microbiome engineering: Current applications and its future. *Biotechnol. J.* **2017**, *12*. [CrossRef] [PubMed]

3.   Arnold, J.W.; Roach, J.; Azcarate-Peril, M.A. Emerging Technologies for Gut Microbiome Research. *Trends Microbiol.* **2016**, *24*, 887–901. [CrossRef] [PubMed]

4.   Siegwald, L.; Touzet, H.; Lemoine, Y.; Hot, D.; Audebert, C.; Caboche, S. Assessment of Common and Emerging Bioinformatics Pipelines for Targeted Metagenomics. *PLoS ONE* **2017**, *12*, e0169563. [CrossRef] [PubMed]

5.   Maier, T.V.; Lucio, M.; Lee, L.H.; VerBerkmoes, N.C.; Brislawn, C.J.; Bernhardt, J.; Lamendella, R.; McDermott, J.E.; Bergeron, N.; Heinzmann, S.S.; et al. Impact of Dietary Resistant Starch on the Human Gut Microbiome, Metaproteome, and Metabolome. *mBio* **2017**, *8*, 1343–1417. [CrossRef] [PubMed]

6.   Heintz-Buschart, A.; May, P.; Laczny, C.C.; Lebrun, L.A.; Bellora, C.; Krishna, A.; Wampach, L.; Schneider, J.G.; Hogan, A.; de Beaufort, C.; et al. Integrated multi-omics of the human gut microbiome in a case study of familial type 1 diabetes. *Nat. Microbiol.* **2016**, *2*, 16180. [CrossRef] [PubMed]

7.   Wilmes, P.; Bond, P.L. Metaproteomics: Studying functional gene expression in microbial ecosystems. *Trends Microbiol.* **2006**, *14*, 92–97. [CrossRef] [PubMed]

8.   Heintz-Buschart, A.; Wilmes, P. Human Gut Microbiome: Function Matters. *Trends Microbiol.* **2017**, *17*, 30251–30252. [CrossRef] [PubMed]

9.   Wilmes, P.; Heintz-Buschart, A.; Bond, P.L. A decade of metaproteomics: Where we stand and what the future holds. *Proteomics* **2015**, *15*, 3409–3417. [CrossRef] [PubMed]

10.   Tanca, A.; Abbondio, M.; Palomba, A.; Fraumene, C.; Manghina, V.; Cucca, F.; Fiorillo, E.; Uzzau, S. Potential and active functions in the gut microbiota of a healthy human cohort. *Microbiome* **2017**, *5*, 79. [CrossRef] [PubMed]

11.   Human Microbiome Project Consortium. A framework for human microbiome research. *Nature* **2012**, *486*, 215–221.

12.   Tanca, A.; Palomba, A.; Fraumene, C.; Pagnozzi, D.; Manghina, V.; Deligios, M.; Muth, T.; Rapp, E.; Martens, L.; Addis, M.F.; et al. The impact of sequence database choice on metaproteomic results in gut microbiota studies. *Microbiome* **2016**, *4*, 51. [CrossRef] [PubMed]

13.   Tanca, A.; Palomba, A.; Deligios, M.; Cubeddu, T.; Fraumene, C.; Biosa, G.; Pagnozzi, D.; Addis, M.F.; Uzzau, S. Evaluating the impact of different sequence databases on metaproteome analysis: Insights from a lab-assembled microbial mixture. *PLoS ONE* **2013**, *8*, e82981. [CrossRef] [PubMed]

14.   Timmins-Schiffman, E.; May, D.H.; Mikan, M.; Riffle, M.; Frazar, C.; Harvey, H.R.; Noble, W.S.; Nunn, B.L. Critical decisions in metaproteomics: Achieving high confidence protein annotations in a sea of unknowns. *ISME J.* **2017**, *11*, 309–314. [CrossRef] [PubMed]

15.   May, D.H.; Timmins-Schiffman, E.; Mikan, M.P.; Harvey, H.R.; Borenstein, E.; Nunn, B.L.; Noble, W.S. An Alignment-Free "Metapeptide" Strategy for Metaproteomic Characterization of Microbiome Samples Using Shotgun Metagenomic Sequencing. *J. Proteome Res.* **2016**, *15*, 2697–2705. [CrossRef] [PubMed]

16.   Tang, H.; Li, S.; Ye, Y. A Graph-Centric Approach for Metagenome-Guided Peptide and Protein Identification in Metaproteomics. *PLoS Comput. Biol.* **2016**, *12*, 1005224. [CrossRef] [PubMed]

17.   Muth, T.; Renard, B.Y.; Martens, L. Metaproteomic data analysis at a glance: Advances in computational microbial community proteomics. *Expert Rev. Proteom.* **2016**, *13*, 757–769. [CrossRef] [PubMed]

18.   Muth, T.; Kolmeder, C.A.; Salojärvi, J.; Keskitalo, S.; Varjosalo, M.; Verdam, F.J.; Rensen, S.S.; Reichl, U.; de Vos, W.M.; Rapp, E.; et al. Navigating through metaproteomics data: A logbook of database searching. *Proteomics* **2015**, *15*, 3439–3453. [CrossRef] [PubMed]

19.   Mesuere, B.; Debyser, G.; Aerts, M.; Devreese, B.; Vandamme, P.; Dawyndt, P. The Unipept metaproteomics analysis pipeline. *Proteomics* **2015**, *15*, 1437–1442. [CrossRef] [PubMed]

20.   Xiong, W.; Brown, C.T.; Morowitz, M.J.; Banfield, J.F.; Hettich, R.L. Genome-resolved metaproteomic characterization of preterm infant gut microbiota development reveals species-specific metabolic shifts and variabilities during early life. *Microbiome* **2017**, *5*, 72. [CrossRef] [PubMed]

21.   Huson, D.H.; Beier, S.; Flade, I.; Górska, A.; El-Hadidi, M.; Mitra, S.; Ruscheweyh, H.J.; Tappu, R. MEGAN Community Edition—Interactive Exploration and Analysis of Large-Scale Microbiome Sequencing Data. *PLoS Comput. Biol.* **2016**, *12*, 1004957. [CrossRef] [PubMed]

22. Muth, T.; Behne, A.; Heyer, R.; Kohrs, F.; Benndorf, D.; Hoffmann, M.; Lehtevä, M.; Reichl, U.; Martens, L.; Rapp, E. The MetaProteomeAnalyzer: A powerful open-source software suite for metaproteomics data analysis and interpretation. *J. Proteome Res.* **2015**, *14*, 1557–1565. [CrossRef] [PubMed]

23. Jagtap, P.D.; Johnson, J.E.; Onsongo, G.; Sadler, F.W.; Murray, K.; Wang, Y.; Shenykman, G.M.; Bandhakavi, S.; Smith, L.M.; Griffin, T.J. Flexible and accessible workflows for improved proteogenomic analysis using the Galaxy framework. *J. Proteome Res.* **2014**, *13*, 5898–5908. [CrossRef] [PubMed]

24. Jagtap, P.D.; Blakely, A.; Murray, K.; Stewart, S.; Kooren, J.; Johnson, J.E.; Rhodus, N.L.; Rudney, J.; Griffin, T.J. Metaproteomic analysis using the Galaxy framework. *Proteomics* **2015**, *15*, 3553–3565. [CrossRef] [PubMed]

25. Afgan, E.; Baker, D.; van den Beek, M.; Blankenberg, D.; Bouvier, D.; Čech, M.; Chilton, J.; Clements, D.; Coraor, N.; Eberhard, C.; et al. The Galaxy platform for accessible, reproducible and collaborative biomedical analyses: 2016 update. *Nucleic Acids Res.* **2016**, *44*, 3–10. [CrossRef] [PubMed]

26. Wilmes, P.; Bond, P.L. The application of two-dimensional polyacrylamide gel electrophoresis and downstream analyses to a mixed community of prokaryotic microorganisms. *Environ. Microbiol.* **2004**, *6*, 911–920. [CrossRef] [PubMed]

27. Klaassens, E.S.; de Vos, W.M.; Vaughan, E.E. Metaproteomics approach to study the functionality of the microbiota in the human infant gastrointestinal tract. *Appl. Environ. Microbiol.* **2007**, *73*, 1388–1392. [CrossRef] [PubMed]

28. Rudney, J.D.; Xie, H.; Rhodus, N.L.; Ondrey, F.G.; Griffin, T.J. A metaproteomic analysis of the human salivary microbiota by three-dimensional peptide fractionation and tandem mass spectrometry. *Mol. Oral Microbiol.* **2010**, *25*, 38–49. [CrossRef] [PubMed]

29. Haange, S.B.; Oberbach, A.; Schlichting, N.; Hugenholtz, F.; Smidt, H.; von Bergen, M.; Till, H.; Seifert, J. Metaproteome analysis and molecular genetics of rat intestinal microbiota reveals section and localization resolved species distribution and enzymatic functionalities. *J. Proteome Res.* **2012**, *11*, 5406–5417. [CrossRef] [PubMed]

30. Jagtap, P.; McGowan, T.; Bandhakavi, S.; Tu, Z.J.; Seymour, S.; Griffin, T.J.; Rudney, J.D. Deep metaproteomic analysis of human salivary supernatant. *Proteomics* **2012**, *12*, 992–1001. [CrossRef] [PubMed]

31. Bastida, F.; Hernández, T.; García, C. Metaproteomics of soils from semiarid environment: Functional and phylogenetic information obtained with different protein extraction methods. *J. Proteom.* **2014**, *101*, 31–42. [CrossRef] [PubMed]

32. Wu, J.; Zhu, J.; Yin, H.; Liu, X.; An, M.; Pudlo, N.A.; Martens, E.C.; Chen, G.Y.; Lubman, D.M. Development of an Integrated Pipeline for Profiling Microbial Proteins from Mouse Fecal Samples by LC-MS/MS. *J. Proteome Res.* **2016**, *15*, 3635–3642. [CrossRef] [PubMed]

33. Kohrs, F.; Wolter, S.; Benndorf, D.; Heyer, R.; Hoffmann, M.; Rapp, E.; Bremges, A.; Sczyrba, A.; Schlüter, A.; Reichl, U. Fractionation of biogas plant sludge material improves metaproteomic characterization to investigate metabolic activity of microbial communities. *Proteomics* **2015**, *15*, 3585–3589. [CrossRef] [PubMed]

34. Bao, Z.; Okubo, T.; Kubota, K.; Kasahara, Y.; Tsurumaru, H.; Anda, M.; Ikeda, S.; Minamisawa, K. Metaproteomic identification of diazotrophic methanotrophs and their localization in root tissues of field-grown rice plants. *Appl. Environ. Microbiol.* **2014**, *80*, 5043–5052. [CrossRef] [PubMed]

35. Colatriano, D.; Ramachandran, A.; Yergeau, E.; Maranger, R.; Gélinas, Y.; Walsh, D.A. Metaproteomics of aquatic microbial communities in a deep and stratified estuary. *Proteomics* **2015**, *15*, 3566–3579. [CrossRef] [PubMed]

36. Young, J.C.; Pan, C.; Adams, R.M.; Brooks, B.; Banfield, J.F.; Morowitz, M.J.; Hettich, R.L. Metaproteomics reveals functional shifts in microbial and human proteins during a preterm infant gut colonization case. *Proteomics* **2015**, *15*, 3463–3473. [CrossRef] [PubMed]

37. Mattarozzi, M.; Manfredi, M.; Montanini, B.; Gosetti, F.; Sanangelantoni, A.M.; Marengo, E.; Careri, M.; Visioli, G. A metaproteomic approach dissecting major bacterial functions in the rhizosphere of plants living in serpentine soil. *Anal. Bioanal. Chem.* **2017**, *409*, 2327–2339. [CrossRef] [PubMed]

38. Jovel, J.; Patterson, J.; Wang, W.; Hotte, N.; O'Keefe, S.; Mitchel, T.; Perry, T.; Kao, D.; Mason, A.L.; Madsen, K.L.; et al. Characterization of the Gut Microbiome Using 16S or Shotgun Metagenomics. *Front. Microbiol.* **2016**, *7*, 459. [CrossRef] [PubMed]

39. Haider, B.; Ahn, T.H.; Bushnell, B.; Chai, J.; Copeland, A.; Pan, C. Omega: An overlap-graph de novo assembler for metagenomics. *Bioinformatics* **2014**, *30*, 2717–2722. [CrossRef] [PubMed]

40. Chatterjee, S.; Stupp, G.S.; Park, S.K.; Ducom, J.C.; Yates, J.R., 3rd; Su, A.I.; Wolan, D.W. A comprehensive and scalable database search system for metaproteomics. *BMC Genom.* **2016**, *17*, 642. [CrossRef] [PubMed]

41. Guo, X.; Li, Z.; Yao, Q.; Mueller, R.S.; Eng, J.K.; Tabb, D.L.; Hervey, W.J., 4th; Pan, C. Sipros Ensemble Improves Database Searching and Filtering for Complex Metaproteomics. *Bioinformatics* **2017**. [CrossRef] [PubMed]

42. Rooijers, K.; Kolmeder, C.; Juste, C.; Doré, J.; de Been, M.; Boeren, S.; Galan, P.; Beauvallet, C.; de Vos, W.M.; Schaap, P.J. An iterative workflow for mining the human intestinal metaproteome. *BMC Genom.* **2011**, *12*, 6. [CrossRef] [PubMed]

43. Kertesz-Farkas, A.; Keich, U.; Noble, W.S. Tandem Mass Spectrum Identification via Cascaded Search. *J. Proteome Res.* **2015**, *14*, 3027–3038. [CrossRef] [PubMed]

44. Rudney, J.D.; Jagtap, P.D.; Reilly, C.S.; Chen, R.; Markowski, T.W.; Higgins, L.; Johnson, J.E.; Griffin, T.J. Protein relative abundance patterns associated with sucrose-induced dysbiosis are conserved across taxonomically diverse oral microcosm biofilm models of dental caries. *Microbiome* **2015**, *3*, 69. [CrossRef] [PubMed]

45. Jagtap, P.; Goslinga, J.; Kooren, J.A.; McGowan, T.; Wroblewski, M.S.; Seymour, S.L.; Griffin, T.J. A two-step database search method improves sensitivity in peptide sequence matches for metaproteomics and proteogenomics studies. *Proteomics* **2013**, *13*, 1352–1357. [CrossRef] [PubMed]

46. Vaudel, M.; Barsnes, H.; Berven, F.S.; Sickmann, A.; Martens, L. SearchGUI: An open-source graphical user interface for simultaneous OMSSA and X!Tandem searches. *Proteomics* **2011**, *11*, 996–999. [CrossRef] [PubMed]

47. Vaudel, M.; Burkhart, J.M.; Zahedi, R.P.; Oveland, E.; Berven, F.S.; Sickmann, A.; Martens, L.; Barsnes, H. PeptideShaker enables reanalysis of MS-derived proteomics data sets. *Nat. Biotechnol.* **2015**, *33*, 22–24. [CrossRef] [PubMed]

48. Huson, D.H.; Weber, N. Microbial community analysis using MEGAN. *Methods Enzymol.* **2013**, *531*, 465–485. [PubMed]

49. Mesuere, B.; Van der Jeugt, F.; Willems, T.; Naessens, T.; Devreese, B.; Martens, L.; Dawyndt, P. High-throughput metaproteomics data analysis with Unipept: A tutorial. *J. Proteom.* **2017**, *17*, 30189–30196. [CrossRef] [PubMed]

50. Mesuere, B.; Willems, T.; Van der Jeugt, F.; Devreese, B.; Vandamme, P.; Dawyndt, P. Unipept web services for metaproteomics analysis. *Bioinformatics* **2016**, *32*, 1746–1748. [CrossRef] [PubMed]

51. Gene Ontology Consortium. The Gene Ontology: Enhancements for 2011. *Nucleic Acids Res.* **2012**, *40*, 559–564.

52. Kanehisa, M.; Furumichi, M.; Tanabe, M.; Sato, Y.; Morishima, K. KEGG: New perspectives on genomes, pathways, diseases and drugs. *Nucleic Acids Res.* **2017**, *45*, 353–361. [CrossRef] [PubMed]

53. Hunter, S.; Apweiler, R.; Attwood, T.K.; Bairoch, A.; Bateman, A.; Binns, D.; Bork, P.; Das, U.; Daugherty, L.; Duquenne, L.; et al. InterPro: The integrative protein signature database. *Nucleic Acids Res.* **2009**, *37*, 211–215. [CrossRef] [PubMed]

54. Huerta-Cepas, J.; Szklarczyk, D.; Forslund, K.; Cook, H.; Heller, D.; Walter, M.C.; Rattei, T.; Mende, D.R.; Sunagawa, S.; Kuhn, M.; et al. eggNOG 4.5: A hierarchical orthology framework with improved functional annotations for eukaryotic, prokaryotic and viral sequences. *Nucleic Acids Res.* **2016**, *44*, 286–293. [CrossRef] [PubMed]

55. Huerta-Cepas, J.; Forslund, K.; Coelho, L.P.; Szklarczyk, D.; Jensen, L.J.; von Mering, C.; Bork, P. Fast Genome-Wide Functional Annotation through Orthology Assignment by eggNOG-Mapper. *Mol. Biol. Evolut.* **2017**, *34*, 2115–2122. [CrossRef] [PubMed]

56. Jones, P.; Binns, D.; Chang, H.Y.; Fraser, M.; Li, W.; McAnulla, C.; McWilliam, H.; Maslen, J.; Mitchell, A.; Nuka, G.; et al. InterProScan 5: Genome-scale protein function classification. *Bioinformatics* **2014**, *30*, 1236–1240. [CrossRef] [PubMed]

57. Binns, D.; Dimmer, E.; Huntley, R.; Barrell, D.; O'Donovan, C.; Apweiler, R. QuickGO: A web-based tool for Gene Ontology searching. *Bioinformatics* **2009**, *25*, 3045–3046. [CrossRef] [PubMed]

58. Supek, F.; Bošnjak, M.; Škunca, N.; Šmuc, T. REVIGO summarizes and visualizes long lists of gene ontology terms. *PLoS ONE* **2011**, *6*, e21800. [CrossRef] [PubMed]

3

MetaGOmics: A Web-Based Tool for Peptide-Centric Functional and Taxonomic Analysis of Metaproteomics Data

Michael Riffle [1,2,*], Damon H. May [1], Emma Timmins-Schiffman [1], Molly P. Mikan [3], Daniel Jaschob [2], William Stafford Noble [1] and Brook L. Nunn [1,*]

[1] Department of Genome Sciences, University of Washington, Seattle, WA 98195, USA; damonmay@uw.edu (D.H.M.); emmats@u.washington.edu (E.T.-S.); william-noble@uw.edu (W.S.N.)

[2] Department of Biochemistry, University of Washington, Seattle, WA 98195, USA; djaschob@uw.edu

[3] Department of Ocean, Earth, and Atmospheric Sciences, Old Dominion University, Norfolk, VA 23529, USA; mmika003@odu.edu

[*] Correspondence: mriffle@uw.edu (M.R.); brookh@uw.edu (B.L.N.)

Abstract: Metaproteomics is the characterization of all proteins being expressed by a community of organisms in a complex biological sample at a single point in time. Applications of metaproteomics range from the comparative analysis of environmental samples (such as ocean water and soil) to microbiome data from multicellular organisms (such as the human gut). Metaproteomics research is often focused on the quantitative functional makeup of the metaproteome and which organisms are making those proteins. That is: What are the functions of the currently expressed proteins? How much of the metaproteome is associated with those functions? And, which microorganisms are expressing the proteins that perform those functions? However, traditional protein-centric functional analysis is greatly complicated by the large size, redundancy, and lack of biological annotations for the protein sequences in the database used to search the data. To help address these issues, we have developed an algorithm and web application (dubbed "MetaGOmics") that automates the quantitative functional (using Gene Ontology) and taxonomic analysis of metaproteomics data and subsequent visualization of the results. MetaGOmics is designed to overcome the shortcomings of traditional proteomics analysis when used with metaproteomics data. It is easy to use, requires minimal input, and fully automates most steps of the analysis—including comparing the functional makeup between samples. MetaGOmics is freely available at https://www.yeastrc.org/metagomics/.

Keywords: metaproteomics; proteomics; bioinformatics; software; data visualization; mass spectrometry; gene ontology

1. Introduction

Recent years have seen tremendous advancements in the availability of high-throughput "omics" technologies for characterizing complex biological samples. These advancements have fueled the growth of meta-omics as a field for characterizing the metagenomes, metatranscriptomes, meta-metabolomes, and metaproteomes of environmental and microbiome samples comprising a taxonomically diverse (often uncharacterized) community of organisms [1–3]. Metagenomics examines questions related to taxonomic composition and genomic architecture of organisms in the sample [4,5]. Meta-metabolomics examines which metabolites are being produced and how those change in response to environmental factors [6]. Meta-transcriptomics aims to use gene expression of mRNA transcripts to track taxonomic and functional abundance [7]. However, transcript and actual protein levels can be poorly correlated because of codon bias, differing rates of protein turnover, and other factors [8,9].

Metaproteomics aims to overcome this by assaying the protein composition directly in order to characterize the taxonomic and functional makeup of a sample at the time of collection [10].

Typically, metaproteomics is carried out by bottom-up, or shotgun, proteomics. Environmental or microbiome samples are digested into peptides (usually with trypsin), the peptides separated by a liquid chromatography column, and then analyzed by inline tandem mass spectrometry. Because these peptides are ionizable and fragment in a predictable manner when using Collision Induced Dissociation (CID), the mass spectrum that results can be interpreted and the peptide sequence can be resolved. The identification of the peptide sequence relies on software (e.g., Comet [11], Sequest [12], Mascot [13], and X! Tandem [14]) that searches the masses of candidate ions against a sequence database. This sequence database may comprise gene products predicted by a metagenomic analysis, annotated reference proteomes of organisms likely to be present (e.g., sequences from NCBI nr [15] for human or specific bacteria), or a combination of the two [16,17]. Lists of tens of 1000's of peptides are generated in a single 60 minute mass spectrometry analysis. The identified peptides are used to predict which proteins from the sequence database are present in the sample [18,19]. Once a final protein list is predicted, the relative abundance of proteins may be estimated using spectral counting, which counts the number of times peptides that matched each protein were observed by the mass spectrometer. Spectral counts are then normalized using one of various methods [20–23]. For example, the Normalized Spectrum Abundance Factor (NSAF) [22] adjusts the spectral counts based on the total number of proteins identified in the sample and the respective lengths of those proteins. Finally, the normalized values, functional annotations and taxonomic assignments for the predicted protein lists are used to ascribe relative abundances to functions and taxa and these abundances are compared between samples.

However, metaproteomics datasets present unique challenges for which traditional protein inference and spectral counting methods, such as NSAF, were not designed. From the list of peptides, the standard protein inference methods predict which proteins are present by considering all identified peptides and correlating them back to the protein sequence database used to search the data. Parsimonious protein inference follows the Occam's razor principle, (e.g., if protein A or B can explain the occurrence of 3 peptides but protein B can also explain the occurrence of 6 other peptides, then it is probably true that protein B is present, not A) [24–26]. As a result of this simplification, protein inference is contentious within the mass spectrometry community, even for single organism studies [27,28]. In the case of microbiomes, this issue becomes far more complex because there are hundreds, possibly thousands, of species contributing to the peptides present. In many cases, when searching mass spectrometry data from microbiomes against a site-specific metaproteome database, a single peptide sequence might be found in more than 1000 of the proteins in that database. Upon close examination, it becomes apparent that that peptide, in many cases, is present in proteins that come from a wide variety of species. So the question becomes, how do you report a protein's presence when the peptide evidence suggests it is from 1000 different species? Deciding to include all proteins matched by peptides will almost certainly include too many and selecting only one to report has inherent bias. Further, because NSAF uses the final list of proteins to normalize the NSAF values for each protein, erroneously including too many or too few proteins may have a significant impact on these values. And, in the case where public databases are used in lieu of metagenomic-derived sequence databases, peptides are matched to proteins that may not be in the sample at all [16,17]. This further confounds NSAF, since the lengths of the proteins in the database are used to normalize the abundance value and the lengths of the protein sequences in the sample may differ from the lengths of proteins in the database. Given that (1) the mass spectrometry assay identifies peptides rather than proteins; (2) we are interested in characterization of functions and taxa rather than proteins; and (3) the list of inferred proteins may be unreliable, we contend that protein inference should be avoided altogether. Instead we pursue a peptide-centric approach.

Several tools for functional and taxonomic analysis of metaproteomes are currently used by the metaproteomics community. Though MEtaGenome ANalyzer, or MEGAN [29], was designed for metagenomics data, it can be used for metaproteomics analysis by substituting the expected genomic BLAST with protein BLAST. The visualization tools for MEGAN rely on RefSeq (NCBI) annotations and were designed for metagenomics data and utilizes full sequence information, such as an assembled gene or gene product (i.e., protein). MetaProteomeAnalyzer (MPA) [30] is a complete metaproteomics pipeline spanning initial database search through final visualization. To deal with the issues associated with protein inference-based analysis of function and taxonomy, MPA collapses highly-redundant protein hits into "metaproteins", whose annotations are derived from its component proteins. MPA then quantifies taxa and functions using spectral counting. Some features of MPA also assume UniProtKB was used to match spectra to peptides, and custom metagenome-derived databases may not be natively supported. Unipept [31], a web-based, peptide-centric metaproteomics application, provides advanced and beautiful data visualization tools but is currently limited to taxonomic analysis and cannot analyze peptides that are not present in UniProtKB.

The web-based MetaGOmics tool is designed to perform functional and taxonomic analysis without the need to predict which proteins are present in the sample. MetaGOmics works at the peptide level, by assigning functional and taxonomic annotations to each peptide based on the annotations of proteins matched in the sequence database used to search the data. Then the relative abundance for each of those annotations is incremented by the relative abundance for that peptide. In the case of spectral counting, this ensures each spectrum increases the spectral count for each functional or taxonomic annotation only once, which may not be the case if we assumed each protein matched by the peptide was in the sample. The MetaGOmics tool requires minimal input, only the sequence database (as a FASTA file) used to search the data and the list of identified peptide sequences and associated relative abundances (e.g., spectral counts). All subsequent processing, including any necessary sequence homology searching, are performed by the MetaGOmics servers. The web application includes tools to visualize, download, and compare data between experiments, including statistical tools for identifying GO annotations with statistically significant changes. MetaGOmics is open-source (https://github.com/metagomics/, Apache 2.0 license) and freely available to use at https://www.yeastrc.org/metagomics/.

2. Methods

2.1. Web Application Implementation

MetaGOmics is implemented as a database-backed web-application for submitting and visualizing results and a series of distributed server-side programs to run sequential parts of the pipeline on other servers as needed. The web application was developed using Java, HTML, CSS, SVG, and Javascript; and designed to run on the Apache Tomcat (http://tomcat.apache.org/) Java servlet container and the Struts application framework (http://struts.apache.org/). The relational database was developed using the MySQL (https://www.mysql.com/) relational database management system. The server-side programs are implemented in Java and execution of the programs is managed by the JobCenter [32] platform for managing distributed computational job execution. All source code for both the web application and server-side programs are available at https://github.com/metagomics.

2.2. MetaGOmics Algorithm

The MetaGOmics algorithm is designed to be applied to spectrum identifications resulting from standard bottom-up shotgun proteomics analysis of metaproteomics samples. The spectrum identifications may be derived from standard analysis workflows, such as those using Comet, Mascot, X! Tandem, or Sequest. There are no requirements that the protein sequence database comprise entries from any particular public database, and may be made from predictions resulting from, for example, metagenomic sequencing. The algorithm takes as input the list of identified peptides, the relative abundance for each peptide (e.g., spectral count), and the protein sequence database (as a FASTA file) used to search the data. Note that the uploaded FASTA file need only contain the proteins matched by any of the uploaded peptides and we recommend trimming very large protein databases in this way.

The first step of the algorithm estimates the relative abundance of protein functions based on the relative frequencies of peptides that can be ascribed to any protein with those functions (Figure 1a). This is done by first matching peptides to unannotated proteins in the FASTA file, where peptides must be N-terminal in the protein or C-terminal to a lysine or arginine (tryptic digestion is assumed) and leucine and isoleucine are treated interchangeably. Those proteins are then annotated by performing Basic Local Alignment Search Tool (BLAST) [33] on these protein sequences against a large, well-annotated protein sequence database (such as UniProtKB [34]) to find Gene Ontology (GO) [35,36] annotations. Then, for each peptide, a non-redundant directed acyclic graph (DAG) is constructed from all of the GO terms for all of the proteins matched by that peptide. Because of the hierarchical nature of GO, the peptide DAG includes both the GO terms with which proteins were directly annotated as well as all ancestor terms (using the "is a" relationship) to the root of the GO DAG. The spectral count for each GO term in this DAG is then incremented by the spectral count of the peptide that gave rise to it. After all peptides are processed, the final spectral count for each GO term is then normalized by dividing its count by the total number of peptide spectrum matches in the search. A table can be produced containing all GO terms found in the sample, the number of times peptides indicating those GO terms were observed (spectral count), and the proportion of all scans in the experiment that were attributable to that GO term (Table 1).

Table 1. Small subset of GO terms, spectral counts, and relative abundance ratio in a hypothetical mass spectrometry experiment.

GO Accession String	GO Aspect	GO Name	Spectral Count	Ratio
GO:0005575	cellular_component	cellular_component	12,217	1
GO:0008150	biological_process	biological_process	12,217	1
GO:0003674	molecular_function	molecular_function	12,217	1
unknownprc	biological_process	unknown biological process	5472	0.45
GO:0005488	molecular_function	binding	4185	0.34
GO:0097159	molecular_function	organic cyclic compound binding	3579	0.29
GO:1901363	molecular_function	heterocyclic compound binding	3579	0.29
GO:0005524	molecular_function	ATP binding	1712	0.14
GO:1901566	biological_process	organonitrogen compound biosynthetic process	1353	0.11
GO:0042026	biological_process	protein refolding	1145	0.09
GO:1990351	cellular_component	transporter complex	200	0.02

The second step of the algorithm estimates the relative taxonomic contribution to the number of peptide identifications for each GO term (Figure 1b). For a given GO term (e.g., "calcium ion binding"), this is done by first collating all the peptides that provided evidence for that GO term. Each of these peptides is matched to proteins in the FASTA database, and the UniProtKB BLAST hits are used to provide taxonomic classifications of the proteins. Then for each peptide, the lowest common ancestor (LCA) is found from all proteins matched by this peptide. This is the most specific taxonomic unit for which we can say this peptide provides unambiguous spectral evidence. For example, if the peptide matches proteins that each were found in species belonging to different orders of class Betaproteobacteria, we identify the class Betaproteobacteria as the LCA for which this peptide provides spectral evidence. We then increment the spectral count for the LCA and all of its parent taxa (e.g., Proteobacteria (phylum) and Bacteria (superkingdom) in the case of Betaproteobacteria (class)) by the spectral count for the peptide. After processing all peptides for a given GO term, a table is produced containing (1) all taxa that provided unambiguous evidence for that GO term; (2) the number of times peptides were observed from that taxon for this GO term; and (3) the proportion of all spectra for this GO term that are attributable to that taxon (Table 2).

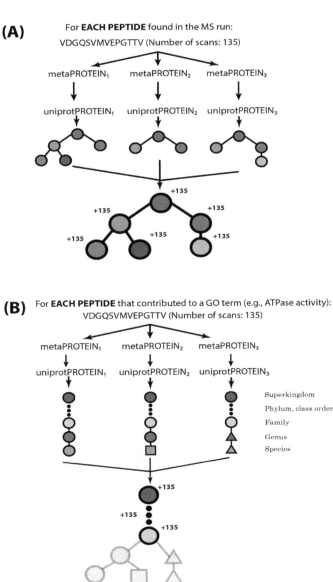

Figure 1. The MetaGOmics algorithm. **(A)** The first phase of the algorithm (functional analysis) examines all peptides identified in a mass spectrometry (MS) experiment. Each peptide is matched to proteins in the FASTA file (metaPROTEIN in figure), those are matched to UniProtKB proteins via BLAST (uniprotPROTEIN in figure), and Gene Ontology (GO) annotations for the UniProtKB proteins are used to create complete GO graphs for each protein containing direct annotations and all ancestor terms. All GO graphs from all proteins matched by a peptide are merged into a single, non-redundant GO graph (the union of the sets), and the spectral count of each term is increased by the spectral count for the peptide. This process is repeated for all peptides in the experiment to obtain final spectral counts for all GO terms; **(B)** The second phase of the algorithm, taxonomic analysis of functions, examines all peptides that are assigned a specific GO term. Each peptide is matched to a FASTA protein (metaPROTEIN in figure), the FASTA proteins are matched to UniProtKB proteins via BLAST (uniprotPROTEIN in figure), and taxonomic annotations for the UniProtKB proteins are used. A taxonomic tree is generated containing the direct taxonomic annotations and all ancestor terms. All taxonomic trees resulting from all matched proteins are merged such that the resulting tree contains only those terms present in all trees (the intersection of the sets). The taxonomic terms have their spectral count increased by the spectral count of the peptide. After all peptides assigned to a GO term are processed, the ratio of the spectral count of each taxonomic term to the total spectral count of the GO term is calculated. This provides the relative, unambiguous contribution (in spectral count) of each taxon to a GO term at any arbitrary level of the taxonomic tree.

Table 2. For a given GO term, the taxa, spectral count, fraction of this GO term's spectral count, and fraction of all spectra in the experiment that could be unambiguously attributed to each respective taxon. E.g., 88% of the spectra for this GO term were attributable to the Bacteroidetes phylum. 3.5% of the spectra in the experiment were attributable to this GO term and the Bacteriodetes phylum.

Taxon Name	Taxonomy Rank	Spectral Count	Ratio of GO	Ratio of Experiment
Bacteria	superkingdom	240	0.88	3.50×10^{-2}
Bacteroidia	class	141	0.52	2.05×10^{-2}
Bacteroidetes	phylum	141	0.52	2.05×10^{-2}
Bacteroidales	order	141	0.52	2.05×10^{-2}
Prevotella	genus	81	0.3	1.18×10^{-2}
Prevotellaceae	family	81	0.3	1.18×10^{-2}
Firmicutes	phylum	41	0.15	5.97×10^{-3}
Lactobacillales	order	33	0.12	4.81×10^{-3}
Lactobacillaceae	family	33	0.12	4.81×10^{-3}
Lactobacillus	genus	33	0.12	4.81×10^{-3}
Bacilli	class	33	0.12	4.81×10^{-3}
Prevotella sp. CAG:873	species	23	0.08	3.35×10^{-3}
Clostridiales	order	6	0.02	8.74×10^{-4}
Clostridia	class	6	0.02	8.74×10^{-4}
Actinobacteria	phylum	5	0.02	7.28×10^{-4}

The third step of the algorithm compares the spectral counts for GO terms between different experiments. This is done by first calculating the log-fold difference (base 2) between the proportions of spectra attributable to a given GO term in two experiments. Because a GO term might only be observed in one of the two experiments and the log-fold change using zero in one of the conditions is undefined, a Laplace correction is performed on the spectral counts for each GO term by adding one to the spectral count for all GO terms and recalculating the proportions. The Laplace correction is equivalent to assigning a uniform prior probability to each GO term. The log-fold change is then calculated between these Laplace-corrected proportions. A *p*-value is calculated using a two-tailed test of proportions that tests the null hypothesis that the proportion for this GO term is the same in the two experiments against the alternative hypothesis that the ratios are different. Multiple hypothesis testing is controlled for using either a Bonferroni correction or a Benjamini-Hochberg adjustment [37,38] (Table 3). At this time, MetaGOmics only supports pairwise comparisons between two experiments, though the data from multiple experiments may be downloaded and compared using any preferred analysis method.

Table 3. For the comparison of two hypothetical MS experiments, a small subset of the GO terms, log-fold changes, and *q*-values for GO terms detected in the two experiments.

GO Name	Fold Change	*q*-Value
outer membrane	1.55	5.27×10^{-106}
cell outer membrane	1.55	5.61×10^{-106}
external encapsulating structure part	1.5	5.64×10^{-102}
membrane	1.14	3.00×10^{-101}
receptor activity	1.47	5.03×10^{-93}
intrinsic component of membrane	1.44	6.37×10^{-88}
integral component of membrane	1.44	6.37×10^{-88}
molecular transducer activity	1.35	4.14×10^{-81}
membrane part	1.01	4.69×10^{-53}
carbohydrate derivative binding	−2.03	1.25×10^{-49}
ribonucleotide binding	−2.03	1.25×10^{-49}
purine ribonucleoside binding	−2.04	5.36×10^{-47}
ribonucleoside binding	−2.04	5.36×10^{-47}
purine ribonucleoside triphosphate binding	−2.04	5.36×10^{-47}

2.3. Specifying Relative Abundance

The required inputs to use MetaGOmics include the protein database used to search the MS data and a peptide list with associated abundance measures (reported as integers). For the sake of simplicity, this manuscript focuses on relative abundances reported as spectral counts. However, MetaGOmics is not limited to spectral counting. This abundance can just as easily be the total area under the curve from the extracted ion chromatogram for each peptide. MetaGOmics performs its analysis in terms of relative abundances in each sample, not absolute abundances. In the case of areas under the curve, the abundance of a given peptide would be considered as the proportion of its area under the curve to the total areas under the curves for all peptides. It is this proportion that would be compared between samples. Examining the changes in functional and taxonomic abundance between samples in terms of changes in absolute abundance is a current focus of development, and functionality specifically designed to handle this case will be added to the MetaGOmics website soon.

2.4. Running BLAST

NCBI BLAST is performed as-needed on behalf of users to find Gene Ontology and taxonomic annotations for proteins in their protein sequence database. Users may select the BLAST database (UniProtKB TrEMBL or Swiss-Prot [39]), E-value cutoff (currently defaults to 1×10^{-10}), and whether or not to only use the top BLAST hit (currently defaults to true). If only the top BLAST hit is used, then all BLAST hits tied for the best score will be used (if the best score is less than or equal to the E-value cutoff). If not only the top BLAST hit is used, then all hits with an E-value meeting the cutoff are used. At the time of this writing, MetaGOmics runs NCBI BLAST 2.6.0+ on two Linux servers, each with 56-core Intel Xeon E5-2697 CPUs.

2.5. Expected Wait Times

The time required to process the initial FASTA upload ranges from a few minutes to a few hours, depending on the size of the FASTA file. Subsequent uploads of the same FASTA file will be instantaneous. To speed up processing, it is highly recommended that this FASTA file be trimmed to include only proteins represented by peptide matches in the results. Please see our GitHub repository for a program to aid in this trimming (https://github.com/metagomics/).

The time required to process uploaded peptide lists and associated spectral counts ranges from a few minutes to more than a day. This time depends on the number of peptides, whether BLAST has already been run on some proteins from this FASTA file from previous analyses, the size of the FASTA file, whether BLAST is searching UniProtKB TrEMBL or Swiss-Prot, and whether or not results beyond the top BLAST hit are being used. Using UniProtKB TrEMBL and retaining more than the top BLAST hit will result in longer processing times.

In all cases, the expected run times may be significantly impacted by user demand. All requests are processed on a first come, first served basis.

2.6. Unknown Gene Ontology Annotations

In the case where no GO annotation is found for a peptide, an "unknown" annotation is created and added as a direct child of the root node of the relevant aspect of the GO DAG. For example, if no annotations were associated with a peptide for the "molecular function" aspect, then an "unknown molecular function" node is added as a child of the "molecular_function" root node. The spectral count for this unknown annotation (and the "molecular_function" root node) is thus increased by the spectral count for that peptide and is included in reported data.

2.7. Analysis of Ocean Metaproteomics Dataset

An example MetaGOmics result set was generated by comparing Bering Strait (BSt) surface water (7 m) samples to Chukchi Sea (CS) bottom water (55.5 m) samples. Briefly, microbial fractions were collected on 0.7 μm filters after passing 15 L of seawater through both 10 μm and 1 μm filters to remove larger eukayrotes. These samples were collected, digested, analyzed on a Q-Exactive-HF, and data were searched according to the methods description in May et al. [40].

Separate results files were created for each of the BSt and CS conditions that contained the combined Comet results for the two replicates from each condition. The BSt and CS results were each post-processed using Percolator [41] version 2.10 with the -X option. Lists of identified peptides and spectral counts were parsed from the resulting Percolator XML files using an in-house script, filtering peptide identifications on a q-value of 0.01 and then using peptide spectrum matches (PSMs) filtered on a q-value of 0.01 for spectral counts. The metaproteome FASTA file (described by May et al.) was filtered using an in-house script to only include protein sequences for which there were peptide matches. Note that this step is not required, but speeds up processing. Both of these scripts are available at our GitHub repository at https://github.com/metagomics/. The resulting list of peptides, associated spectral counts, and filtered FASTA file were uploaded to the MetaGOmics server, with settings to use UniProtKB TrEMBL, an E-value cutoff of 1×10^{-10}, and only keep the top hit as BLAST settings. The results of the re-analysis are available at https://www.yeastrc.org/metagomics/ocean.

3. Results and Discussion

3.1. Web Application

The MetaGOmics web application is available at https://www.yeastrc.org/metagomics/. The user is presented with a simple interface for creating an initial context for the analysis of metaproteomics data (Figure 2a). To create this context the user (1) uploads a FASTA file containing the protein sequences to which peptides identified in any of the experiments are to be matched (e.g., the FASTA file used to search the data); (2) selects a database against which proteins will be searched by BLAST (i.e., UniProtKB/Swiss-Prot or UniProtKB/TrEMBL); (3) chooses which BLAST matches should be considered; and (4) enters an email address where they may be contacted when results are ready.

Upon submitting this form, a unique URL is created where users may request analysis of peptide lists, visualize results, and download text reports from different experiments based on the submitted FASTA file and requested BLAST settings (Figure 2b). This URL contains an unguessable hash string as part of the URL that serves to secure the data. Users will receive an email containing a link to the private URL after submitting the FASTA file. This URL should be retained or recorded. The links do not expire.

To submit data, the user selects "Upload Peptide Count List" and submits a tab-delimited text file containing the peptide list and associated spectral counts from a given experiment. Submitting this file initiates the sequential running of multiple server-side programs to perform the analysis. Once the analysis is complete, the user will receive an email notification and another link to the private URL, which allows for downloading and visualizing the result of this analysis. Users may submit the results of as many experiments as they would like to be analyzed using this FASTA file and BLAST settings, such as the results of each condition or replicates.

To view the results, users click the "Download GO Analysis" button next to the nickname they gave a set of results. This opens an overlay where users may choose to download the results as a text report or as an image of the GO DAG. The text report contains the GO term accession number, GO aspect, GO name, spectral count, and proportion of all spectra in the experiment for each GO term found in the experiment (spectral ratio). The image contains a graphical representation of the GO terms in their hierarchical structure, labeled with their spectral count, spectral ratio, and shaded according to the spectral ratio. The images may be downloaded in portable network graphics (PNG) or scalable vector graphics (SVG) formats.

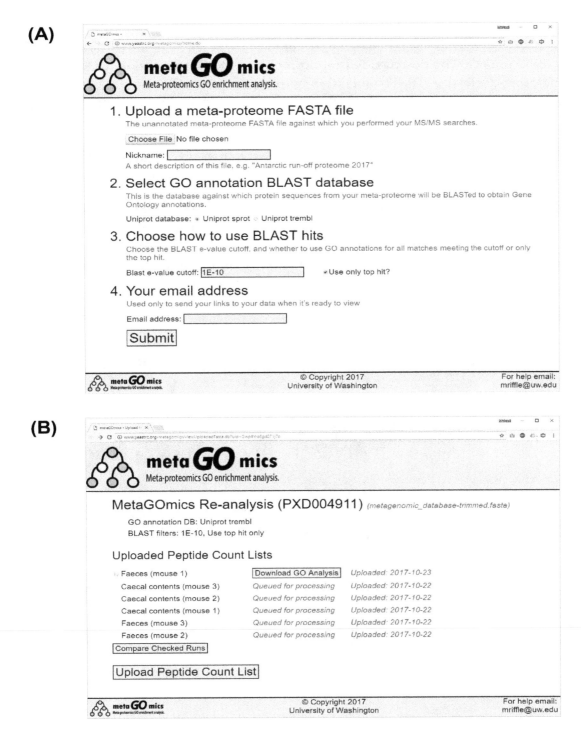

Figure 2. Screenshots from the MetaGOmics web application. (**A**) A user fills out an initial form to create a context for MetaGOmics analysis. The user (1) uploads a FASTA file containing a database of protein sequences to which peptides should be matched, (2) selects a BLAST database to use for protein annotations, (3) chooses cutoffs for the BLAST hits, and (4) enters an email address to be notified when processing is complete; (**B**) After a user submits the form in part (**A**), a unique URL is created for a page where a user may perform MetaGOmics analysis using the desired FASTA file, BLAST database, and BLAST cutoff settings for all uploaded data. To upload data for analysis, the user clicks "Upload Peptide Count List" to upload a text file containing peptide sequences and spectral counts. Each row under "Uploaded Peptide Count Lists" shows each requested analysis and its current status. Upon completion, users may click the "Download GO Analysis" button to download the results as text reports or images. Two analyses may be compared by clicking the checkbox next to two rows and clicking "Compare Checked Runs." Comparisons may also be downloaded as text reports or as images.

If multiple experiments have been analyzed, users may check the box next to any two and click the "Compare Checked Runs" to download a text report or images comparing the ratios of GO terms in the two experiments. The downloaded report includes the GO accession, GO aspect, GO name, ratios from both experiments, PSM counts from both experiments, the log (base 2) fold difference in the ratios, corrected p-values, and q-values (Benjamini-Hochberg adjustment) for all GO terms found in either experiment. Instead of all GO terms from either experiment, the images contain a "trimmed" depiction of the identified GO DAG that is color coded according to statistical significance (Figure 3). Trimming is accomplished by iteratively removing all leaves from the DAG that have a q-value greater than 0.01. The result is a DAG where all leaves have a q-value less than or equal to 0.01; note that parents of these nodes may have non-significant q-values. This is done to preserve the structure of the DAG, including the relative level of specificity of GO terms in the DAG, while still removing much of the noise of insignificant results. GO terms with higher ratios in the second experiment are shaded yellow, and terms with lower ratios are shaded blue, where the intensity of each color depends on the significance of the q-value. Grey terms are not statistically significant. Each GO term is labeled with its name, accession number, log fold change, and q-value. The images may be downloaded as PNG or SVG.

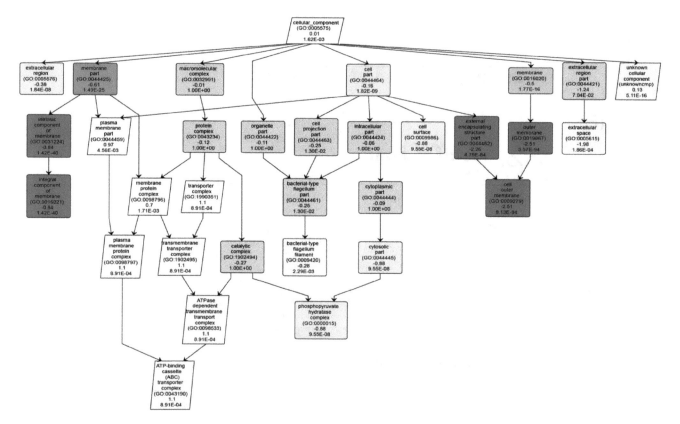

Figure 3. An example GO graph generated by comparing two experiments using the MetaGOmics server. Each node is a GO term, with lines indicating edges between those nodes in the GO structure. Each node is labeled with the name of the GO term, the log fold change, and the q-value. Nodes with a positive log-fold change are shaped as parallelograms, and shaded yellow—where darker shades of yellow indicate more significant q-values. Nodes with a negative log-fold change are shaped as rectangles, with shades of blue indicating q-value significance. Grey terms are not statistically significant. In this example, GO terms with the "cellular component" aspect were compared. The ratio of spectra in the second experiment matching proteins that localized to the outer cell membrane, extracellular space, phosphopyruvate hydralase complex, and integral component of the membrane were significantly reduced. Whereas, the ratio of spectra matching proteins with an unknown cellular component and ATP-binding (ABC) transporter complex were increased.

3.2. Example Analysis: Ocean Metaproteomics

An example dataset was generated via a MetaGOmics analysis of the ocean metaproteomics dataset described in May et al. [40] (see methods) (Figure 4). These data compare the relative abundance of peptides attributable to GO terms between Bering Strait surface water (7 m) samples to Chukchi Sea bottom water (55.5 m) samples. A highly-filtered set of results from the analysis is presented in Table 4, where only the statistically significant leaves of the resulting GO DAG are presented. The surface samples have a higher relative abundance of peptides relating to metabolism, translation, GTP utilization, and photosynthesis. Where, the bottom samples have a higher relative abundance of peptides relating to metal binding, protein refolding, DNA repair, and ATP utilization. These results are consistent with what would generally be expected in surface and sea bottom ocean samples. These results may be viewed in their entirety at https://www.yeastrc.org/metagomics/ocean.

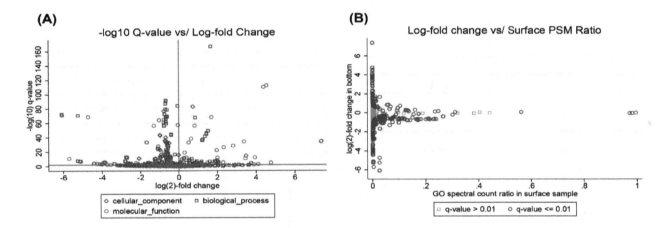

Figure 4. GO term statistics produced by a MetaGOmics analysis comparing ocean surface to bottom water samples from May et al. All data from the analysis are available at https://www.yeastrc.org/metagomics/ocean. (**A**) Volcano plot depicting the negative log (base 10) of the q-value versus the log (base 2) fold change for all GO terms found in either sample. A horizontal reference line is added for a q-value cutoff of 0.01. A vertical reference line is added for no change. Each point is a GO term and has been colored and re-shaped according to its GO aspect; (**B**) A scatter plot depicting the log (base 2) fold change from the surface sample to the bottom sample for each GO term versus that GO term's spectral count ratio in the surface sample. Each GO term has been colored either lavender (not statistically significant) or red (q-value ≤ 0.01).

Table 4. Up to the top 10 leaf GO terms with a q-value $\leq$ 0.01 for positive and negative log-fold changes comparing ocean water samples from BSt (Surface) to CS (Bottom) from May et al. Shown are the name of the GO term, the log-fold change from surface to bottom samples, and the q-value resulting from the Benjamini-Hochberg adjustment.

Biological Process

Higher in Ocean Surface Water			Higher in Ocean Bottom Water		
GO Term	log Change	q-Value	GO Term	log Change	q-Value
D-xylose transport	-6.09	3.49×10^{-73}	protein refolding	1.66	1.01×10^{-167}
translation	-0.77	2.43×10^{-57}	chromosome condensation	1.52	3.69×10^{-50}
translational elongation	-1.11	9.60×10^{-26}	DNA repair	1.61	1.82×10^{-7}
transcription anti-termination	-2.79	5.66×10^{-8}	dephosphorylation	2.04	6.62×10^{-7}
fatty acid biosynthetic process	-1.21	6.42×10^{-8}	de novo' pyrimidine nucleobase biosynthetic process	3.74	4.53×10^{-5}
GTP biosynthetic process	-5.12	7.30×10^{-8}	RNA phosphodiester bond hydrolysis, exonucleolytic	1	5.52×10^{-5}
UTP biosynthetic process	-5.12	7.30×10^{-8}	mRNA catabolic process	0.93	1.69×10^{-4}
CTP biosynthetic process	-5.12	7.30×10^{-8}	7,8-dihydroneopterin 3'-triphosphate biosynthetic process	3.09	4.48×10^{-3}
tricarboxylic acid cycle	-3.84	4.41×10^{-8}	response to cadmium ion	1.45	7.42×10^{-3}
cell division	-1.07	1.18×10^{-5}			

Molecular Function

Higher in Ocean Surface Water			Higher in Ocean Bottom Water		
GO Term	log Change	q-Value	GO Term	log Change	q-Value
monosaccharide binding	-6.07	6.71×10^{-72}	histidine ammonia-lyase activity	4.54	1.93×10^{-113}
receptor activity	-1.03	5.77×10^{-65}	unfolded protein binding	0.83	2.85×10^{-57}
structural constituent of ribosome	-0.68	1.12×10^{-34}	nitrate reductase activity	7.35	4.13×10^{-35}
DNA-directed RNA polymerase activity	-1.68	4.06×10^{-34}	heme binding	3.38	4.23×10^{-35}
translation elongation factor activity	-1.09	7.50×10^{-25}	ATP binding	0.51	6.87×10^{-34}
GTP binding	-0.92	6.40×10^{-17}	4 iron, 4 sulfur cluster binding	2.33	3.28×10^{-27}
GTPase activity	-0.92	8.64×10^{-7}	prephenate dehydratase activity	3.94	4.38×10^{-13}
nucleoside diphosphate kinase activity	-5.11	1.28×10^{-7}	selenium binding	2	8.97×10^{-13}
tRNA binding	-0.87	3.00×10^{-6}	4-phytase activity	3.45	7.83×10^{-73}
acetyl-CoA carboxylase activity	-3.91	3.46×10^{-6}	formate dehydrogenase (NAD+) activity	1.48	9.79×10^{-6}

Table 4. *Cont.*

Cellular Component

Higher in Ocean Surface Water			Higher in Ocean Bottom Water		
GO Term	log Change	q-Value	GO Term	log Change	q-Value
cell outer membrane	−0.98	1.17×10^{-43}	cytoplasm	0.74	8.53×10^{-84}
intracellular	−0.71	1.09×10^{-36}	bacterial-type flagellum filament	3.49	1.47×10^{-15}
ribosome	−0.64	1.78×10^{-33}	bacterial-type flagellum	2.01	1.53×10^{-8}
integral component of membrane	−0.98	9.77×10^{-24}	unknown cellular component	0.07	1.04×10^{-6}
thylakoid	−2.08	1.07×10^{-11}	ATP-binding cassette (ABC) transporter complex	0.6	1.04×10^{-5}
large ribosomal subunit	−1.21	1.54×10^{-11}	cytosolic small ribosomal subunit	3.66	9.19×10^{-3}
acetyl-CoA carboxylase complex	−3.84	4.22×10^{-6}			
plasma membrane	−0.29	3.37×10^{-4}			
pyruvate dehydrogenase complex	−3.99	7.77×10^{-4}			
proton-transporting ATP synthase complex, catalytic core F(1)	−0.37	1.36×10^{-3}			

3.3. Interpreting Results With Many "Unknown" GO Annotations

It is important to pay close attention to the "unknown" GO terms when interpreting changes in relative abundance for GO terms between samples, particularly more general GO terms. A peptide will receive an "unknown" GO annotation for two reasons: there were no BLAST matches for that peptide that met the E-value cutoff or there were no GO annotations for any of the UniProtKB proteins matched via BLAST for the respective GO aspect (molecular function, biological process, or cellular component). In either of these cases, the spectral count of the "unknown" GO term for the respective GO aspect is increased by the spectral count for the peptide. However, it is important to consider that if we did know the true annotations for the unknown peptides, it is likely that many of them would be annotated with a molecular function, biological process, or cellular component that falls under the most general GO terms.

For example, a peptide has a spectral count of 100, but no GO annotation for molecular function can be found. The spectral count for the "unknown molecular function" is increased by 100. However, if the GO annotation were known, it is quite likely it would fall under "binding", "catalytic activity", "structural molecule activity", or some other general term for molecular function. In experiments with a large ratio of unknown molecular functions, it is likely that the ratios of the general GO terms for molecular functions are really higher than reported. As such, interpretations of changes in relative abundance of general GO terms should be treated with care when there is a large ratio of "unknown" GO annotations in one of the experiments.

3.4. Interpreting Taxonomic Changes

When a taxonomic tree is calculated for a peptide, only the taxa that are unambiguously matched by that peptide have their spectral counts incremented. For example, if a peptide matches two proteins with the same species, the taxonomic node for that species and all of its ancestors (genus, family, order, and so on to kingdom) have their spectral counts increased by the spectral count for that peptide. If a peptide matches two proteins with different species, but the same genus, we consider that genus to be the most specific taxonomic unit for which this peptide provides unambiguous evidence. Consequently, if a peptide matches multiple proteins in different families, but with the same class, then that class is the most specific taxonomic unit for this peptide. Only the spectral counts for the most specific, shared taxonomic unit for a peptide (and all of its ancestors to the root) are increased by that peptide.

The ramifications of this approach are that more specific levels of the taxonomic tree (i.e., species and genus) may have fewer total spectral counts than more general levels of the taxonomic tree. A peptide that matches a very taxonomically homogeneous set of proteins is able to provide unambiguous evidence for more specific taxa than a peptide that matches a very taxonomically diverse set of proteins. This may be seen in the data by summing the fraction of a GO term's spectral count that is unambiguously attributable to all the taxa at different levels in the taxonomic tree. For example, when summing the fraction at a general taxonomic level (e.g., order), the fractions associated with all the orders may approach one. This means that nearly all the peptides provided unambiguous taxonomic evidence to at least the order level of the taxonomic tree. On the other hand, if we pick a more granular level (e.g., genus), then the fraction may be considerably lower, such as 0.25. This would indicate that only 25% of the spectra could be unambiguously assigned to a specific genus. Each step up in the taxonomic tree (to a more general term) will have a summed fraction greater than or equal to the step below.

We recommend that when interpreting the taxonomic contributions to a GO term's spectral count, that the most specific taxonomic level that sums nearly to one be chosen. This is the most specific level of unambiguous taxonomic inference that makes use of most of the spectra.

3.5. Current Usage

An early implementation of the functional analysis and experimental comparison steps of the MetaGOmics algorithm (implemented then as in-house scripts) was used to perform a GO-based functional analysis for Timmins-Schiffman, et al. (2017) [17]. This study aimed to characterize the effect of the chosen protein sequence database on the peptide yield and biological inferences that are made from environmental metaproteomics data. It was found that using the metagenome-derived metaproteome yielded a larger number of confident peptide identifications versus protein sequences constructed from existing public databases. It was also found that the choice of sequence database had a profound effect on the functional annotation of the experiment, which could lead to profoundly different biological conclusions.

4. Conclusions

As the field of metaproteomics grows, standardized approaches to data analysis must be established to allow comparisons between treatments (e.g., gut microbiomes) or environmental locations (e.g., ocean transects). In addition to the research field having a desire to compare microbiome proteomes between treatments or sites, temporal shifts in community structure and function are paramount to creating accurate models for predicting efficacy of treatments or, in the case of the environment, tracking the effects of global climate change or anthropogenic perturbations. The complexity of peptide sequence assignment in mixed community proteomics cannot be simplified by assuming protein inference yields accurate depictions of the community.

Here we present MetaGOmics, an algorithm and web application for peptide-centric functional and taxonomic analysis of metaproteomics samples, and comparisons between those samples. The MetaGOmics algorithm is designed to overcome drawbacks implicit in a protein-centric approach, in which spectral counting of proteins is used as the basis for functional analysis. Because MetaGOmics requires as input only a list of peptide sequences and associated abundances, the method works with data from any shotgun proteomics pipeline. The web application includes tools for submitting data, viewing results, and downloading reports. MetaGOmics is open-source (https://github.com/metagomics/) and free to use at https://www.yeastrc.org/metagomics/.

Acknowledgments: This work is supported by (NSF-OCE 1633939) awarded to B.L. Nunn and W.S. Noble, as well as by the University of Washington's Proteomics Resource (UWPR95794). M. Mikan was supported by NSF-OCE grant 1636045.

Author Contributions: M.R. designed the algorithm, developed the web application and processing pipeline, and wrote the paper. D.H.M. provided statistical input, and implemented code necessary for testing. E.T.-S. and M.P.M. provided scientific input and generated data for algorithm development and testing. D.J. wrote code to implement data processing in JobCenter. W.S.N. helped direct the project and provided statistical and algorithmic direction. B.L.N. conceived of and directed the project.

References

1. Sunagawa, S.; Coelho, L.P.; Chaffron, S.; Kultima, J.R.; Labadie, K.; Salazar, G.; Djahanschiri, B.; Zeller, G.; Mende, D.R.; Alberti, A.; et al. Structure and function of the global ocean microbiome. *Science* **2015**, *348*, 1261359. [CrossRef] [PubMed]

2. Group, N.H.W.; Peterson, J.; Garges, S.; Giovanni, M.; McInnes, P.; Wang, L.; Schloss, J.A.; Bonazzi, V.; McEwen, J.E.; Wetterstrand, K.A.; et al. The NIH human microbiome project. *Genome Res.* **2009**, *19*, 2317–2323. [CrossRef] [PubMed]

3. Morris, R.M.; Nunn, B.L.; Frazar, C.; Goodlett, D.R.; Ting, Y.S.; Rocap, G. Comparative metaproteomics reveals ocean-scale shifts in microbial nutrient utilization and energy transduction. *ISME J.* **2010**, *4*, 673–685. [CrossRef] [PubMed]

4.   Oulas, A.; Pavloudi, C.; Polymenakou, P.; Pavlopoulos, G.A.; Papanikolaou, N.; Kotoulas, G.; Arvanitidis, C.; Iliopoulos, I. Metagenomics: Tools and insights for analyzing next-generation sequencing data derived from biodiversity studies. *Bioinform. Biol. Insights* **2015**, *9*, 75–88. [CrossRef] [PubMed]

5.   Quince, C.; Walker, A.W.; Simpson, J.T.; Loman, N.J.; Segata, N. Shotgun metagenomics, from sampling to analysis. *Nat. Biotechnol.* **2017**, *35*, 833–844. [CrossRef] [PubMed]

6.   Jones, O.A.H.; Maguire, M.L.; Griffin, J.L.; Dias, D.A.; Spurgeon, D.J.; Svendsen, C. Metabolomics and its use in ecology. *Austral Ecol.* **2013**, *38*, 713–720. [CrossRef]

7.   Bashiardes, S.; Zilberman-Schapira, G.; Elinav, E. Use of metatranscriptomics in microbiome research. *Bioinform. Biol. Insights* **2016**, *10*, 19–25. [CrossRef] [PubMed]

8.   Haider, S.; Pal, R. Integrated analysis of transcriptomic and proteomic data. *Curr. Genom.* **2013**, *14*, 91–110. [CrossRef] [PubMed]

9.   Maier, T.; Guell, M.; Serrano, L. Correlation of mrna and protein in complex biological samples. *FEBS Lett.* **2009**, *583*, 3966–3973. [CrossRef] [PubMed]

10.   Petriz, B.A.; Franco, O.L. Metaproteomics as a complementary approach to gut microbiota in health and disease. *Front. Chem.* **2017**, *5*. [CrossRef] [PubMed]

11.   Eng, J.K.; Jahan, T.A.; Hoopmann, M.R. Comet: An open-source ms/ms sequence database search tool. *Proteomics* **2013**, *13*, 22–24. [CrossRef] [PubMed]

12.   Eng, J.K.; McCormack, A.L.; Yates, J.R. An approach to correlate tandem mass spectral data of peptides with amino acid sequences in a protein database. *J. Am. Soc. Mass Spectrom.* **1994**, *5*, 976–989. [CrossRef]

13.   Perkins, D.N.; Pappin, D.J.C.; Creasy, D.M.; Cottrell, J.S. Probability-based protein identification by searching sequence databases using mass spectrometry data. *Electrophoresis* **1999**, *20*, 3551–3567. [CrossRef]

14.   Craig, R.; Beavis, R.C. Tandem: Matching proteins with tandem mass spectra. *Bioinformatics* **2004**, *20*, 1466–1467. [CrossRef] [PubMed]

15.   Coordinators, N.R. Database resources of the national center for biotechnology information. *Nucleic Acids Res.* **2017**, *45*, D12–D17.

16.   Tanca, A.; Palomba, A.; Fraumene, C.; Pagnozzi, D.; Manghina, V.; Deligios, M.; Muth, T.; Rapp, E.; Martens, L.; Addis, M.F.; et al. The impact of sequence database choice on metaproteomic results in gut microbiota studies. *Microbiome* **2016**, *4*, 51. [CrossRef] [PubMed]

17.   Timmins-Schiffman, E.; May, D.H.; Mikan, M.; Riffle, M.; Frazar, C.; Harvey, H.R.; Noble, W.S.; Nunn, B.L. Critical decisions in metaproteomics: Achieving high confidence protein annotations in a sea of unknowns. *ISME J.* **2017**, *11*, 309–314. [CrossRef] [PubMed]

18.   Nesvizhskii, A.I.; Aebersold, R. Interpretation of shotgun proteomic data: The protein inference problem. *Mol. Cell. Proteom.* **2005**, *4*, 1419–1440. [CrossRef] [PubMed]

19.   Rappsilber, J.; Mann, M. What does it mean to identify a protein in proteomics? *Trends Biochem. Sci.* **2002**, *27*, 74–78. [CrossRef]

20.   Griffin, N.M.; Yu, J.Y.; Long, F.; Oh, P.; Shore, S.; Li, Y.; Koziol, J.A.; Schnitzer, J.E. Label-free, normalized quantification of complex mass spectrometry data for proteomic analysis. *Nat. Biotechnol.* **2010**, *28*, 83–89. [CrossRef] [PubMed]

21.   Ishihama, Y.; Oda, Y.; Tabata, T.; Sato, T.; Nagasu, T.; Rappsilber, J.; Mann, M. Exponentially modified protein abundance index (empai) for estimation of absolute protein amount in proteomics by the number of sequenced peptides per protein. *Mol. Cell. Proteom.* **2005**, *4*, 1265–1272. [CrossRef] [PubMed]

22.   Paoletti, A.C.; Parmely, T.J.; Tomomori-Sato, C.; Sato, S.; Zhu, D.X.; Conaway, R.C.; Conaway, J.W.; Florens, L.; Washburn, M.P. Quantitative proteomic analysis of distinct mammalian mediator complexes using normalized spectral abundance factors. *Proc. Natl. Acad. Sci. USA* **2006**, *103*, 18928–18933. [CrossRef] [PubMed]

23.   Zhang, Y.; Wen, Z.H.; Washburn, M.P.; Florens, L. Refinements to label free proteome quantitation: How to deal with peptides shared by multiple proteins. *Anal. Chem.* **2010**, *82*, 2272–2281. [CrossRef] [PubMed]

24.   Li, Y.F.G.; Arnold, R.J.; Li, Y.X.; Radivojac, P.; Sheng, Q.H.; Tang, H.X. A bayesian approach to protein inference problem in shotgun proteomics. *J. Comput. Biol.* **2009**, *16*, 1183–1193. [CrossRef] [PubMed]

25.   Nesvizhskii, A.I.; Keller, A.; Kolker, E.; Aebersold, R. A statistical model for identifying proteins by tandem mass spectrometry. *Anal. Chem.* **2003**, *75*, 4646–4658. [CrossRef] [PubMed]

26.   Zhang, B.; Chambers, M.C.; Tabb, D.L. Proteomic parsimony through bipartite graph analysis improves accuracy and transparency. *J. Proteome Res.* **2007**, *6*, 3549–3557. [CrossRef] [PubMed]

27. Serang, O.; Moruz, L.; Hoopmann, M.R.; Kall, L. Recognizing uncertainty increases robustness and reproducibility of mass spectrometry-based protein inferences. *J. Proteome Res.* **2012**, *11*, 5586–5591. [CrossRef] [PubMed]

28. Audain, E.; Uszkoreit, J.; Sachsenberg, T.; Pfeuffer, J.; Liang, X.; Hermjakob, H.; Sanchez, A.; Eisenacher, M.; Reinert, K.; Tabb, D.L.; et al. In-depth analysis of protein inference algorithms using multiple search engines and well-defined metrics. *J. Proteom.* **2017**, *150*, 1701–1782. [CrossRef] [PubMed]

29. Huson, D.H.; Auch, A.F.; Qi, J.; Schuster, S.C. Megan analysis of metagenomic data. *Genome Res.* **2007**, *17*, 377–386. [CrossRef] [PubMed]

30. Muth, T.; Behne, A.; Heyer, R.; Kohrs, F.; Benndorf, D.; Hoffmann, M.; Lehteva, M.; Reichl, U.; Martens, L.; Rapp, E. The metaproteomeanalyzer: A powerful open-source software suite for metaproteomics data analysis and interpretation. *J. Proteome Res.* **2015**, *14*, 1557–1565. [CrossRef] [PubMed]

31. Mesuere, B.; Devreese, B.; Debyser, G.; Aerts, M.; Vandamme, P.; Dawyndt, P. Unipept: Tryptic peptide-based biodiversity analysis of metaproteome samples. *J. Proteome Res.* **2012**, *11*, 5773–5780. [CrossRef] [PubMed]

32. Jaschob, D.; Riffle, M. Jobcenter: An open source, cross-platform, and distributed job queue management system optimized for scalability and versatility. *Source Code Biol. Med.* **2012**, *7*, 8. [CrossRef] [PubMed]

33. Altschul, S.F.; Gish, W.; Miller, W.; Myers, E.W.; Lipman, D.J. Basic local alignment search tool. *J. Mol. Biol.* **1990**, *215*, 403–410. [CrossRef]

34. Chen, C.; Huang, H.; Wu, C.H. Protein bioinformatics databases and resources. *Methods Mol. Biol.* **2017**, *1558*, 33–39.

35. Ashburner, M.; Ball, C.A.; Blake, J.A.; Botstein, D.; Butler, H.; Cherry, J.M.; Davis, A.P.; Dolinski, K.; Dwight, S.S.; Eppig, J.T.; et al. Gene ontology: Tool for the unification of biology. The gene ontology consortium. *Nat. Genet.* **2000**, *25*, 25–29. [CrossRef] [PubMed]

36. The Gene Ontology Consortium. Expansion of the gene ontology knowledgebase and resources. *Nucleic Acids Res.* **2017**, *45*, D331–D338.

37. Benjamini, Y.; Hochberg, Y. Controlling the false discovery rate—A practical and powerful approach to multiple testing. *J. Roy. Stat. Soc. B Methodol.* **1995**, *57*, 289–300.

38. Yekutieli, D.; Benjamini, Y. Resampling-based false discovery rate controlling multiple test procedures for correlated test statistics. *J. Stat. Plan. Inference* **1999**, *82*, 171–196. [CrossRef]

39. O'Donovan, C.; Martin, M.J.; Gattiker, A.; Gasteiger, E.; Bairoch, A.; Apweiler, R. High-quality protein knowledge resource: Swiss-prot and trembl. *Brief. Bioinform.* **2002**, *3*, 275–284. [CrossRef] [PubMed]

40. May, D.H.; Timmins-Schiffman, E.; Mikan, M.P.; Harvey, H.R.; Borenstein, E.; Nunn, B.L.; Noble, W.S. An alignment-free "metapeptide" strategy for metaproteomic characterization of microbiome samples using shotgun metagenomic sequencing. *J. Proteome Res.* **2016**, *15*, 2697–2705. [CrossRef] [PubMed]

41. Kall, L.; Canterbury, J.D.; Weston, J.; Noble, W.S.; MacCoss, M.J. Semi-supervised learning for peptide identification from shotgun proteomics datasets. *Nat. Methods* **2007**, *4*, 923–925. [CrossRef] [PubMed]

4

Biochemical and Computational Insights on a Novel Acid-Resistant and Thermal-Stable Glucose 1-Dehydrogenase

Haitao Ding [1,*], Fen Gao [2], Yong Yu [1] and Bo Chen [1]

[1] Key Laboratory for Polar Science of State Oceanic Administration, Polar Research Institute of China, Shanghai 200136, China; yuyong@pric.org.cn (Y.Y.); chenbo@pric.org.cn (B.C.)

[2] East China Sea Fisheries Research Institute, Shanghai 200090, China; gaofen2011@163.com

* Correspondence: dinghaitao@pric.org.cn

Academic Editor: Quan Zou

Abstract: Due to the dual cofactor specificity, glucose 1-dehydrogenase (GDH) has been considered as a promising alternative for coenzyme regeneration in biocatalysis. To mine for potential GDHs for practical applications, several genes encoding for GDH had been heterogeneously expressed in *Escherichia coli* BL21 (DE3) for primary screening. Of all the candidates, GDH from *Bacillus* sp. ZJ (BzGDH) was one of the most robust enzymes. BzGDH was then purified to homogeneity by immobilized metal affinity chromatography and characterized biochemically. It displayed maximum activity at 45 °C and pH 9.0, and was stable at temperatures below 50 °C. BzGDH also exhibited a broad pH stability, especially in the acidic region, which could maintain around 80% of its initial activity at the pH range of 4.0–8.5 after incubating for 1 hour. Molecular dynamics simulation was conducted for better understanding the stability feature of BzGDH against the structural context. The in-silico simulation shows that BzGDH is stable and can maintain its overall structure against heat during the simulation at 323 K, which is consistent with the biochemical studies. In brief, the robust stability of BzGDH made it an attractive participant for cofactor regeneration on practical applications, especially for the catalysis implemented in acidic pH and high temperature.

Keywords: *Bacillus*; glucose 1-dehydrogenase; acid-resistant; thermal-stable; molecular dynamics simulation

1. Introduction

NAD(P)-dependent glucose 1-dehydrogenase (GDH, EC 1.1.1.47) is an oxidoreductase present in various organisms and involved in glucose metabolic pathways, catalyzing the oxidation of D-glucose to D-glucono-1,5-lactone while simultaneously reducing NAD(P) to NAD(P)H [1–6]. As a member of the short-chain dehydrogenases/reductases family (SDRs), GDH is a tetrameric protein consisting of four identical subunits, which shares similar overall folding and oligomeric architecture with those of its homologous counterparts [7,8]. Due to the dual cofactor specificity, high activity, easy preparation, and cheap substrate, GDH has been widely used in biocatalysis [9–11], bioremediation [12], biosensors [13], and biofuel cells [14].

Biocatalysis has been considered as a powerful tool for the pharmaceutical and fine chemical synthetic processes due to the chemo-, regio-, and stereo-selectivity of enzymes [15]. However, because many kinds of industrial enzymes are cofactor-dependent, the enzymatic synthesis is limited by the considerable expenses of the cofactors. To tackle the issue of manufacturing expense on biocatalysis, several cofactor regeneration approaches have been proposed, of which the enzymatic regeneration method has been considered as an effective technique [16]. Due to the activity toward both NAD and NADP, GDH has been proposed as a promising candidate for coenzyme regeneration [17,18], compared with other oxidoreductases such as formate dehydrogenase [19], alcohol dehydrogenase [20], glucose-6-phosphate dehydrogenase [21], and phosphite dehydrogenase [22].

Although GDHs from various microorganisms have been employed as coenzyme regenerators for biocatalysis [9–11], new enzymes with robust stability against broad temperature and pH range are still preferred. In this work, a novel NAD(P)-dependent glucose 1-dehydrogenase from *Bacillus* sp. ZJ (BzGDH), with considerable acidic tolerance and thermal stability, has been extensively characterized through biochemical experiments. In contrast to previously reported acid-resistant GDHs, including GDH from *Bacillus thuringiensis* M15 (BtGDH) [3], *Bacillus* sp. G3 (BgGDH) [23], and *Bacillus cereus* var. *mycoides* (BcGDH) [24], BzGDH exhibited superior thermal stability to its homologous counterparts. To better understand this remarkable feature that distinguishes BzGDH from other acid-resistant GDHs, molecular dynamic (MD) simulation was conducted to investigate the conformational flexibility and fluctuations of BzGDH over time and spatial scales. Analysis of the trajectory shows that BzGDH is stable and can maintain its overall structure against heat during the simulation at 323 Kelvin (K), which is in accordance with the biochemical studies.

2. Results and Discussion

2.1. Sequence Analysis

The gene *bzgdh* encodes a peptide consisting of 261 amino acids with a predicted molecular weight of 28 kDa and a theoretical isoelectric point of 5.4. Significant Pfam-A matches [25] revealed that BzGDH was affiliated to adh_short_C2 family (PF13561, Enoyl-(Acyl carrier protein) reductase), which belonged to the FAD/NAD(P)-binding Rossmann fold superfamily (CL0063), as well as other GDHs. BzGDH also shared the conserved coenzyme-binding GXXXGXG motif (14–20) and catalytic triad (Ser145/Tyr158/Lys162) with other GDHs. In addition, amino acid substitutions mostly occurred at the N-terminus of GDHs (Figure 1), indicating that the N-terminal sequence is less conservative than the C-terminal sequence, which played critical roles in substrate recognition. Phylogenetic analysis showed that these GDHs diverged into two clusters, and BzGDH belonged to the sub-branch consisting of BtGDH, BcGDH, and BgGDH (Figure 2), of which all exhibited acidic resistance in previous studies, suggesting that these four GDHs might originate from the same ancestral sequences.

```
              1        10        20        30        40        50        60
BgGDH     MYSDLEGKVVVITGSASGLGRAMGVRFAREKAKVVINYRSRESEANDVL.EEIKKVGGEAIAVKGDVTVE
BtGDH     MYSDLEGKVVVITGSASGLGRAMGVRFAAEKAKVVINYRSRESEANDVL.EEIKKVGGEAIAVKGDVTVE
BzGDH     MYSDLAGKVVVITGSATGLGRAMGVRFAKEKAKVVINYRSRESEANDVL.EEIKKVGGEAIAVKGDVTVE
BcGDH     MYSDLEGKVVVITGSATGLGRAMGVRFAKEKAKVVINYRSRESEANDVL.EEIKKVGGEAIAVKGDVTVE
BmGDHB    MYKDLEGKVVVITGSSTGLGKSMAIRFATEKAKVVNYRSKEDEANSVLEEEIKKVGGEAIAVKGDVTVE
BmGDH     MYKDLEGKVVVITGSSTGLGKSMAIRFATEKAKVVVNYRSKEDEANSVL.EEIKKVGGEAIAVKGDVTVE
BmGDHI    MYKDLEGKVVVITGSSTGLGKSMAIRFATEKAKVVVNYRSKEEEANSVL.EEIKKVGGEAIAVKGDVTVE
LsGDH     MYSDLEGKVIVITGASTGLGKAMALRFGEBKAKVIVNFRSDENEANAVV.EGVKKAGGDAIAVKGDVTVE
BaGDH     MYTDLKGKVVAITGASSGLGKAMAIRFGQEQAKVVVNYYSNEKDAQTVK.EEIQKAGGEAVIIQGDVTKE
BsGDH     MYPDLKGKVVAITGAASGLGKAMAIRFGKEQAKVVINYYSNKQDPNEVK.EEVIKAGGEAVVVQGDVTKE
BmGDHIII  MYTDLKDKVVVITGGSTGLGRAMAVRFGQEEAKVINYYNNEEEALDAK.KEVEEAGGQAIIVQGDVTKE
BmGDHIV   MYTDLKDKVVVITGGSTGLGRAMAVRFGQEEAKVVINYYNNEEEALDAK.KEVEEAGGQAIIVQGDVTKE
BmGDHA    MYTDLKDKVVVITGGSTGLGRAMAVRFGQEEAKVVINYYNNEEEALDAK.KEVEEAGGQAIIVQGDVTKE
BmGDHII   MYTDLKDKVVVVTGGSKGLGRAMAVRFGQEQSKVVVNYRSNEEEALEVK.KEIEEAGGQAIIVRGDVTKE
```

```
              70        80        90       100       110       120       130
BgGDH     SDVVNLIQSAVKEFGTLDVMINNAGIENAVPSHEMPLEDWNRVINTNLTGAFLGSREAIKYFVEHDIKGS
BtGDH     SDVANLIQSAVKEFGTLDVMINNAGIENAVPSHEMPLEDWNRVINTNLTGAFLGSREAIKYFVEHDIKGS
BzGDH     SDVANLIQSAVKEFGTLDVMINNAGIENAVPSHEMPLEDWNRVINTNLTGAFLGSREAIKYFVEHDIKGS
BcGDH     ADVVNLIQSAVKEFGTLDVMINNAGIENAVPSHEMPLEDWNRVIHTNLTGAFLGSREAIKYFVEHDIKGS
BmGDHB    SDVINLVQSAIKEFGKLDVMINNAGMENPVSSHEMSLSDWNKVIDTNLTGAFLGSREAIKYFVENDIKGT
BmGDH     SDVINLVQSAIKEFGKLDVMINNAGLENPVSSHEMSLSDWNKVIDTNLTGAFLGSREAIKYFVENDIKGT
BmGDHI    SDVINLVQSSIKEFGKLDVMINNAGMENPVSSHEMSLSDWNKVIDTNLTGAFLGSREAIKYFVENDIKGT
LsGDH     EDVINLVQTAVNKFGTLDVMINNAGIENPVASHEMPLSDWNRVINTNLTGAFLGSREAIKYFVENDIKGS
BaGDH     EDVKNIVQTAVKEFGTLDVMINNAGMENPVQSHEMPLKDWNKVINTNLTGAFLGSREAIKYYVENDIQGN
BsGDH     EDVKNIVQTAIKEFGTLDIMINNAGLENPVPSHEMPLKDWDKVIGTNLTGAFLGSREAIKYFVENDIKGN
BmGDHIII  EDVVNLVQTAIKEFGTLDVMINNAGVENPVPSHELSLDNWNKVIDTNLTGAFLGSREAIKYFVENDIKGN
BmGDHIV   EDVVNLVQTAIKEFGTLDVMINNAGVENPVPSHELSLDNWNKVIDTNLTGAFLGSREAIKYFVENDIKGN
BmGDHA    EDVVNLVQTAIKEFGTLDVMINNAGVENPVPSHELSLDNWNKVIDTNLTGAFLGSREAIKYFVENDIKGN
BmGDHII   EDVVNLVETAVKEFGSLDVMINNAGVENPVPSHELSLENWNQVIDTNLTGAFLGSREAIKYFVENDIKGN
```

Figure 1. *Cont.*

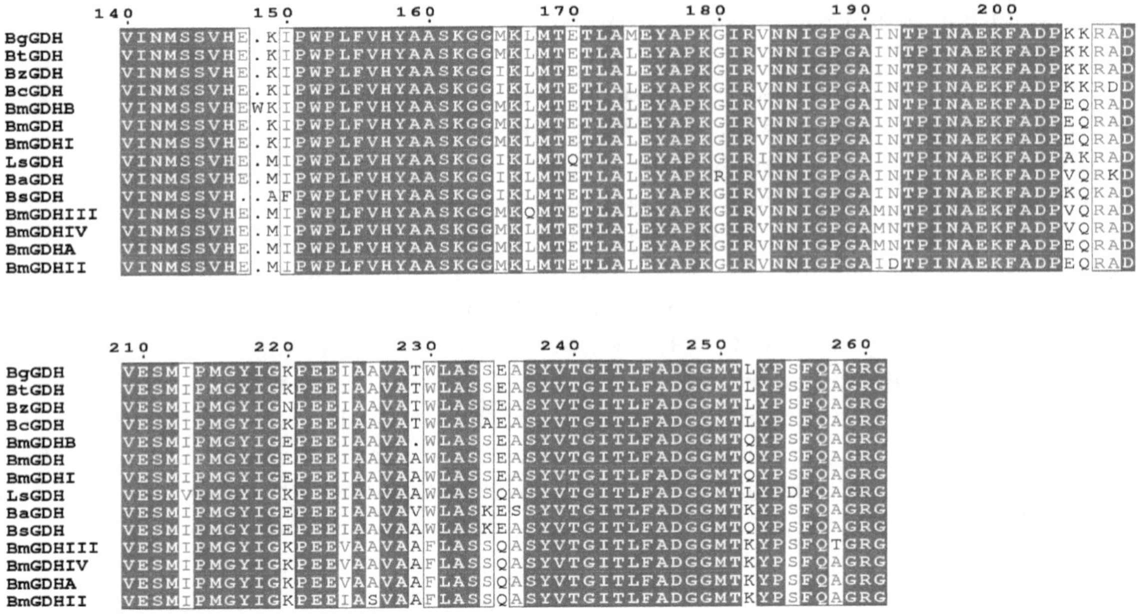

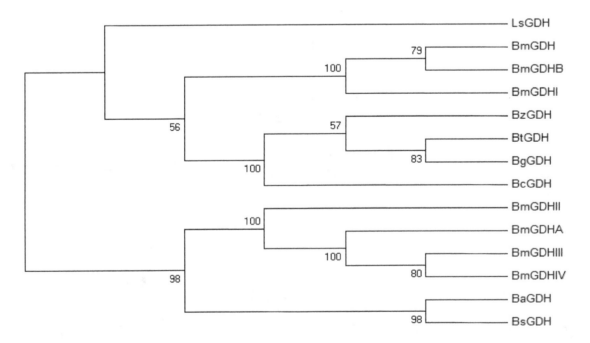

Figure 1. Multiple alignment of the primary structure of glucose 1-dehydrogenases (GDHs). Identical residues and conserved substitutions are shaded red and enveloped by rectangles, respectively. GDHs from *Bacillus* sp. ZJ (this study), *Bacillus megaterium* IWG3 [5], *Lysinibacillus sphaericus* G10 [2], *Bacillus cereus* var. *mycoides* [24], *Bacillus* sp. G3 [23], *Bacillus amyloliquefaciens* SB5 [1], *Bacillus thuringiensis* M15 [26], and *Bacillus subtilis* W168 [27] are abbreviated as BzGDH, BmGDH, LsGDH, BcGDH, BgGDH, BaGDH, BtGDH, and BsGDH, respectively. BmGDHA and BmGDHB are from *Bacillus megaterium* M1286 [6]. BmGDHI, BmGDHII, BmGDHIII, and BmGDHIV are from *Bacillus megaterium* IAM1030 [4,5]. Alignment of multiple protein sequences was conducted by using the Clustal X 2.0 program [28] and rendered by ESPript [29].

Figure 2. Unrooted phylogenetic tree of GDHs. The phylogenetic tree was constructed using the neighbor joining method [30] in MEGA7 software [31], with a bootstrap test of 1000 replicates. The evolutionary distances were computed using the Poisson correction method [32] and are in the units of the number of amino acid substitutions per site.

2.2. Heterologous Expression and Purification

The specific activity of the purified BzGDH was 194 ± 2 U·mg^{-1} at 25 °C using NAD (nicotinamide adenine dinucleotide) as a cofactor. SDS-PAGE (sodium dodecyl sulfate polyacrylamide gel electrophoresis) analysis showed a homogeneous band corresponding to 30 kDa (Figure 3). By using gel filtration chromatography through a Zorbax Bio-series GF-450 column, the molecular weight of the native BzGDH was estimated to be 120 kDa. These results indicated that BzGDH was a homo-tetramer composed of four identical subunits, as well as other NAD-dependent GDHs derived from *Bacillus* [1–6,23,24,26,27].

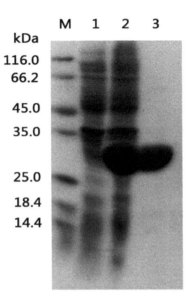

Figure 3. SDS-PAGE (sodium dodecyl sulfate polyacrylamide gel electrophoresis) analysis of total cell lysate and the purified enzyme. Lane M: protein molecular weight marker. Lane 1: uninduced total cell lysate of BzGDH. Lane 2: induced total cell lysate of BzGDH. Lane 3: purified BzGDH.

2.3. Effects of pH and Temperature on the Activity and Stability

BzGDH exhibited activity at a wide pH range from 4.0 to 10.5, and displayed maximum activity at pH 9.0 in Tris-HCl buffer among all buffers. Actually, the chemical composition of the sodium citrate, sodium phosphate, and Tris-HCl buffers, showed no significant influence on the specific activity of the enzyme, and the differences in the specific activity are mainly caused by the change of the pH of the solution. However, a significant decrease of specific activity was observed in the glycine-NaOH buffer at pH values of 8.5 and 9.0 when compared to those of the same pH values of the Tris-HCl buffer, indicating that glycine might inhibit the activity of BzGDH. Surprisingly, the optimum pH was determined as 9.5 in Glycine-NaOH buffer (Figure 4a), which is inconsistent with the maximum activity pH of 9.0. A reasonable explanation for this discrepancy is that the observed activity of the enzyme is not only affected by the pKa of its catalytic residues which played critical roles on the activity, but is influenced by the stability of the enzyme which might be unstable at its optimum pH (Figure 4b), and is even sometimes affected by the chemicals in the buffer such as glycine in this case. In regards to its pH stability, BzGDH was stable over a broad pH range, especially in the acidic region, which could maintain around 80% of its initial activity in the pH range of 4.0–8.5 after incubating for 1 hour (Figure 4b).

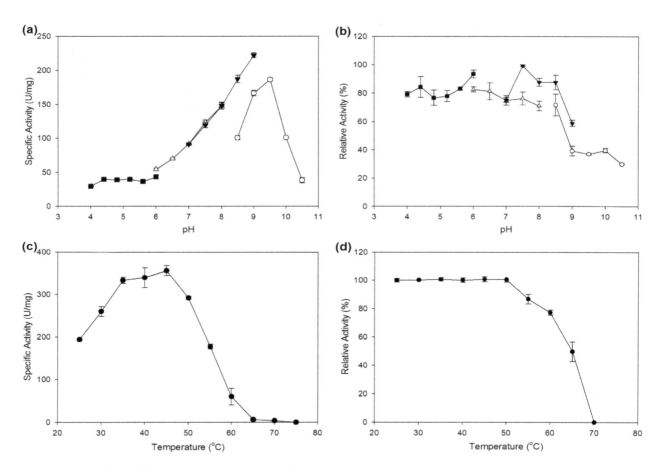

Figure 4. Effects of pH and temperature on the activity and stability of BzGDH. (**a**) Effect of pH on the activity of BzGDH; (**b**) Effect of pH on the stability of BzGDH. (■) pH 4.0–6.0, 100 mM sodium citrate buffer; (Δ) pH 6.0–8.0, 100 mM sodium phosphate buffer; (▼) pH 7.0–9.0, 100 mM Tris-HCl buffer; (○) pH 8.5–10.5, 100 mM glycine-NaOH buffer; (**c**) Effect of temperature on the activity of BzGDH; (**d**) Effect of temperature on the stability of BzGDH.

As demonstrated in Figure 4c, the optimum catalytic temperature of BzGDH was determined as 45 °C. The activity of BzGDH decreased linearly from 45 to 65 °C and could not be measurable at 75 °C. In consistent with its higher optimum reaction temperature, the recombinant enzyme also possessed good thermal stability, which was stable after incubation at temperatures below 50 °C for 30 min and still maintained 50% of its initial activity after incubation at 65 °C for 30 min (Figure 4d). BzGDH exhibited superior thermal stability to its homologous counterparts, BgGDH [23] and BcGDH [24], which were almost completely inactivated after incubation at 50 °C without any protective agent.

Since stability is an indispensable characteristic for the utilization of enzymes in real life, the considerable stability of BzGDH against both heat and acid made it a very promising candidate in practical application in harsh conditions.

2.4. Substrate Specificity and Steady-State Kinetics

As shown in Table 1, the substrate spectrum of BzGDH was similar to that of BcGDH. However, both BzGDH and BcGDH displayed stricter substrate specificity toward various sugars than that of BgGDH, especially for galactose and mannose, indicating that BzGDH could be a potential diagnostic reagent for blood glucose measurement as well as BcGDH.

The steady-state kinetic constants of BzGDH were determined by using a nonlinear fitting plot (Table 2). Although BzGDH had similar k_{cat} values for both NAD and NADP, the K_m value for NADP was 5.6-fold higher than that for NAD, indicating that BzGDH preferred NAD rather than NADP as the cofactor. The cofactor preference of BzGDH resembled that of BmGDHIII, BmGDHIV [4], and BtGDH [3], while BmGDH, BmGDHI, BmGDHII [5], and BgGDH [23] preferred NADP.

Table 1. Substrate specificity of GDHs.

Substrate	Relative Activity (%) [1]		
	BzGDH	BgGDH [23]	BcGDH [24]
D-glucose	100	100	100
D-galactose	6.8	22.0	7.3
D-mannose	3.2	7.1	4.4
D-fructose	0.9	0.6	0
D-xylose	6.1	6.4	6.0
D-arabinose	0	0.2	0
D-maltose	10.0	13.0	11.0
D-lactose	3.1	2.6	5.2
D-sucrose	0.9	6.3	2.51

[1] The activities are expressed relative to those for D-glucose.

Table 2. Kinetic constants of BzGDH.

Substrate/Cofactor	K_m (mM)	k_{cat} (s^{-1}) [1]	k_{cat}/K_m (mM^{-1}·s^{-1})
D-glucose	17.126 ± 0.946	87.844 ± 1.362	5.129
NAD	0.072 ± 0.009	84.521 ± 2.175	1166.294
NADP	0.404 ± 0.088	73.960 ± 2.677	182.978

[1] The values of k_{cat} were calculated for one subunit.

2.5. Homology Modeling and Electrostatic Potential Analysis

The quaternary structure of BzGDH was constructed by SWISS-MODEL [33] and evaluated by ProSA-web [34] and PROCHECK [35]. Both of the Z-score and Ramachandran plot statistics indicated that the dimensional structure of BzGDH (Figure 5a) had been modeled reasonably (Table 3). To investigate the electrostatic potential of BzGDH, the model of BzGDH was subjected to the software APBS [36] and PyMOL (The PyMOL Molecular Graphics System, Version 1.7 Schrödinger, LLC. available online: http://pymol.org/), to generate the electrostatic potential molecular surface. As shown in Figure 5, the contact surfaces of subunits AB, AC, and AD circled by black ellipses are mainly constituted by non-polar amino acid residues and are surrounded by acidic amino acid residues. The non-polar areas can maintain their electrically neutral state in either acidic or alkaline solutions, whereas the acidic areas would be negatively charged in alkaline solutions, leading to the mutual repulsion between subunits. Therefore, the acid-resistance of BzGDH could be explained by the electrostatic potential of contact surfaces between subunits, as well as BcGDH [24].

Table 3. Evaluation of models generated by homology modeling.

Model	Z-Score [1]	Ramachandran Plot [2]			
		Most Favored (%)	Additional Allowed (%)	Generously Allowed (%)	Disallowed (%)
1GEE	−8.87	91.2	8.0	0.9	0
BzGDH	−8.80	89.1	10.5	0.4	0

[1] Calculated by ProSA-web [34]; [2] Calculated by PROCHECK [35].

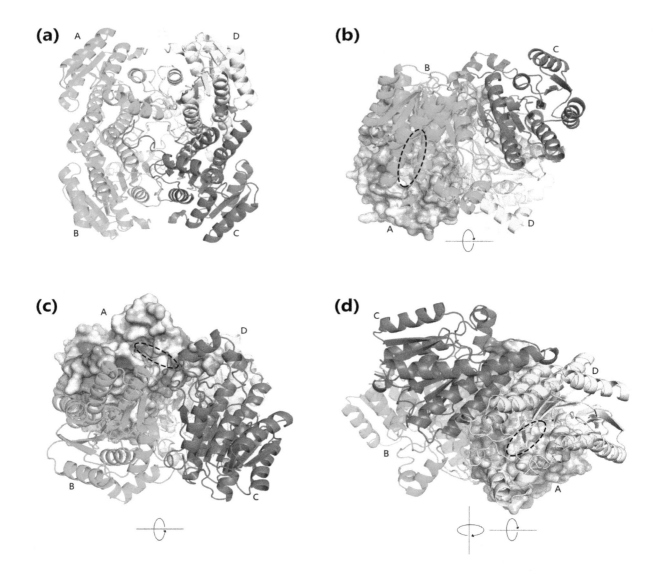

Figure 5. ±5 kT/e electrostatic potential surface of BzGDH. (**a**) Tetrameric structure of BzGDH; (**b**) Electrostatic potential surface representation of the interface between subunits A and B; (**c**) Electrostatic potential surface representation of the interface between subunits A and C; (**d**) Electrostatic potential surface representation of the interface between subunits A and D. Subunits ABCD were labeled using the corresponding capital letters nearby, respectively. Positive, negative, and neutral electrostatic potential surfaces are rendered by blue, red, and white, respectively. The non-polar regions of the contact surfaces of subunits AB, AC, and AD were circled by dashed ellipses.

2.6. Global Structure Stability

To study the stability and mobility of BzGDH, the model was subjected to a 20-ns MD simulation at 323 K. The stability of BzGDH was analyzed by the all-atom and backbone-atom root mean square deviation (RMSD), respectively, both of which increased from the beginning of the simulation and reached an equilibrium state at about 10 ns (Figure 6a), suggesting no significant structural changes for BzGDH during the simulation. In addition, the radius of gyration, the hydrogen bonds of intra-protein, and the solvent accessible surface area (SASA) of BzGDH all displayed steadily dynamic changes against time (Figure 6b–d), further confirming the stable global behavior of BzGDH during the simulation at 323 K.

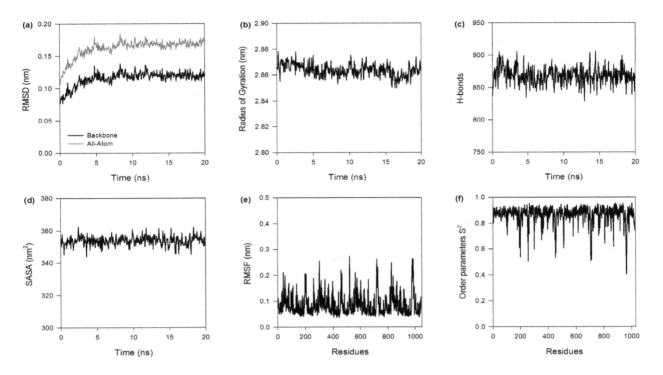

Figure 6. Dynamic changes of BzGDH in the molecular dynamics (MD) simulation. (**a**) All-atom and backbone-atom root mean square deviation (RMSD) as functions of time; (**b**) Radius of gyration as a function of time; (**c**) Hydrogen bonds as a function of time. Hydrogen bonds were detected by GROMACS (GROningen MAchine for Chemical Simulations) with default geometrical criterion, which defined both the donor-acceptor distance (≤ 0.35 nm) and the hydrogen-donor-acceptor angle ($\leq 30\ °C$); (**d**) Solvent accessible surface areas (SASA) as a function of time; (**e**) Root mean square fluctuation (RMSF) as a function of residue numbers; (**f**) N-H generalized order parameter S^2 as a function of residue numbers.

2.7. Structure Flexibility

The conformational flexibility of BzGDH was assessed using the root mean square fluctuation (RMSF) of C-alpha (Cα) atoms per residue. Generally, regular secondary structure regions display tiny fluctuations with small RMSF values during the simulation, whereas prominent fluctuations with large RMSF are observed for irregular secondary structure regions such as terminal or loop regions, which often bear certain function of proteins. As shown in Figure 6e, regions involved in coenzyme binding (39–55) and substrate binding (190–210) of each subunit are more flexible with large RMSF values than other regions. The RMSF values were converted to B-factors using the equation:

$$\text{B-factor} = (8 \times \pi^2 \times \text{RMSF}^2)/3 \tag{1}$$

to visualize global structure rigidity and flexibility of BzGDH. As shown in Figure 7, most regions of BzGDH are rigid, except for the aforementioned flexible regions, indicating that the enzyme is stable during the simulation at 323 K.

In addition to the observation of RMSF, the bond-specific fluctuations in protein structure can further be captured by the Lipari–Szabo order parameter S^2 [37], which provide an intuitive description of the amplitude of spatial restriction of the internal motions of the bond vectors on a fast timescale from picosecond to nanosecond (ps-ns). More specifically, S^2 represents the component of the H-X bond vector autocorrelation function which is dissipated by global molecular tumbling, while $(1 - S^2)$ characterizes the bond vector orientational disorder arising from internal motion occurring more rapidly than the molecular tumbling. The S^2 order parameter can range from 0 to 1, with 1 corresponding to a rigid bond vector (completely restricted) and 0 corresponding to the highest degree of disorder for a bond vector (completely isotropic). Higher order parameters (0.85) were observed in the regions of secondary structure, while unstructured regions showed lower order parameters (0.4–0.6).

The order parameter S^2 of the main chain N-H bonds of BzGDH has been calculated based on the equilibrium MD trajectories. The average value of the order parameter S^2, over all residues, is 0.86 for BzGDH. The most flexible region that showed lower S^2 of each subunit is the substrate binding domain, with residues Lys 179, Gly 180, Arg 182, Asn 184, Asn 185, Ala 190, Asn 196, and Asp 202 involved, indicating that these residues exhibit considerable disorder on the ps-ns timescale. Similarly, residues Gln 257, Ala 258, and Gly 259 in the C-terminal region of the protein have low order parameters, also implying that this region is disordered on the ps-ns timescale. Indeed, the order parameter revealed that these regions are flexible on the ps–ns timescale, with the fluctuations functioning to allow substrate access to and release of products from the active site. The results of the computation of the order parameters are in considerable agreement with the RMSF profiles, with the greatest flexibility occurring in loop regions, while other secondary structural elements are more constrained.

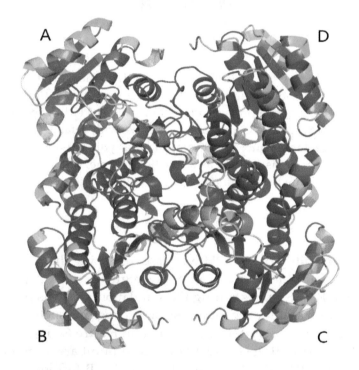

Figure 7. Cartoon representation of BzGDH shaded according to the B-factors (temperature factor) of each residue. Subunits ABCD were labeled using the corresponding capital letters nearby. The structure was shaded from the blue to red spectrum along with the increase of B-factor values from 3.98 to 193.74.

2.8. Essential Dynamics

To reveal the concerted fluctuations of BzGDH over time and spatial scales, essential dynamics (ED) is employed to extract information from sampled conformations over the molecular dynamics trajectory [38]. Practically, the essential dynamics of a protein is obtained by performing principal component analysis (PCA), which is a multivariate statistical technique involving diagonalization of the covariance matrix (Figure 8) constructed from atomic displacements of Cα atoms, to reduce the number of dimensions required to describe protein dynamics and yield a set of eigenvectors that provide information about collective motions of the protein [39].

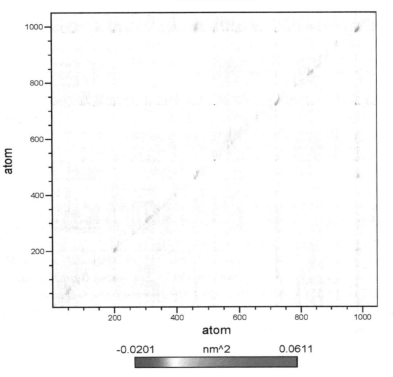

Figure 8. Covariance analysis of the atomic fluctuation of BzGDH in the MD (molecular dynamics) simulation. The correlation matrix is computed using the C-α Cartesian coordinates. The collective motions between pairs of residues are represented as red for correlated, white for uncorrelated, and blue for anti-correlated motions, respectively. The amplitude of fluctuation was represented by the color depth.

The eigenvectors represent the directions of motion, and the corresponding eigenvalues represent the amount of motion along each eigenvector, where larger eigenvalues describe motions on larger spatial scales. Generally, the first 10–20 eigenvectors are enough to capture the principal motions of the protein and describe more than 90% of all cumulative protein fluctuations [40]. However, it can be seen that only 14.3% of the total Cα motion can be explained by the first two eigenvectors, even the first 20 eigenvectors merely contribute for 51.2% of the total Cα motion from Figure 9a. This shows that most of the internal motions of BzGDH are not confined within a subspace of small dimension, and no obvious collective motion of the backbone of BzGDH is observed from the MD simulation performed at 323 K, reflecting that the enzyme can maintain its overall structure against heat, which is in accordance with the biochemical experiments.

Figure 9b shows the trajectory projected on the plane defined by the first two principal eigenvectors. The trajectories filled most of the expected ranges, suggesting the deficiency of a coupled force field, which leads to independent motions. The trajectories were projected onto the individual eigenvectors against time to further investigate the motion along the eigenvector directions. It is clear from Figure 10 that the fluctuations of the first six eigenvectors are relatively large, whereas those of the subsequent eigenvectors become successively flat, indicating that the motions belong to the last four eigenvectors have reached their equilibrium fluctuation, which cannot be used to describe the motions of the system. Due to the limitation of hardware, such simulations may not capture the essential motions related to function at much longer timescales. Improvements in computational power will fill the gap between reality and simulation.

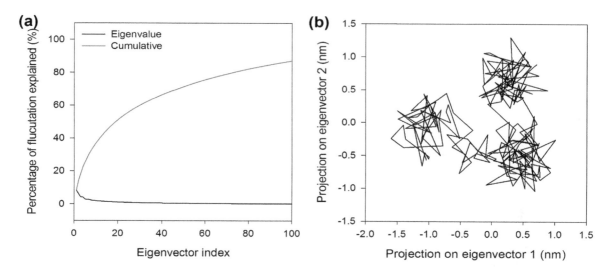

Figure 9. Principal component analysis of BzGDH in the MD simulation. (**a**) Relative cumulative deviation up to the first 100 eigenvectors provided by the essential dynamics analysis performed on the Cα atoms of BzGDH; (**b**) Projections of the trajectory on the plane defined by the first two principal eigenvectors. Horizontal axis: atomic displacement along the first eigenvector. Vertical axis: atomic displacement along the second eigenvector.

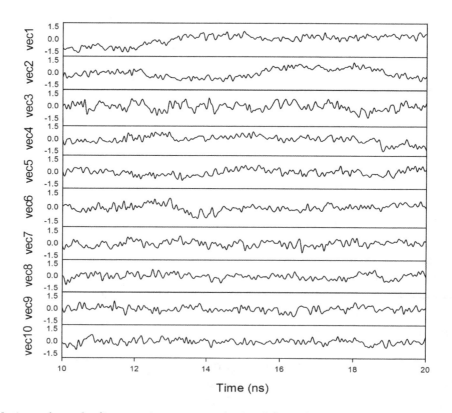

Figure 10. Motions along the first ten eigenvectors obtained from the Cα coordinates' covariance matrix.

3. Materials and Methods

3.1. Strains, Plasmids, and Chemicals

Strain *Bacillus* sp. ZJ isolated from the soil near Yuhangtang River in Hangzhou, China, was used as a source for retrieving glucose 1-dehydrogenase. *Escherichia coli* (*E. coli*) DH5α and BL21 (DE3), expression vector pET28a (+), and Ni-NTA resin were purchased from Invitrogen. Taq

DNA polymerase, PrimeSTAR HS DNA Polymerase, T4 DNA ligase, *Nde*I, and *Bam*HI were purchased from TaKaRa. Genomic DNA, plasmid and gel extraction kits were purchased from Axygen. All other chemicals were of analytical grade.

3.2. Cloning of the Bzgdh Gene and Sequence Analysis

The gene *bzgdh* was amplified by using genomic DNA of *Bacillus* sp. ZJ as a template with the forward primer 5′-GGAATTCCATATGTATAGTGATTTAGCAGG-3′ and the reverse primer 5′-CGGGATCCTATTACCCACGCCCAGC-3′, which carried cutting sites of *Nde*I and *Bam*HI (underlined), respectively. The amplified fragments were digested with *Nde*I and *Bam*HI simultaneously, then purified by using a gel extraction kit prior to ligate with the pre-digested vector pET-28a (+). The recombinant plasmid harboring gene *bzgdh* was transformed into competent cells of *E. coli* DH5α for sequencing.

Homologous searches in GenBank were performed using the BLAST server (available online: http://blast.ncbi.nlm.nih.gov). Alignment of multiple protein sequences was conducted by using the Clustal X 2.0 program [28] and rendered by ESPript [29]. The phylogenetic tree was constructed using the neighbor-joining method in MEGA7 [31], with a bootstrap test of 1000 replicates.

The nucleotide sequence for GDH of *Bacillus* sp. ZJ was deposited in GenBank under accession number KJ701281.

3.3. Expression and Purification of Recombinant BzGDH

The recombinant plasmid was transformed into competent cells of *E. coli* BL21 (DE3) for expression. The recombinant cells were cultivated in Luria-Bertani broth containing 50 μg kanamycin·mL^{-1} at 37 °C with a shaking speed of 250 rpm. The expression of recombinant protein was induced by adding 0.5 mM of IPTG to the medium when the OD_{600} of the culture reached 0.5–0.8, followed by another 12 h incubation at 25 °C with a shaking speed of 200 rpm. The cells were harvested by centrifugation at 10,000× g for 10 min at 4 °C and were washed with the binding buffer (50 mM NaH_2PO_4, 500 mM NaCl, 20 mM imidazole, pH 8.0), and then lysed by ultrasonication. The cell debris was removed by centrifugation at 15,000× g for 30 min at 4 °C, and then the supernatant was loaded onto a column containing pre-equilibrated Ni-NTA resin. The column was washed with binding buffer and subsequently eluted with elution buffer (50 mM NaH_2PO_4, 500 mM NaCl, 250 mM imidazole, pH 8.0). The eluted enzyme was desalted and concentrated by ultrafiltration and stored at −80 °C in 25 mM sodium phosphate buffer (pH 6.5) with 30% of glycerol contained. The protein concentration was determined by Bradford's method using bovine serum albumin as the reference.

Denaturing discontinuous polyacrylamide gel electrophoresis was performed on a 5% stacking gel and a 12% separating gel. The native molecular weight of GDH was determined by size-exclusion chromatography according to the protocol of the manufacture (Zorbax Bio-series GF-450, Agilent, Santa Clara, CA, USA), using lysozyme (14.3 kDa), chicken ovalbumin (45 kDa), bovine serum albumin fraction V (67 kDa), and goat IgG (150 kDa) as standards.

3.4. Enzyme Activity Assay

Glucose dehydrogenase activity was determined by assaying the absorbance of NADH at 340 nm in 100 mM sodium phosphate (pH 8.0) containing 200 mM glucose and 1 mM NAD at 25 °C. All measurements were conducted in triplicate. One unit of enzyme activity was defined as the amount of the enzyme that catalyzed the formation of 1 μmol of NADH per minute.

3.5. Effects of pH and Temperature on the Activity and Stability of BzGDH

The optimum pH of BzGDH was measured at pH ranging from 4.0 to 10.5 at 25 °C. The effect of pH on the stability of BzGDH was determined by measuring the residual activity after incubating BzGDH in buffers with different pH values for one hour at 25 °C.

The optimal temperature of BzGDH was determined at different temperatures (25–75 °C) in phosphate buffer at pH 7.0. The thermal stability of BzGDH was assayed by measuring the residual activity after incubating BzGDH at different temperatures (25–75 °C) in phosphate buffer at pH 7.0 for 30 min.

3.6. Substrate Specificity of BzGDH

The substrate specificity of BzGDH was determined by the aforementioned enzyme activity assay, except that glucose was replaced by sucrose, lactose, maltose, xylose, galactose, mannose, fructose, and arabinose, respectively.

3.7. Steady-State Kinetics of BzGDH

In order to obtain the kinetic constants for the coenzyme, 200 μM of glucose was employed as the substrate and 0.01 to 0.2 mM NAD and NADP were used as the coenzymes, respectively. For analysis of the kinetics for glucose, 1 mM NAD was used as a cofactor, 1 to 200 mM glucose was used as the substrate. GDH activity was measured as described above. The kinetic constants were determined by using a nonlinear fitting of the Michaelis-Menten equation:

$$v = (V_{max} \times [S])/(K_m + [S]) \tag{2}$$

where [S] is the concentration of the cofactor or substrate, K_m is the Michaelis constants for the cofactor or substrate, v is the reaction velocity, and V_{max} is the maximum reaction velocity. The turnover number k_{cat} was calculated by the equation:

$$V_{max} = k_{cat} \times [E] \tag{3}$$

where [E] is the concentration of the enzyme.

3.8. Homology Modeling and Electrostatic Potential of BzGDH

The crystal structure of glucose 1-dehydrogenase from *Bacillus megaterium* IWG3 (PDB code: 1GEE, 1.60 Å) [41], which shares 88.12% identity with BzGDH, was served as the template for homology modeling of BzGDH. The three-dimensional model of BzGDH was constructed by using the SWISS-MODEL [33]. Precise evaluation of the model quality was performed using ProSA-web [34] and PROCHECK [35]. The structure for electrostatics calculations was prepared by PDB2PQR [42] using the AMBER force field and assigned protonation states at pH 7.0. The electrostatic potential of BzGDH was calculated by APBS [36] using the linearized Poisson-Boltzmann equation (lpbe) at 298 K with the monovalent ion concentration of 0.1 M. The dielectric constants of protein and solvent were set as 2.0 and 78.0, respectively. The electrostatic potential molecular surface was represented by PyMOL.

3.9. Molecular Dynamic Simulations of BzGDH

The constructed model of BzGDH was subjected to the software package GROMACS 5.0.2 [43], with the AMBER99SB [44] force field adopted, for molecular dynamics simulations. The model was first placed into the center of a virtual cubic box with a side length of 11.049 nm and solvated with 39,486 TIP3P water molecules. The pH condition was 7.0 according to the ionization state of the protein with a charge of −20, and twenty Na^+ ions were added to the water box as counter ions to neutralize the negative charge of the entire system. Bond lengths were constrained by the LINCS algorithm to ensure covalent bonds to maintain their correct lengths during the simulation. Energy minimization of the system was conducted using the steepest descent algorithm for 5000 steps, followed by a 500-ps equilibration simulation with harmonic position restraints on the heavy protein atoms to equilibrate the solvent molecules around the protein. Subsequently, a 2-ns simulation without position restraints was conducted to equilibrate the entire system. Finally, the production simulation was performed for 20 ns at the target temperature. All simulations were performed under the NPT ensemble with periodic boundary conditions and a time step of 2 fs. The temperature of the system was kept at 323 K using the v-rescale method, and the pressure was kept at 1 bar using the Parrinello-Rahman method.

According to the RMSD profile of BzGDH, trajectories that reached the equilibrium state (10–20 ns) were used for further analysis. Principal component analysis was conducted to identify the direction and amplitude of the most prominent characteristics of the motions of BzGDH along the simulation trajectory. Generalized order parameters S^2, employed as a measure of the degree of spatial restriction of motion, were also calculated for the N-H bonds of BzGDH.

4. Conclusions

In this study, a novel NAD(P)-dependent glucose 1-dehydrogenase from *Bacillus* sp. ZJ has been extensively characterized, with remarkable acidic tolerance and thermal stability. To better understand the stability feature of BzGDH against the structural context, molecular dynamics simulation was conducted to investigate the conformational flexibility and fluctuations of BzGDH over time and spatial scales. Analysis of the trajectory shows that BzGDH is stable and can maintain its overall structure against heat during the simulation at 323 K, which is in accordance with the biochemical studies. In brief, the robust stability of BzGDH made it a promising participant for cofactor regeneration in practical applications, especially for catalysis implemented in acidic pH and high temperature.

Acknowledgments: This study was supported by the Chinese Polar Environment Comprehensive Investigation and Assessment Program (CHINARE-01-05, CHINARE-04-02, CHINARE-02-01), Open Fund of Key Laboratory of Biotechnology and Bioresources Utilization of Dalian Minzu University (KF2015009), Youth Innovation Fund of Polar Science (201602), and the National Natural Science Foundation of China (31200599).

Author Contributions: Haitao Ding conceived and designed the experiments; Haitao Ding and Fen Gao performed the experiments; Haitao Ding, Fen Gao, Yong Yu, and Bo Chen analyzed the data; Haitao Ding, Yong Yu, and Bo Chen contributed reagents/materials/analysis tools; Haitao Ding wrote the paper.

References

1. Pongtharangkul, T.; Chuekitkumchorn, P.; Suwanampa, N.; Payongsri, P.; Honda, K.; Panbangred, W. Kinetic properties and stability of glucose dehydrogenase from *Bacillus amyloliquefaciens* SB5 and its potential for cofactor regeneration. *AMB Express* **2015**, *5*, 1–12. [CrossRef] [PubMed]
2. Ding, H.; Du, Y.; Liu, D.; Li, Z.; Chen, X.; Zhao, Y. Cloning and expression in *E. coli* of an organic solvent-tolerant and alkali-resistant glucose 1-dehydrogenase from *Lysinibacillus sphaericus* G10. *Bioresour. Technol.* **2011**, *102*, 1528–1536. [CrossRef] [PubMed]
3. Boontim, N.; Yoshimune, K.; Lumyong, S.; Moriguchi, M. Purification and characterization of D-glucose dehydrogenase from *Bacillus thuringiensis* M15. *Ann. Microbiol.* **2004**, *54*, 481–492.
4. Nagao, T.; Mitamura, T.; Wang, X.H.; Negoro, S.; Yomo, T.; Urabe, I.; Okada, H. Cloning, nucleotide sequences, and enzymatic properties of glucose dehydrogenase isozymes from *Bacillus megaterium* IAM1030. *J. Bacteriol.* **1992**, *174*, 5013–5020. [CrossRef] [PubMed]
5. Mitamura, T.; Urabe, I.; Okada, H. Enzymatic properties of isozymes and variants of glucose dehydrogenase from *Bacillus megaterium*. *Eur. J. Biochem.* **1989**, *186*, 389–393. [CrossRef] [PubMed]
6. Heilmann, H.J.; Magert, H.J.; Gassen, H.G. Identification and isolation of glucose dehydrogenase genes of *Bacillus megaterium* M1286 and their expression in *Escherichia coli*. *Eur. J. Biochem.* **1988**, *174*, 485–490. [CrossRef] [PubMed]
7. Nishioka, T.; Yasutake, Y.; Nishiya, Y.; Tamura, T. Structure-guided mutagenesis for the improvement of substrate specificity of *Bacillus megaterium* glucose 1-dehydrogenase IV. *FEBS J.* **2012**, *279*, 3264–3275. [CrossRef] [PubMed]
8. Yamamoto, K.; Kurisu, G.; Kusunoki, M.; Tabata, S.; Urabe, I.; Osaki, S. Crystal structure of glucose dehydrogenase from *Bacillus megaterium* IWG3 at 1.7 A resolution. *J. Biochem.* **2001**, *129*, 303–312. [CrossRef] [PubMed]

9. Liu, Z.Q.; Ye, J.J.; Shen, Z.Y.; Hong, H.B.; Yan, J.B.; Lin, Y.; Chen, Z.X.; Zheng, Y.G.; Shen, Y.C. Upscale production of ethyl (S)-4-chloro-3-hydroxybutanoate by using carbonyl reductase coupled with glucose dehydrogenase in aqueous-organic solvent system. *Appl. Microbiol. Biotechnol.* **2015**, *99*, 2119–2129. [CrossRef] [PubMed]

10. Zhang, R.; Zhang, B.; Xu, Y.; Li, Y.; Li, M.; Liang, H.; Xiao, R. Efficient (R)-phenylethanol production with enantioselectivity-alerted (S)-carbonyl reductase II and NADPH regeneration. *PLoS ONE* **2013**, *8*, e83586. [CrossRef] [PubMed]

11. Gao, C.; Zhang, L.; Xie, Y.; Hu, C.; Zhang, Y.; Li, L.; Wang, Y.; Ma, C.; Xu, P. Production of (3S)-acetoin from diacetyl by using stereoselective NADPH-dependent carbonyl reductase and glucose dehydrogenase. *Bioresour. Technol.* **2013**, *137*, 111–115. [CrossRef] [PubMed]

12. Gao, F.; Ding, H.; Shao, L.; Xu, X.; Zhao, Y. Molecular characterization of a novel thermal stable reductase capable of decoloration of both azo and triphenylmethane dyes. *Appl. Microbiol. Biot.* **2015**, *99*, 255–267. [CrossRef] [PubMed]

13. Liang, B.; Lang, Q.; Tang, X.; Liu, A. Simultaneously improving stability and specificity of cell surface displayed glucose dehydrogenase mutants to construct whole-cell biocatalyst for glucose biosensor application. *Bioresour. Technol.* **2013**, *147*, 492–498. [CrossRef] [PubMed]

14. Yan, Y.M.; Yehezkeli, O.; Willner, I. Integrated, Electrically Contacted NAD(P)+-Dependent Enzyme–Carbon Nanotube Electrodes for Biosensors and Biofuel Cell Applications. *Chem. A Eur. J.* **2007**, *13*, 10168–10175. [CrossRef] [PubMed]

15. Bornscheuer, U.; Huisman, G.; Kazlauskas, R.; Lutz, S.; Moore, J.; Robins, K. Engineering the third wave of biocatalysis. *Nature* **2012**, *485*, 185–194. [CrossRef] [PubMed]

16. Liu, W.; Wang, P. Cofactor regeneration for sustainable enzymatic biosynthesis. *Biotechnol. Adv.* **2007**, *25*, 369–384. [CrossRef] [PubMed]

17. Zhao, H.; van der Donk, W.A. Regeneration of cofactors for use in biocatalysis. *Curr. Opin. Biotechnol.* **2003**, *14*, 583–589. [CrossRef] [PubMed]

18. Weckbecker, A.; Gröger, H.; Hummel, W. Regeneration of nicotinamide coenzymes: Principles and applications for the synthesis of chiral compounds. In *Biosystems Engineering I*; Wittmann, C., Krull, R., Eds.; Springer: Berlin, Germany, 2010; Volume 120, pp. 195–242.

19. Sheng, B.; Zheng, Z.; Lv, M.; Zhang, H.; Qin, T.; Gao, C.; Ma, C.; Xu, P. Efficient production of (R)-2-hydroxy-4-phenylbutyric acid by using a coupled reconstructed d-lactate dehydrogenase and formate dehydrogenase system. *PLoS ONE* **2014**, *9*, e104204. [CrossRef] [PubMed]

20. Lo, H.C.; Fish, R.H. Biomimetic NAD+ Models for Tandem Cofactor Regeneration, Horse Liver Alcohol Dehydrogenase Recognition of 1, 4-NADH Derivatives, and Chiral Synthesis. *Angew. Chem. Int. Ed.* **2002**, *41*, 478–481. [CrossRef]

21. Lee, W.H.; Chin, Y.W.; Han, N.S.; Kim, M.D.; Seo, J.H. Enhanced production of GDP-L-fucose by overexpression of NADPH regenerator in recombinant *Escherichia coli*. *Appl. Microbiol. Biot.* **2011**, *91*, 967–976. [CrossRef] [PubMed]

22. Johannes, T.W.; Woodyer, R.D.; Zhao, H.M. Efficient regeneration of NADPH using an engineered phosphite dehydrogenase. *Biotechnol. Bioeng.* **2007**, *96*, 18–26. [CrossRef] [PubMed]

23. Chen, X.; Ding, H.; Du, Y.; Lin, H.; Li, Z.; Zhao, Y. Cloning, expression and characterization of a glucose dehydrogenase from *Bacillus* sp. G3 in *Escherichia coli*. *Afr. J. Microbiol. Res.* **2011**, *5*, 5882–5888.

24. Wu, X.; Ding, H.; Ke, L.; Xin, Y.; Cheng, X. Characterization of an acid-resistant glucose 1-dehydrogenase from *Bacillus cereus* var. mycoides. *Romanian Biotechnol. Lett.* **2012**, *17*, 7540–7548.

25. Finn, R.D.; Coggill, P.; Eberhardt, R.Y.; Eddy, S.R.; Mistry, J.; Mitchell, A.L.; Potter, S.C.; Punta, M.; Qureshi, M.; Sangrador-Vegas, A.; et al. The Pfam protein families database: Towards a more sustainable future. *Nucleic. Acids Res.* **2016**, *44*, D279–D285. [CrossRef] [PubMed]

26. Boontim, N.; Yoshimune, K.; Lumyong, S.; Moriguchi, M. Cloning of D-glucose dehydrogenase with a narrow substrate specificity from *Bacillus thuringiensis* M15. *Ann. Microbiol.* **2006**, *56*, 237–240. [CrossRef]

27. Ramaley, R.F.; Vasantha, N. Glycerol protection and purification of *Bacillus subtilis* glucose dehydrogenase. *J. Biol. Chem.* **1983**, *258*, 12558–12565. [PubMed]

28. Larkin, M.A.; Blackshields, G.; Brown, N.P.; Chenna, R.; McGettigan, P.A.; McWilliam, H.; Valentin, F.; Wallace, I.M.; Wilm, A.; Lopez, R.; et al. Clustal W and Clustal X version 2.0. *Bioinformatics* **2007**, *23*, 2947–2948. [CrossRef] [PubMed]

29. Gouet, P.; Robert, X.; Courcelle, E. ESPript/ENDscript: Extracting and rendering sequence and 3D information from atomic structures of proteins. *Nucleic Acids Res.* **2003**, *31*, 3320–3323. [CrossRef] [PubMed]

30. Saitou, N.; Nei, M. The neighbor-joining method: A new method for reconstructing phylogenetic trees. *Mol. Biol. Evol.* **1987**, *4*, 406–425. [PubMed]

31. Kumar, S.; Stecher, G.; Tamura, K. MEGA7: Molecular Evolutionary Genetics Analysis version 7.0 for bigger datasets. *Mol. Biol. Evol.* **2016**, *33*, 1870–1874. [CrossRef] [PubMed]

32. Zuckerkandl, E.; Pauling, L. Evolutionary divergence and convergence in proteins. *Evol. Genes Proteins* **1965**, *97*, 97–166.

33. Biasini, M.; Bienert, S.; Waterhouse, A.; Arnold, K.; Studer, G.; Schmidt, T.; Kiefer, F.; Gallo Cassarino, T.; Bertoni, M.; Bordoli, L.; et al. SWISS-MODEL: Modelling protein tertiary and quaternary structure using evolutionary information. *Nucleic Acids Res.* **2014**, *42*, W252–W258. [CrossRef] [PubMed]

34. Wiederstein, M.; Sippl, M.J. ProSA-web: Interactive web service for the recognition of errors in three-dimensional structures of proteins. *Nucleic Acids Res.* **2007**, *35*, W407–W410. [CrossRef] [PubMed]

35. Laskowski, R.A.; MacArthur, M.W.; Moss, D.S.; Thornton, J.M. PROCHECK: A program to check the stereochemical quality of protein structures. *J. Appl. Crystallogr.* **1993**, *26*, 283–291. [CrossRef]

36. Baker, N.A.; Sept, D.; Joseph, S.; Holst, M.J.; McCammon, J.A. Electrostatics of nanosystems: Application to microtubules and the ribosome. *Proc. Natl. Acad. Sci. USA* **2001**, *98*, 10037–10041. [CrossRef] [PubMed]

37. Lipari, G.; Szabo, A. Model-free approach to the interpretation of nuclear magnetic resonance relaxation in macromolecules. 1. Theory and range of validity. *J. Am. Chem. Soc.* **1982**, *104*, 4546–4559. [CrossRef]

38. Amadei, A.; Linssen, A.; Berendsen, H.J. Essential dynamics of proteins. *Proteins Struct. Funct. Bioinform.* **1993**, *17*, 412–425. [CrossRef] [PubMed]

39. Balsera, M.A.; Wriggers, W.; Oono, Y.; Schulten, K. Principal component analysis and long time protein dynamics. *J. Phys. Chem.* **1996**, *100*, 2567–2572. [CrossRef]

40. Berendsen, H.J.; Hayward, S. Collective protein dynamics in relation to function. *Curr. Opin. Struc. Biol.* **2000**, *10*, 165–169. [CrossRef]

41. Baik, S.H.; Michel, F.; Aghajari, N.; Haser, R.; Harayama, S. Cooperative effect of two surface amino acid mutations (Q252L and E170K) in glucose dehydrogenase from *Bacillus megaterium* IWG3 on stabilization of its oligomeric state. *Appl. Environ. Microbiol.* **2005**, *71*, 3285–3293. [CrossRef] [PubMed]

42. Dolinsky, T.J.; Nielsen, J.E.; McCammon, J.A.; Baker, N.A. PDB2PQR: An automated pipeline for the setup of Poisson–Boltzmann electrostatics calculations. *Nucleic Acids Res.* **2004**, *32*, W665–W667. [CrossRef] [PubMed]

43. Pronk, S.; Páll, S.; Schulz, R.; Larsson, P.; Bjelkmar, P.; Apostolov, R.; Shirts, M.R.; Smith, J.C.; Kasson, P.M.; van der Spoel, D. GROMACS 4.5: A high-throughput and highly parallel open source molecular simulation toolkit. *Bioinformatics* **2013**, *29*, 845–854. [CrossRef] [PubMed]

44. Hornak, V.; Abel, R.; Okur, A.; Strockbine, B.; Roitberg, A.; Simmerling, C. Comparison of multiple Amber force fields and development of improved protein backbone parameters. *Proteins Struct. Funct. Bioinform.* **2006**, *65*, 712–725. [CrossRef] [PubMed]

5

Relationship of Triamine-Biocide Tolerance of *Salmonella enterica* Serovar Senftenberg to Antimicrobial Susceptibility, Serum Resistance and Outer Membrane Proteins

Bożena Futoma-Kołoch [1,*], Bartłomiej Dudek [1], Katarzyna Kapczyńska [2], Eva Krzyżewska [2], Martyna Wańczyk [1], Kamila Korzekwa [1], Jacek Rybka [2], Elżbieta Klausa [3] and Gabriela Bugla-Płoskońska [1,*]

[1] Department of Microbiology, Institute of Genetics and Microbiology, University of Wrocław, 51-148 Wrocław, Poland; bartlomiej.dudek@uwr.edu.pl (B.D.); kamila.korzekwa@uwr.edu.pl (K.K.); wanczyk.martyna@gmail.com (W.M.)
[2] Department of Immunology of Infectious Diseases, Hirszfeld Institute of Immunology and Experimental Therapy, Polish Academy of Sciences, 53-114 Wrocław, Poland; katarzyna.kapczynska@iitd.pan.wroc.pl (K.K.); eva.krzyzewska@iitd.pan.wroc.pl (E.K.); rybka@iitd.pan.wroc.pl (J.R.)
[3] Regional Centre of Transfusion Medicine and Blood Bank, 50-345 Wrocław, Poland; e.klausa@rckik.wroclaw.pl
*   Correspondence: bozena.futoma-koloch@uwr.edu.pl (B.F.-K.); gabriela.bugla-ploskonska@uwr.edu.pl (G.B.-P.)

Abstract: A new emerging phenomenon is the association between the incorrect use of biocides in the process of disinfection in farms and the emergence of cross-resistance in *Salmonella* populations. Adaptation of the microorganisms to the sub-inhibitory concentrations of the disinfectants is not clear, but may result in an increase of sensitivity or resistance to antibiotics, depending on the biocide used and the challenged *Salmonella* serovar. Exposure of five *Salmonella enterica* subsp. *enterica* serovar Senftenberg (*S.* Senftenberg) strains to triamine-containing disinfectant did not result in variants with resistance to antibiotics, but has changed their susceptibility to normal human serum (NHS). Three biocide variants developed reduced sensitivity to NHS in comparison to the sensitive parental strains, while two isolates lost their resistance to serum. For *S.* Senftenberg, which exhibited the highest triamine tolerance (6 × MIC) and intrinsic sensitivity to 22.5% and 45% NHS, a downregulation of flagellin and enolase has been demonstrated, which might suggest a lower adhesion and virulence of the bacteria. This is the first report demonstrating the influence of biocide tolerance on NHS resistance. In conclusion, there was a potential in *S.* Senftenberg to adjust to the conditions, where the biocide containing triamine was present. However, the adaptation did not result in the increase of antibiotic resistance, but manifested in changes within outer membrane proteins' patterns. The strategy of bacterial membrane proteins' analysis provides an opportunity to adjust the ways of infection treatments, especially when it is connected to the life-threating bacteremia caused by *Salmonella* species.

Keywords: *Salmonella*; biocide; serum; antimicrobial resistance; molecular biology; outer membrane protein analysis

1. Introduction

Cross-resistance to antibiotics of bacteria exposed to disinfectants (biocides) is an increasing problem for public health as cross-resistant phenotypes of microorganisms could potentially develop into life-threatening infections. The main reasons for increasing microbial resistance to disinfectants are mistakes during the disinfecting process itself, using chemicals that are not designed for specific microbiological pollution, inaccurate cleaning of surfaces with biocides (which causes high levels of organic matter and biofilm formation) or applying too low concentrations of biocides [1]. The implementation of hygiene supervision and standardization of the use of antibiotics and disinfectants seem to be a promising way to improve public safety [2]. There is still a lack of understanding of the mode of action of the biocides, especially when used at low or sub-inhibitory concentrations. A single exposure to some biocides has been found to be insufficient to select for multidrug-resistant (MDR) strains; however, repeated, sub-inhibitory exposure to biocides does result in the selection of MDR bacteria [3]. MDR is a major problem in the treatment of infections caused in humans by *Salmonella* isolates. It has also been noted that the drug resistance was found more frequently in the internal farm environment than in the external environment [2]. It is interesting that, although cross-resistance between biocides and antibiotics is often described for biocide-resistant mutants [2,4,5], increased susceptibility for some antimicrobials has been observed [6,7]. Moreover, resistance levels can also differ even between *Salmonella* serovars [7]. In two recent studies, we demonstrated that growth of *Salmonella enterica* subsp. *enterica* serovars Enteritidis, Typhimurium, Virchow and Zanzibar isolated from human fecal samples with sub-inhibitory concentrations of farm disinfectants containing dodecylamine (triamine) led to increased isolation of multiple antibiotic-resistant strains [8,9]. The antimicrobial efficacy of commercially-manufactured disinfectant substances (represented by quaternary ammonium salts (QAC) and QAC combined with other additives) were tested against *Salmonella* Enteritidis strains by others [7,10].

QAC and triamine-containing disinfectants are effective against many Gram-positive and Gram-negative bacteria. The antibacterial effect is caused by an increase in the permeability of the bacterial cell membrane, which leads to an osmotic imbalance and an outpouring of cytoplasm [11]. A blend containing dodecylamine-based structure was designed for the cleaning and disinfection of workplaces and devices that come into contact with food and in veterinary hygiene to disinfect animal houses (manufacturer's data, Amity International). Exposure and further *Salmonella* adaptation to biocides may result in modification of cell envelope (an activity of efflux systems), virulence or motility [7]. It may also include various alterations of chemotaxis pathways and protein synthesis [1,12]. Several proteins have been found to be differentially expressed between biocide-tolerant variants and their parental counterparts. Recently, we have suggested that the resistance of the *S.* Typhimurium disinfectant (dodecylamine) variant to ciprofloxacin and cefotaxime was connected to the 55-kDa surface protein repression [8]. Moreover, *S.* Typhimurium and *S.* Enteritidis dodecylamine-tolerant isolates produced more surface proteins in the range of 30–40 kDa, which probably were porins OmpC (36 kDa), OmpF (35 kDa) and OmpD (34 kDa) [9]. Exposure of *Salmonella* cells to disinfectants can induce the expression of the AcrAB-TolC efflux system [13]; but after single exposure, MDR strains were not found and probably, this is not the primary mechanism of biocide tolerance generation [6]. Additionally, SugE protein is implicated in QAC resistance and is frequently found in *Salmonella* isolates of clinical and animal origin [14,15].

The majority of *Salmonella* infections result in a mild, self-limiting, gastrointestinal illness and usually do not require antimicrobial treatment. In some cases, *Salmonella* infection can develop to bacteremia in a minor subset of patients [16]. In the situation of severe enteric disease, or when *Salmonella* invades and causes bloodstream infection, therapy with antimicrobials is essential and can be life-saving. Infections with antimicrobial-resistant *Salmonella* strains resistant to first-line treatments, i.e., fluoroquinolones and cephalosporins, may cause treatment failure. There is a lack of studies regarding the susceptibility of biocide-tolerant bacteria to normal human serum (NHS), so the present work is the first study in which these aspects of bacterial virulence are discussed. Outer membrane proteins (OMPs) are described as surface virulence factors necessary for bacterial adaptation to human immune response [17]. Therefore, they have been analyzed in our research in the context of resistance to complement-mediated killing. The aim of the present study was to assess the in vitro antimicrobial effects of triamine on *S. enterica* subsp. *enterica* serovar Senftenberg sensitive to antibiotics using both MIC (minimal inhibitory concentration) and MBC (minimal bactericidal concentration) in correlation with susceptibility to NHS and OMPs patterns. Understanding the mechanisms of the individual and cross-resistance of bacteria may provide reliable clues for the design of more effective antimicrobials.

2. Results

2.1. Salmonella enterica Tolerance to the Biocide Formulation

In this paper, the biocide formulation containing active substances triamine, ethanol, cationic surfactant and nonionic surfactant (Amity International) was used in the experiments of the generation of disinfectant-tolerant bacteria. Five *S.* Senftenberg (*Salmonella* Senftenberg) strains (131, 132, 133, 134, 135) were exposed to the disinfectant in Luria-Bertani (LB) liquid medium (Table 1). We found that the threshold for the bacterial growth was the concentration of the biocide of $8 \times$ MIC (Minimal Inhibitory Concentration) in the LB medium, which was lethal for all tested microorganisms. The strains were grown in LB supplemented with the biocide used in the concentrations of $4 \times$ MIC (131, 132, 134) or $6 \times$ MIC (133, 135). After 25 days of incubation in LB containing the biocidal formulation and the following 10 days of the stability test (incubation in LB broth), the cultures were subjected to *Salmonella* spp. detection, because of the possible contamination with other microorganisms during extended passages. The isolates before and after the stability test were identified as *Salmonella* spp. on Brilliant Green agar plates as red to pink-white colonies with a red zone.

Table 1. Generation of triamine-tolerant *Salmonella* Senftenberg (*S.* Senftenberg) variants.

Time of Incubation	Concentration of Biocide	S. Senftenberg Strain				
		131	132	133	134	135
1-day preculture in LB broth	none	+	+	+	+	+
7 days in Luria-Bertani (LB) broth	0.5 × MIC	+ 0.05	+ 0.2	+ 0.05	+ 0.05	+ 0.1
Gradient 4 × 4 days in LB broth	0.75 × MIC	+ 0.075	+ 0.3	+ 0.075	+ 0.075	+ 0.15
	1.0 × MIC	+ 0.1	+ 0.4	+ 0.1	+ 0.1	+ 0.2
	1.25 × MIC	+ 0.125	+ 0.5	+ 0.125	+ 0.125	+ 0.25
	1.5 × MIC	+ 0.15	+ 0.6	+ 0.15	+ 0.15	+ 0.3
1 day in LB broth	2 × MIC	+ 0.2	+ 0.8	+ 0.2	+ 0.2	+ 0.4
	4 × MIC	+ 0.4	+ 1.6	+ 0.4	+ 0.4	+ 0.8
	6 × MIC	− 0.6	− 2.4	+ 0.6	− 0.6	+ 1.2
	8 × MIC	− 0.8	− 3.2	− 0.8	− 0.8	− 1.6
Identification on Brilliant Green	from the highest MIC (where growth was observed)	+	+	+	+	+
Stability test 10 days in LB broth	none	+	+	+	+	+
Identification on Brilliant Green	none	+	+	+	+	+

Definitions of abbreviations: (+) the growth of bacteria in broth supplemented with the biocide seen as the turbidity of the tubes contents or the presence of the colonies typical for *Salmonella* bacteria on Brilliant Green Agar; (−) no growth; concentrations of the biocide (μL/mL) are also shown. MIC, minimal inhibitory concentration.

Salmonella variants were tested for MIC determination before and after the 10-day stability test (incubation in LB) to verify if the feature of tolerance to the biocide is stable or not. As can be seen in Table 2, MIC values were getting higher in the case of *S.* Senftenberg strains (131 [bST], 131 [aST], 132 [bST], 133 [bST], 133 [aST], 134 [bST], 134 [aST], 135 [bST]) in comparison to their wild-type counterparts. Except for the *S.* Senftenberg 131 strain, in almost all tested variants cultures, MICs were decreased after the stability test, to almost the same level as it was at the beginning of the experiments. Additional MBC comparison showed that MBCs were equal to MICs for four wild tested strains: *S.* Senftenberg (131, 132, 133, 134), but not for *S.* Senftenberg 135. It was also interesting to verify if MBC levels were maintained after the test of the stability of the tolerant phenotypes. It was demonstrated that MBC did not change (strain 132 ^{aST}) or was even slightly higher (131 ^{aST}, 133 ^{aST}, 134 ^{aST}, 135 ^{aST}) in comparison to the MBC value estimated for the wild-type strains. In general, both parameters MIC and MBC increased as the effect of bacterial adaptation to the increasing concentration of the biocide containing triamine.

Table 2. MIC and MBC values of the triamine-containing disinfectant for *Salmonella* Senftenberg strains.

Test	S. Senftenberg Strains														
	131	131 ^{bST}	131 ^{aST}	132	132 ^{bST}	132 ^{aST}	133	133 ^{bST}	133 ^{aST}	134	134 ^{bST}	134 ^{aST}	135	135 ^{bST}	135 ^{aST}
MIC (μL/mL)	0.1	0.4	0.4	0.4	1.6	0.2	0.1	0.6	0.4	0.1	0.4	0.2	0.2	0.8	0.2
MBC (μL/mL)	0.1	nt	0.4	0.4	nt	0.4	0.1	nt	0.8	0.1	nt	0.2	0.4	nt	0.8

Definitions of abbreviations: MIC, minimal inhibitory concentration; MBC, minimal bactericidal concentration; nt, not tested; bST, before the test of stability; aST, after the test of stability.

2.2. Antibiotic Susceptibility Profiling

The obtained results showed that the passages of *S.* Senftenberg strains in medium containing disinfectant did not change the susceptibility pattern to antibiotics. The wild-type strains, as well as their biocide variants, were sensitive to ciprofloxacin (CIP, 5 μg), co-trimoxazole (STX, 25 μg), cefotaxime (CTX, 5 μg), amoxicillin/clavulanic acid (AMX 30; 20/10 μg) and ampicillin (AMP, 10 μg).

2.3. Bactericidal Activity of Human Serum against S. Senftenberg Variants Tolerant to the Disinfectant

As C3 is a crucial protein in the activation of the serum complement cascade, standard analysis of C3 protein level in NHS used for experiments was performed. C3 concentration in NHS was 1470 mg/L, which was in the range of standard values (970–1576 mg/L for males and 1032–1495 for females).

Bactericidal activity of diluted NHS (22.5%, 45%) was performed on *Salmonella* wild-type strains, as well as on their biocide variants. The average number of colony forming units (CFU/mL) was calculated from the colonies grown on the agar plates from the volume of 10 μL of bacterial suspensions. Between zero and 20 colonies were achieved. Two mechanisms of bacterial susceptibility to the antibacterial activity of serum were observed. Three variants of *Salmonella* strains (131 ^{bST}, 131 ^{aST}, 133 ^{aST}, 134 ^{bST}, 134 ^{aST}) were found to become resistant in T_1 or T_2 to NHS in comparison to the sensitive parental strains, while two biocide variants (132 ^{bST}, 132 ^{aST}, 135 ^{bST}, 135 ^{aST}), lost their resistance to serum (wild-type strains were resistant) (Tables S1 and S2). In detail, the survival rate estimated for two serovars (131 and 133) increased from 10.8% (131) to 55.6% (131 ^{bST}) at T_1 in 45% NHS and to 85.3% in the case of the variant obtained after the test of stability (131 ^{aST}). Survival of bacteria increased from 0.3% (parent strain 133) to 385.7% after 15 min of incubation in 45% NHS and from 15.0–103.0% at the same time in 22.5% serum. The third strain, which exhibited resistance to NHS, was 134 ^{bST}, which multiplied before the test of stability in 22.5% at T_2 (survival changed from 7.2% to 128.6%), as well as after the test of stability (76.9% survival, 134 ^{aST}), at the same time, in comparison to the parent strain. In the higher concentration of serum of 45%, the same strain became resistant, as its survival raised from 5.5–70.0% at T_2. It was interesting also to compare the feature of resistance between strains before the test of stability and after that. It was helpful to determine if the phenotype of resistance in variants was stable even if the cultures did not have any contact with the disinfectant during 10 days of incubation in LB not supplemented with the biocide. It has been observed that after the test of stability, the resistance rose (131 ^{aST}, 132 ^{aST}, 133 ^{aST}, 134 ^{aST}), or the resistance was maintained (131 ^{aST}, 134 ^{aST}), or vanished (132 ^{aST}), depending on the time of incubation and the serum dilution. When the bactericidal activity of NHS was heat inhibited (HIS, control of experiments), bacterial cells proliferated very intensively, and all of the tested strains were resistant to 22.5% NHS (Table S1) and 45% NHS (Table S2). Regarding our results and reports of other research groups, showing that resistance to the bactericidal activity of serum is determined by OMPs [17–20], the next stage of research focused on the analysis of the protein profiles of OMPs in the context of unknown OMP-dependent tolerance to the biocide.

2.4. Analysis of the Two-Dimensional (2-DE) Profiles of Isolated Membrane Proteins

We applied a proteomic approach using the 2-DE and mass spectrometry analysis for the identification of specific proteins that could be involved in the phenomenon of biocide and NHS resistance of the *S.* Senftenberg 133 strain. We have chosen for this analysis *S.* Senftenberg 133 as the only strain that was primarily sensitive to NHS and belonging to the group of the highest triamine tolerance (6 × MIC). Protein spots on 2-DE were visualized within the molecular weight (MW) range of 10–250 kDa and isoelectric points (pI) of 4–7. The comparative protein pattern analysis of *S.* Senftenberg strains resistant to triamine and NHS showed differences in the presence of some proteins (Figure 1), from which four were described in detail (Table 3). MWs of identified OMPs were distributed in the range of 37.49–89.52 kDa. These research spots were distributed in the range of pI of 4.85–8.48. The detailed MASCOT search results are provided as Supporting Information. It has been noted that flagellar protein FliC (Spot 1, Figure 1), as well as enolase (Spot 2) were present in lower quantities in the biocide-tolerant variant in comparison to the wild-type parent strain. In contrast, two identified molecules, chemotaxis response regulator protein-glutamate methylesterase (Spot 3), and outer membrane protein assembly factor (Spot 4), were overproduced in the *S.* Senftenberg biocide-serum-resistant isolate, although the molecular mass of Spot 4 from 2-DE does not reflect the mass of the identified protein from the database (89.252 kDa), suggesting the degradation of the protein during the preparation process.

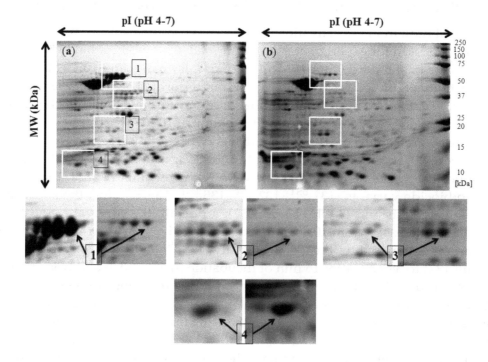

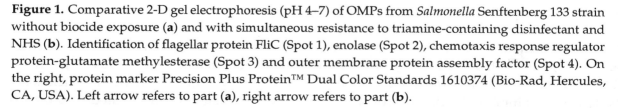

Figure 1. Comparative 2-D gel electrophoresis (pH 4–7) of OMPs from *Salmonella* Senftenberg 133 strain without biocide exposure (**a**) and with simultaneous resistance to triamine-containing disinfectant and NHS (**b**). Identification of flagellar protein FliC (Spot 1), enolase (Spot 2), chemotaxis response regulator protein-glutamate methylesterase (Spot 3) and outer membrane protein assembly factor (Spot 4). On the right, protein marker Precision Plus Protein™ Dual Color Standards 1610374 (Bio-Rad, Hercules, CA, USA). Left arrow refers to part (**a**), right arrow refers to part (**b**).

Table 3. Identification of isolated proteins from *Salmonella* Senftenberg 133 with resistance to both triamine-containing biocide (6 × MIC) and normal human serum (NHS).

Spots	Identified Proteins	Gene Symbols	Molecular Weight (kDa)	pI	Expression
1	Flagellin (FliC)	*fliC*	52.081	4.85	downregulated
2	Enolase	*eno*	45.628	5.25	downregulated
3	Chemotaxis response regulator protein-glutamate methylesterase	*cheB*	37.498	8.48	upregulated
4	Outer membrane protein assembly factor BamA	*bamA*	89.525	4.92	upregulated

3. Discussion

Salmonella enterica serovars continue to be among the most important foodborne pathogens worldwide due to the significant human rates of illness reported. Public concern for the appearance of resistant zoonotic pathogens such as *Salmonella* strains to many antibiotics is challenging the poultry industry to find successful means of control [21]. The increasing use of biocides in farming, food production, hospital settings and the home is contributing to the selection of antibiotic-resistant strains [3]. Within several years, it has also been documented that biocide-resistant *Salmonella* mutants demonstrated reduced susceptibility to antibiotics or, differently, the exposure of these microorganisms to the disinfectants has not changed their sensitivity to antimicrobials. Shengzhi et al. [2] showed that 109 *Salmonella* strains were co-resistant to antibiotic and disinfectant. In inquiring research, Whitehead and co-workers [3] isolated mutants able to survive challenge with "in-use" concentrations of biocides after one exposure using fluorescence-activated cell sorting (FACS). These mutants were multidrug resistant and overexpressed the AcrEF efflux pump and MarA, demonstrating that biocide exposure can select for mutants with a broad, low-level antibiotic resistance. Working on *S.* Typhimurium phage type 104 (DT104) Majtánova and Majtán indicated that isolate 5551/99 represented the multiresistant phenotype, resistant to ampicillin, chloramphenicol, streptomycin and tetracycline, but the second isolate 577/99 was sensitive to all antibiotics tested [7]. Others also observed increased susceptibility of *Salmonella* for some drugs. In vitro exposure to a quaternary ammonium disinfectant containing formaldehyde and glutaraldehyde (QACFG) and triclosan led to the selection of *S.* Typhimurium cells with reduced susceptibility to several antibiotics. This was associated with overexpression of the AcrAB efflux pump and accompanied with reduced invasiveness [22]. Strains used in our study, despite the tolerance to biocide, were sensitive to antibiotics, such as ciprofloxacin, co-trimoxazole, cefotaxime, amoxicillin/clavulanic acid and ampicillin. Overall, the issue of bacterial cross-resistance needs to be clarified, but in this paper, the main characteristic that was chosen for testing was *Salmonella* sensitivity to serum.

It has been suggested that the involvement of common general responses in biocide-tolerant mutants includes several alterations in metabolic and chemotaxis pathways, protein synthesis, cell envelope or regulation of pathogenicity islands. Unlike what has been commonly reported, overexpression of AcrAB-like pumps did not seem to be the primary mechanism involved in biocide tolerance. QACs are widely used in different settings, including the food industry as a hard-surface disinfectant, antiseptic and in foaming hand sanitizers [6]. It has been known that QACs are the membrane-active agents with the target site predominantly at the cytoplasmic membrane of bacteria. Although it was found that the antibacterial efficacy of substances containing QACs with other additives was high against *S.* Enteritidis strain 85/01, it was possible to select isolates resistant to these compounds [10]. Repeated passages of *Arcobacter* spp. in a medium with a low concentration of the disinfectant Incidur, containing cationic surfactant benzalkonium chloride, increased their initial resistance to 1.5–3.5×, depending on the bacterial species or origin [5]. In our study, following several rounds of in vitro variants' selection using increasing concentrations of triamine-containing disinfectant, *S.* Senftenberg isolates developed the biocide tolerance phenotype, with a four-fold or six-fold increase in the MIC. The test of stability relied on the incubation of the variants for 10 consecutive days in fresh LB medium without the addition of the biocide. Determination of MICs helped to conclude that the exposure of the tested strains to the triamine-containing blend resulted in increased tolerance immediately after the end of the generation of mutants that was before the stability test. However, after 10 days of incubation in non-stressful conditions, the bacteria became more sensitive to the disinfectant (Table 2). This is an optimistic phenomenon considering the public health, since tolerance to triamine was not stable. The question remains which conditions may favor stable tolerance to the biocides. The possible explanation is the presence of proteins or organic materials that reduce disinfectant activity and contribute to biofilm formation [23]. Quorum sensing is known to contribute to antibiotic resistance in *Salmonella* [24], but its role in biocide tolerance is not understood. In out further investigations, the growth of the bacteria was inhibited using the concentration of 0.04% (strains 131, 134), 0.16% (strain 132), 0.06% (strain 133) and 0.12% (strain 135). It is important to note that the bacteria were able to adapt to the increasing concentrations of the biocidal formulation, as has been previously shown [8,9].

The ability of human pathogens to survive in serum is another feature worth determining. *Salmonella* infections can result in uncomplicated diarrhea in most cases, but can lead to invasive disease [25]. Unfortunately, the mechanism of bacterial survival in NHS is not entirely understood. Considering *Salmonella* spp. surface antigens' composition, it has been shown that long-chain LPS [26], O-antigen (O-Ag) [27], Vi capsules [28], OMPs [17,29] or the presence of fimbriae on the cell surface are virulence factors necessary for bacterial adaptation to human immune response. Recent investigations by Dudek et al. [20] revealed that sensitive *S*. Enteritidis strains possessed a high level of flagellar hook-associated protein 2 (FliD). Furthermore, others showed that O-Ag capsule-deficient mutants produced exclusively phase I flagellin (FliC) [27]. In this paper, we demonstrate that the triamine tolerant mutants displayed changes in their susceptibility profile to a diluted NHS (22.5% and 45%) when compared to their isogenic, wild-type parental strains. To our knowledge, this is the first report demonstrating the influence of biocide-tolerant phenotype to NHS-resistant pattern. *Salmonella* after repeated exposure to the biocide did not become resistant to antibiotics, but have developed resistance to NHS (Table 4). Hypothesizing, if it came to systemic infection by the bacteria with a cross-resistance to antibacterials and reduced susceptibility to serum, it would have produced treatment failure, because of an inadequate dose of a drug.

After the revision of the literature information on the role of membrane proteins in biocide or antibiotics tolerance, it can be summarized that exposure to triclosan has been associated with an upregulation of AcrAB, a major efflux system [23]. Moreover, *Salmonella* can survive challenge with in-use concentrations of some biocides; this is due to de-repression of the AcrEF efflux system, and these mutants were MDR [3]. They also included SugE, classically implicated in QACs resistance and frequently found in *Salmonella* isolates of clinical and animal origin. In this study, we compared the proteomic profile of the *S*. Senftenberg 133 variant (133 [bST]) with the reduced susceptibility to triamine and NHS with its isogenic biocide-tolerant counterpart. We have chosen for this analysis *S*. Senftenberg 133, because it was the only one primarily sensitive to HS, belonging to the group of the highest triamine tolerance (6 × MIC). Intrinsic susceptibility of the tested serovar was essential for evaluation of 2DE analysis since sensitive *Salmonella enterica* serovars were shown to possess higher levels of flagellar hook-associated protein 2 (FliD) [20]. In our analysis, even though the variant was tolerant to the disinfectant and was sensitive to antibiotics, we have observed four distinct changes in protein patterns related to flagellin (FliC), outer membrane protein assembly factor, chemotaxis response regulator protein-glutamate methylesterase and enolase. Downregulation of flagellin and enolase factor might suggest a lower pathogenicity, including adhesion and invasion of the host cells. On the other hand, over-production of chemotaxis response regulator protein and outer membrane protein assembly factor in *S*. Senftenberg 133 [bST] could compensate a loss of motility. It has to be stressed that enolase is described as the multifunctional bacterial protein with the unique function of the receptor to human plasminogen. The enolase/plasminogen system is one of the mechanisms facilitating the invasiveness of pathogens, and it plays also an important role in the development of tumor tissues [30]. It seems that the tested biocide might weaken the motility-dependent virulence of *S*. Senftenberg.

Table 4. Collective phenotypic characteristic of the tested *Salmonella* Senftenberg strains and their biocide-tolerant variants.

No.	Strain	Maximal Tolerance to Biocide (See Table 1)	MIC (See Table 2)	RP in 22.5% NH (See Table S1)	RP in 45% NHS (See Table S2)	Comments
131 131[bST] 131[aST]	S. Senftenberg	4 × MIC	0.1 higher higher	R in T_1 R in T_1 R in T_1 and T_2	S R in T_1 R in T_1 and T_2	Resistance of the variants is maintained
133 133[bST] 133[aST]	S. Senftenberg	6 × MIC	0.1 higher higher	S S R in T_1	S S R in T_1	Resistance of the variant is maintained
135 135[bST] 135[aST]	S. Senftenberg	6 × MIC	0.2 the same	R in T_1 and T_2 S S	R in T_1 and T_2 S S	Resistance of the variants to NHS is lost in both serum concentrations
132 132[bST] 132[aST]	S. Senftenberg	4 × MIC	0.4 higher lower	R in T_1 R in T_2 R in T_1	R in T_1 S S	Resistance of the variants to NHS is lost in higher serum concentration
134 134[bST] 134[aST]	S. Senftenberg	4 × MIC	0.1 higher higher	R in T_1 R in T_1 and T_2 R in T_1 and T_2	R in T_1 R in T_1 R in T_1 and T_2	Resistance of the variants is maintained

Definitions of abbreviations: MIC, minimal inhibitory concentration; NHS, normal human serum; HIS, heat-inactivated normal human serum; RP, resistance pattern; S, sensitive; R, resistant; bST, before the test of stability; aST, after the test of stability.

In summary, there is much potential in *Salmonella* spp. to adjust to hostile environments, where the biocide containing triamine is present; however, the adaptation of the bacteria to the sub-inhibitory disinfectants' concentrations does not always result in the increase of antibiotic resistance. In cases of reduced sensitivity of bacteria to antimicrobials, a good idea would be the use of different disinfectants alternately to minimize the risk of cross-resistance and developing of MDR phenotypes.

4. Materials and Methods

4.1. Bacterial Strains

Salmonella enterica subsp. *enterica* serovar Senftenberg strains were isolated from poultry food samples in the period of November–December 2014 at the LAB-VET Veterinary Diagnostic Laboratory (Tarnowo Podgórne, Poland) by the procedures approved by Polish Centre for Accreditation. Bacterial species were serotyped in the National Serotype *Salmonella* Centre (Gdańsk, Poland). Strains used in this study were as follows: *S.* Senftenberg 131; *S.* Senftenberg 132; *S.* Senftenberg 133; *S.* Senftenberg 134; *S.* Senftenberg 135. Strains originated from the collection of the Department of Microbiology at the University of Wrocław (Wrocław, Poland). Variants before the test of stability were marked as bST and after the test of stability as aST.

4.2. Disinfectants and Antibiotics

Disinfectant: commercial biocide formulation contained: triamine, 2-aminoethanol, cationic surfactants, nonionic surfactants, potassium carbonate (Amity International, Barnsley, UK) (Table 5). Antibiotics: ciprofloxacin (CIP), co-trimoxazole (STX), cefotaxime (CTX), amoxicillin/clavulanic acid (AMX 30) and ampicillin (AMP) (Thermo Fisher Scientific, Waltham, MA, USA).

4.3. Antimicrobial Susceptibility

Parent *S.* Senftenberg strains and their variants were tested with the broth microdilution method to determine MIC and MBC of the biocides according to Andrews et al. [31] with minor modifications. In short, biocide concentrations were prepared in Mueller-Hinton broth (Merck, Kenilworth, NJ, USA) as follows: 204.8, 102.4, 51.2, 25.6, 12.8, 6.4, 3.2, 1.6, 0.8, 0.4, 0.2, 0.1, 0.05, 0.025 μL/mL in U-bottom microtitration plates (Medlab, Raszyn, Poland). The adjustment of the bacterial precultures suspension to the density of the 0.5 McFarland standard was performed. Next, the inoculum was adjusted so that 10^4 CFU/mL per spot were applied into the wells. The plates were incubated at 37 °C, and finally, MICs were estimated as the lowest concentration of biocide at which there was no visible growth. Either MBC was calculated. MBC was the lowest concentration that demonstrated a significant reduction (such as 99.9%) in CFU/mL when compared to the MIC dilution. The testing of bacterial susceptibility to antibiotics was done using disc diffusion and the E-test method. Data interpretation was performed according to the European Committee for Antimicrobial Susceptibility Testing (EUCAST) epidemiological cut-off values and clinical breakpoints [32]. The tests were repeated three times, including appropriate controls.

Table 5. Veterinary industry and healthcare environment biocide formulations used in this study (according to the manufacturers' instructions).

Active Substances	Recommended Contact Time	Experimental Contact Time (See Table 1)	Recommended Working Concentration	Experimental Working Concentration	Mechanisms of Action Against Bacteria
triamine, 2-aminoethanol, cationic surfactants, nonionic surfactants	5–10 min	24 days	(2.5%) 2.5 mL/100 mL	From 5 µL/100 mL (0.005%) to 320 µL/100 mL (0.32%)	penetration of outer membrane of bacterial cell disrupting of RNA of the microorganism preventing of replication of DNA

4.4. Isolation of Biocide Tolerant Variants and the Stability of Their Phenotypes

Isolation (generation) of variants from each culture of *Salmonella* was done according to Ricci et al. [33] and Karatzas et al. [22] (Table 1). The test was performed as previously described [8,9]. One-day precultures of the wild-type strains of *Salmonella* were exposed to subinhibitory concentrations of the disinfectant ($0.5 \times$ MIC) relevant to 0.05, 0.1, 0.2 μL/mL in dependence of the strain for 7 days; gradually increasing concentrations of the same substance (4 days for each concentration $0.75 \times$ MIC, $1.0 \times$ MIC, $1.25 \times$ MIC, $1.5 \times$ MIC); one-day incubation in LB broth (Merck) containing a 2-fold, 4-fold and 6-fold increase in biocides MICs; and ten days of incubation in LB broth, in the absence of the disinfectant to test the stability of the phenotypes. The tests of the stability of phenotypes were done on the cultures from the highest possible MICs. Three replicates of each concentration were used. Typical *Salmonella* colonies from the agar plates were transferred into sterile saline to set the density of 0.5 in McFarland standard (2×10^8 cells). Then, an inoculum was created by suspending of 0.1 mL of the culture in 10 mL of saline. Next, 9.8 mL of LB medium, 0.1 mL of bacterial suspension and 0.1 mL of a given concentration of the biocide were pipetted into a sterile tube. The concentration of the biocide for each test depended on the value of the MIC estimated at the beginning of the experiments. The cultures were incubated at 37 °C for 24 h in a shaking water bath. The cultures of the bacteria were revitalized every day through the collecting of 0.1 mL bacterial suspension from the previous day's incubation and transferring into fresh LB medium. The whole experiment, to obtain the variants tolerant to triamine-containing disinfectant, took 35 days. To confirm the presence of *Salmonella* spp. in the study, the cultures of the bacteria were inoculated onto Brilliant Green Lab-Agar (Biocorp, Warszawa, Poland).

4.5. Serum

NHS was obtained from Regional Centre of Transfusion Medicine and Blood Bank, Wrocław, Poland. This was conducted according to the principles expressed in the Law on public service of the blood of 20 May 2016 and in the Directive 2002/98/EC of the European Parliament and of the Council of 27 January 2003, establishing standards of quality and safety for the collection, testing, processing, storage and distribution of human blood and blood components. Blood samples were collected into aseptic tubes with clot activator and with gel for serum separation. The samples were then stored at room temperature (RT) for 30 min. After that time, the samples were centrifuged for 5 min at $4000 \times g$. Only the serum samples without hemolysis and lipemia were used for experiments. The serum samples were collected, pooled and kept frozen (-70 °C) for a period no longer than 3 months. A suitable volume of serum was thawed immediately before use. Each portion was used only once.

4.6. Serum C3 Concentration

The C3 concentration in the pool of NHS was quantified by a radial immunodiffusion test Human Complement C3&C4 "Nl" Bindarid™ Radial Immunodiffusion Kit (The Binding Site, Birmingham, UK). C3 protein is thought to be the most important component of the C system [34]. NHS with the proper concentration of C3 glycoprotein (between 970 and 1576 mg/L) was used for these studies.

4.7. Serum Susceptibility Assay

The bactericidal activity of normal human serum (NHS) was determined as described previously [35] with minor modifications. It was performed in sterile polystyrene U-bottom microtitration plates (Medlab, Raszyn, Polska). *S.* Senftenberg strains and their biocide variants before and after the test of stability were subjected to the challenge of 22.5% and 45% NHS. Serum decomplemented by heating at 56 °C for 30 min (heat-inactivated normal human serum (HIS)) was used as a control [36]. After overnight incubation in LB medium (Merck, Kenilworth, NJ, USA), bacteria (500 μL) in their early exponential phase were collected by centrifugation ($1500 \times g$ for 20 min at 4 °C).

The pellet was suspended in 3 mL of phosphate-buffered saline (PBS) (POCH, Gliwice, Poland) and then diluted in the same saline to produce a suspension of approximately 10^7 cells/mL. The volume of 20 μL of bacterial suspension and 180 μL of active or inactivated serum were transferred into the wells of the plates. Each concentration of the serum was loaded in triplicate. Finally, each well contained about 2×10^5 of the bacterial cells. The mixtures were incubated at 37 °C for 0, 15 and 30 min (T_0, T_1 and T_2, respectively) on a laboratory shaker with rotation at 20 rpm. Appropriate dilutions in the volume of 10 μL were then spread in triplicate onto nutrient agar plates (Biocorp, Warszawa, Poland) and incubated at 37 °C for 24 h. The average number of CFU/mL was calculated from the replicate plate counts. The survival rate for T_1 and T_2 was calculated as a percentage of the cell count for T_0 (set at 100%). Strains with survival rates below 50% were considered susceptible to the bactericidal action of NHS, while those with survival rates above 50% were described as resistant to NHS. Each test was performed three times.

4.8. Outer Membrane Proteins Isolation and Preparation

The isolation of OMPs was performed according to Murphy and Bartos [37] with minor modifications [20,38]. Bacterial strains were cultured overnight at 37 °C in 25 mL LB medium (Merck, Kenilworth, NJ, USA). The cells from the overnight culture were harvested ($1500 \times g$ at 4 °C for 20 min) and suspended in 1.25 mL 1 M sodium acetate (POCH, Gliwice, Polska) with 1 mM β-mercaptoethanol (Merck). Subsequently, 11.25 mL water solution containing 5% (w/v) Zwittergent Z 3–14 (Merck, Kenilworth, NJ, USA) and 0.5 M $CaCl_2$ (POCH) were added. This mixture was stirred at room temperature for 1h. To precipitate nucleic acids, 3.13 mL of 96% (v/v) cold ethanol (POCH) were added very slowly. The mixture was then centrifuged at $17,000 \times g$ at 4 °C for 10 min. The proteins were precipitated from the supernatant by the addition of 46.75 mL of 96% (v/v) cold ethanol and centrifuged at $17,000 \times g$ at 4 °C for 20 min. The pellet was left to dry at RT and then suspended in 1.5 mL 50 mM Trizma-Base (Merck) buffer, pH 8.0 containing 0.05% (w/v) Zwittergent Z 3–14 and 10 mM EDTA (Merck) and stirred at room temperature for 1 h. The solution was kept at 4 °C overnight. Insoluble material was removed by centrifugation at $12,000 \times g$ at 4 °C for 10 min, with OMPs present in the supernatant. Total protein concentration was measured using a commercial BCA Protein Assay Kit (Thermo Fisher Scientific, Waltham, MA, USA).

4.9. Two-Dimensional Gel Electrophoresis

The OMPs were separated with 4–7 pH immobilized gradient strips (IPG 7 cm) (Bio-Rad, Hercules, CA, USA). 2-DE was carried out with the Mini-PROTEAN Tetra Cell System (Bio-Rad). Isoelectric focusing (IEF) was conducted by a stepwise increase of voltage as follows: 250 V, 20 min (linear); 4000 V, 120 min (linear); and 4000 V (rapid); until the total volt-hours reached 14 kVh. IPG strips were then loaded onto the top of 1-mm slabs comprised of a 9% polyacrylamide stacking gel and 12.5% polyacrylamide separating gel, using 0.5% agarose (Bio-Rad) with bromophenol blue dye in the running buffer. Electrophoresis was performed at 4 °C with constant power (3 W) until the dye front reached the bottom [39–41]. The protein spots were visualized by Coomassie Brilliant Blue (Bio-Rad). Band patterns were visualized under white light and photographed using Gel Doc™ EZ System (Bio-Rad). Image spots of proteins were analyzed by PDQuest software 8.0.1 (Bio-Rad) [20].

4.10. In-Gel Protein Digestion and MS Protein Identification

After isolation, 2-DE separation and staining with the Coomassie Brilliant Blue method, protein spots of interest were excised from the gel and subjected to the in-gel tryptic digestion according to the method described by Shevchenko et al. [42]. Briefly, after destaining (100 mM NH_4HCO_3/acetonitrile, 1:1, v/v), reduction (10 mM dithiothreitol in 100 mM NH_4HCO_3) and alkylation (55 mM iodoacetamide in 100 mM NH_4HCO_3), a suitable volume of 13 ng/μL trypsin in 10 mM ammonium bicarbonate containing 10% (v/v) acetonitrile was added to the excised gel spot cut into cubes. The obtained peptides were extracted from the gel, concentrated and desalted with the Pierce C18 tip and subjected to mass spectrometry analysis using the MALDI-TOF ultrafleXtreme instrument (Bruker, Bremen, Germany). Ten milligrams per milliliter of α-cyano-4-hydroxycinnamic acid (Bruker) in acetonitrile/0.1% TFA in H_2O (1:1, v/v) were used as the eluent of peptides from the Pierce C18 tip directly on an MALDI plate. Spectra were acquired in positive reflector mode, averaging 2000 laser shots per MALDI-TOF spectrum. OMPs identification was achieved using a bioinformatics platform (ProteinScape 3.0., Bruker) and MASCOT (Matrix Science, 2.3.02) as a search engine to search protein sequence databases (NCBI, Swiss-Prot, date of access 10/03/2017) using the peptide mass fingerprinting method. All solvents used for digestion, MS preparation and analysis were of LC-MS grade and purchased from Merck Millipore (Billerica, MA, USA). Ammonium bicarbonate eluent additive for LC-MS, dithiothreitol and iodoacetamide were from Sigma-Aldrich (Saint Louis, MO, USA). Sequencing-grade modified trypsin was obtained from Promega (Madison, WI, USA).

Acknowledgments: The authors thank Jarosław Wilczyński from LAB-VET Veterinary Diagnostic Laboratory in Tarnowo Podgórne (Poland) for *Salmonella* strains. We also direct special thanks to Dr. Katarzyna Guz-Regner and Dr. Aleksandra Pawlak for their surveillance during performing preliminary serum bactericidal assays by Martyna Wańczyk. The present work was financed by the European Union under the framework of the European Social Fund No. BPZ.506.50.2012.MS and University of Wrocław Grant No. 1213/M/IGM/15. Publication was also supported by Wrocław Centre of Biotechnology, program The Leading National Research Centre (KNOW, Krajowy Narodowy Ośrodek Wiodący) for the years 2014–2018. The funding agencies had no direct role in the conduct of the study, the collection, management, statistical analysis and interpretation of the data, preparation nor approval of the manuscript.

Author Contributions: Bożena Futoma-Kołoch, Jacek Rybka and Gabriela Bugla-Płoskońska conceived of and designed the experiments. Bożena Futoma-Kołoch obtained funding from The European Social Fund No. BPZ.506.50.2012.MS. Bożena Futoma-Kołoch and Gabriela Bugla-Płoskońska obtained funding from University of Wrocław Grant No. 1213/M/IGM/15. Bożena Futoma-Kołoch, Bartłomiej Dudek, Katarzyna Kapczyńska, Eva Krzyżewska, Kamila Korzekwa and Martyna Wańczyk performed the experiments. Bożena Futoma-Kołoch, Bartłomiej Dudek, Katarzyna Kapczyńska, Eva Krzyżewska and Gabriela Bugla-Płoskońska analyzed the data. Bożena Futoma-Kołoch, Bartłomiej Dudek, Jacek Rybka, Elżbieta Klausa and Gabriela Bugla-Płoskońska contributed reagents/materials/analysis tools. Bożena Futoma-Kołoch wrote the paper. Gabriela Bugla-Płoskońska and Jacek Rybka provided study supervision. All co-authors revised and approved the final manuscript.

Abbreviations

aST	After the test of stability
bST	Before the test of stability
CFU	Colony-forming units
h	Hours
HIS	Heat-inactivated normal human serum
MDR	Multidrug-resistant
MIC	Minimal inhibitory concentration
MBC	Minimal bactericidal concentration
NHS	Normal human serum
OMPs	Outer membrane proteins
RP	Resistance pattern
QAC	Quaternary ammonium salts
R	Resistant
S	Sensitive
2-DE	Two-dimensional gel electrophoresis

References

1. Ortega Morente, E.; Fernández-Fuentes, M.A.; Grande Burgos, M.J.; Abriouel, H.; Pérez Pulido, R.; Gálvez, A. Biocide tolerance in bacteria. *Int. J. Food Microbiol.* **2013**, *162*, 13–25. [CrossRef] [PubMed]
2. Shengzhi, Y.; Guoyan, W.; Mei, L.; Wenwen, D.; Hongning, W.; Likou, Z. Antibiotic and disinfectant resistance of *Salmonella* isolated from egg production chains. *Yi Chuan* **2016**, *38*, 948–956. [PubMed]

3.   Whitehead, R.N.; Overton, T.W.; Kemp, C.L.; Webber, M.A. Exposure of *Salmonella enterica* serovar Typhimurium to high level biocide challenge can select multidrug resistant mutants in a single step. *PLoS ONE* **2011**, *6*, e22833. [CrossRef] [PubMed]

4.   Gadea, R.; Glibota, N.; Pérez Pulido, R.; Gálvez, A.; Ortega, E. Adaptation to biocides cetrimide and chlorhexidine in bacteria from organic foods: Association with tolerance to other antimicrobials and physical stresses. *J. Agric. Food Chem.* **2017**, *65*, 1758–1770. [CrossRef] [PubMed]

5.   Silha, D.; Silhová, L.; Vytrasová, J. Survival of selected bacteria of *Arcobacter* genus in disinfectants and possibility of acquired secondary resistance to disinfectants. *J. Microb. Biotech. Food Sci.* **2016**, *5*, 326–329. [CrossRef]

6.   Curiao, T.; Marchi, E.; Grandgirard, D.; León-Sampedro, R.; Viti, C.; Leib, S.L.; Baquero, F.; Oggioni, M.R.; Martinez, J.L.; Coque, T.M. Multiple adaptive routes of *Salmonella enterica* Typhimurium to biocide and antibiotic exposure. *BMC Genom.* **2016**, *17*, 491–507. [CrossRef] [PubMed]

7.   Majtánová, L.; Majtán, V. Effect of disinfectants on surface hydrophobicity and mobility in *Salmonella enterica* serovar Typhimurium DT104. *Ceska Slov. Farm.* **2003**, *52*, 141–147. [PubMed]

8.   Futoma-Kołoch, B.; Ksiażczyk, M.; Korzekwa, K.; Migdał, I.; Pawlak, A.; Jankowska, M.; Kędziora, A.; Dorotkiewicz-Jach, A.; Bugla-Płoskońska, G. Selection and electrophoretic characterization of *Salmonella enterica* subsp. *enterica* biocide variants resistant to antibiotics. *Pol. J. Vet. Sci.* **2015**, *18*, 725–732.

9.   Futoma-Kołoch, B.; Ksiażczyk, M. The risk of *Salmonella* resistance following exposure to common disinfectants: An emerging problem. *Biol. Int.* **2013**, *53*, 54–66.

10.   Majtán, V.; Majtánová, L. Effect of disinfectants on the metabolism of *Salmonella enterica* serovar enteritidis. *Folia Microbiol.* **2003**, *48*, 643–648. [CrossRef]

11.   Uzer Celik, E.; Tunac, A.T.; Ates, M.; Sen, B.H. Antimicrobial activity of different disinfectants against cariogenic microorganisms. *Braz. Oral Res.* **2016**, *30*, e125. [CrossRef] [PubMed]

12.   Chapman, J.S. Biocide resistance mechanisms. *Int. Biod. Biodegr.* **2003**, *51*, 133–138. [CrossRef]

13.   Randall, L.P.; Coles, S.W.; Coldham, N.G.; Penuela, L.G.; Mott, A.C.; Woodward, M.J.; Piddock, L.J.V.; Webber, M. Commonly used farm disinfectants can select for mutant *Salmonella enterica* serovar Typhimurium with decreased susceptibility to biocides and antibiotics without compromising virulence. *J. Antimicrob. Chemother.* **2007**, *60*, 1273–1280. [CrossRef] [PubMed]

14.   Son, M.S.; Del Castilho, C.; Duncalf, K.A.; Carney, D.; Weiner, J.H.; Turner, R.J. Mutagenesis of SugE, a small multidrug resistance protein. *Biochem. Biophys. Res. Commun.* **2003**, *312*, 914–921. [CrossRef] [PubMed]

15.   Zou, L.; Meng, J.; McDermott, P.F.; Wang, F.; Yang, Q.; Cao, G.; Hoffmann, M.; Zhao, S. Presence of disinfectant resistance genes in *Escherichia coli* isolated from retail meats in the USA. *J. Antimicrob. Chemother.* **2014**, *69*, 2644–2649. [CrossRef] [PubMed]

16.   Kariuki, S.; Gordon, M.A.; Feasey, N.; Parry, C.M. Antimicrobial resistance and management of invasive *Salmonella* disease. *Vaccine* **2015**, *33*, C21–C29. [CrossRef] [PubMed]

17.   Futoma-Kołoch, B.; Bugla-Płoskońska, G.; Sarowska, J. Searching for outer membrane proteins typical of serum-sensitive and serum-resistant phenotypes of *Salmonella*. In *Salmonella-Distribution, Adaptation, Control Measures, and Molecular Technologies*; Annous, B.A., Ed.; InTech: Rijeka, Croatia, 2012; pp. 265–290.

18.   Riva, R.; Korhonen, T.K.; Meri, S. The outer membrane protease PgtE of *Salmonella enterica* interferes with the alternative complement pathway by cleaving factors B and H. *Front. Microbiol.* **2015**, *6*, 63. [CrossRef] [PubMed]

19.   Futoma-Kołoch, B.; Godlewska, U.; Guz-Regner, K.; Dorotkiewicz-Jach, A.; Klausa, E.; Rybka, J.; Bugla-Płoskońska, G. Presumable role of outer membrane proteins of *Salmonella* containing sialylated lipopolysaccharides serovar Ngozi, sv. Isaszeg and subspecies *arizonae* in determining susceptibility to human serum. *Gut Pathog.* **2015**, *7*, 18. [CrossRef] [PubMed]

20.   Dudek, B.; Krzyżewska, E.; Kapczyńska, K.; Rybka, J.; Pawlak, A.; Korzekwa, K.; Klausa, E.; Bugla-Płoskońska, G. Proteomic analysis of outer membrane proteins from *Salmonella* Enteritidis strains with different sensitivity to human serum. *PLoS ONE* **2016**, *11*, e0164069. [CrossRef] [PubMed]

21.   McDonnell, G.; Russell, A.D. Antiseptics and disinfectants: Activity, action, and resistance. *Clin. Microbiol. Rev.* **1999**, *12*, 147–179. [PubMed]

22.   Karatzas, K.A.G.; Randall, L.P.; Webber, M.; Piddock, L.J.V.; Humphrey, T.J.; Woodward, M.J.; Coldham, N.G.

Phenotypic and proteomic characterization of MAR variants of *Salmonella enterica* serovar Typhimurium selected following exposure to disinfectants. *Appl. Environ. Microbiol.* **2007**, *74*, 1508–1516. [CrossRef] [PubMed]

23. Condell, O.; Sheridan, Á.; Power, K.A.; Bonilla-Santiago, R.; Sergeant, K.; Renault, J.; Burgess, C.; Fanning, S.; Nally, J.E. Comparative proteomic analysis of *Salmonella* tolerance to the biocide active agent triclosan. *J. Proteom.* **2012**, *75*, 4505–4519. [CrossRef] [PubMed]

24. Roy, V.; Adams, B.L.; Bentley, W.E. Developing next generation antimicrobials by intercepting AI-2 mediated quorum sensing. *Enz. Microb. Technol.* **2011**, *49*, 113–123. [CrossRef] [PubMed]

25. Müderris, T.; Ürkmez, F.Y.; Küçüker, Ş.A.; Sağlam, M.F.; Yılmaz, G.R.; Güner, R.; Güleşen, R.; Açıkgöz, Z.C. Bacteremia caused by ciprofloxacin-resistant *Salmonella* serotype Kentucky: A case report and the review of literature. *Mikrobiyol. Bull.* **2016**, *50*, 598–605.

26. Bravo, D.; Silva, C.; Carter, J.A.; Hoare, A.; Alvarez, S.A.; Blondel, C.J.; Zaldívar, M.; Valvano, M.A.; Contreras, I. Growth-phase regulation of lipopolysaccharide O-antigen chain length influences serum resistance in serovars of *Salmonella*. *J. Med. Microbiol.* **2008**, *57*, 938–946. [CrossRef] [PubMed]

27. Marshall, J.M.; Gunn, J.S. The O-antigen capsule of *Salmonella enterica* serovar Typhimurium facilitates serum resistance and surface expression of FliC. *Infect. Immun.* **2015**, *83*, 3946–3959. [CrossRef] [PubMed]

28. Hart, P.J.; O'Shaughnessy, C.M.; Siggins, M.K.; Bobat, S.; Kingsley, R.A.; Goulding, D.A.; Crump, J.A.; Reyburn, H.; Micoli, F.; Dougan, G.; et al. Differential killing of *Salmonella enterica* serovar Typhi by antibodies targeting Vi and lipopolysaccharide O:9 antigen. *PLoS ONE* **2016**, *11*, e0145945. [CrossRef] [PubMed]

29. Wanga, W.; Jeffery, C.J. An analysis of surface proteomics results reveals novel candidates for intracellular/surface moonlighting proteins in bacteria. *Mol. BioSyst.* **2016**, *12*, 1420–1431. [CrossRef] [PubMed]

30. Seweryn, E.; Pietkiewicz, J.; Szamborska, A.; Gamian, A. Enolase on the surface of prockaryotic and eukaryotic cells is a receptor for human plasminogen. *Post. Hig. Med. Dośw.* **2007**, *61*, 672–682.

31. Andrews, J.M. Determination of minimum inhibitory concentrations. *J. Antimicrob. Chemother.* **2001**, *48* (Suppl. 1), 5–16. [CrossRef] [PubMed]

32. European Committee on Antimicrobial Susceptibility Testing. 2015, pp. 1–78. Available online: http://www.eucast.org/.

33. Ricci, V.; Tzakas, P.; Buckley, A.; Piddock, L.J.V. Ciprofloxacin-resistant *Salmonella enterica* serovar Typhimurium strains are difficult to select in the absence of AcrB and TolC. *Antimicrob. Agents Chemother.* **2006**, *50*, 38–42. [CrossRef] [PubMed]

34. Futoma-Kołoch, B.; Godlewska, U.; Bugla-Płoskońska, G.; Pawlak, A. Bacterial cell surface—Place where C3 complement activation occurs. 13th Conference Molecular biology in diagnostics of infectious diseases and biotechnology. *Diag. Mol.* **2012**, 120–123.

35. Bugla-Płoskońska, G.; Rybka, J.; Futoma-Kołoch, B.; Cisowska, A.; Gamian, A.; Doroszkiewicz, W. Sialic acid-containing lipopolysaccharides of *Salmonella* O48 strains-potential role in camouflage and susceptibility to the bactericidal effect of normal human serum. *Microb. Ecol.* **2010**, *59*, 601–613. [CrossRef] [PubMed]

36. Doroszkiewicz, W. Mechanism of antigenic variation in *Shigella flexneri* bacilli. IV. Role of lipopolysaccharides and their components in the sensitivity of *Shigella flexneri* 1b and its Lac+ recombinant to killing action of serum. *Arch. Immunol. Ther. Exp.* **1997**, *45*, 235–242.

37. Murphy, T.F.; Bartos, L.C. Surface-exposed and antigenically conserved determinants of outer membrane proteins of *Branhamella catarrhalis*. *Infect. Immun.* **1989**, *57*, 2938–2941. [PubMed]

38. Bugla-Płoskońska, G.; Korzeniowska-Kowal, A.; Guz-Regner, K. Reptiles as a source of *Salmonella* O48-clinically important bacteria for children: The relationship between resistance to normal cord serum and outer membrane protein patterns. *Microb. Ecol.* **2011**, *61*, 41–51. [CrossRef] [PubMed]

39. O'Farrell, P.H. High resolution two-dimensional electrophoresis of proteins. *J. Biol. Chem.* **1975**, *250*, 4007–4021. [PubMed]

40. Bednarz-Misa, I.; Serek, P.; Dudek, B.; Pawlak, A.; Bugla-Płoskońska, G.; Gamian, A. Application of zwitterionic detergent to the solubilization of *Klebsiella pneumoniae* outer membrane proteins for two-dimensional gel electrophoresis. *J. Microbiol. Methods* **2014**, *107*, 74–79. [CrossRef] [PubMed]

41. Bugla-Płoskońska, G.; Futoma-Kołoch, B.; Skwara, A.; Doroszkiewicz, W. Use of zwitterionic type of detergent in isolation of *Escherichia coli* O56 outer membrane proteins improves their two-dimensional electrophoresis (2-DE). *Pol. J. Microbiol.* **2009**, *58*, 205–209. [PubMed]
42. Shevchenko, A.; Tomas, H.; Havlis, J.; Olsen, J.V.; Mann, M. In-gel digestion for mass spectrometric characterization of proteins and proteomes. *Nat. Protoc.* **2006**, *1*, 2856–2860. [CrossRef] [PubMed]

6

PSFM-DBT: Identifying DNA-Binding Proteins by Combing Position Specific Frequency Matrix and Distance-Bigram Transformation

Jun Zhang and Bin Liu *

School of Computer Science and Technology, Harbin Institute of Technology Shenzhen Graduate School, Shenzhen 518055, China; junzhangcs@foxmail.com
* Correspondence: bliu@hit.edu.cn

Abstract: DNA-binding proteins play crucial roles in various biological processes, such as DNA replication and repair, transcriptional regulation and many other biological activities associated with DNA. Experimental recognition techniques for DNA-binding proteins identification are both time consuming and expensive. Effective methods for identifying these proteins only based on protein sequences are highly required. The key for sequence-based methods is to effectively represent protein sequences. It has been reported by various previous studies that evolutionary information is crucial for DNA-binding protein identification. In this study, we employed four methods to extract the evolutionary information from Position Specific Frequency Matrix (PSFM), including Residue Probing Transformation (RPT), Evolutionary Difference Transformation (EDT), Distance-Bigram Transformation (DBT), and Trigram Transformation (TT). The PSFMs were converted into fixed length feature vectors by these four methods, and then respectively combined with Support Vector Machines (SVMs); four predictors for identifying these proteins were constructed, including PSFM-RPT, PSFM-EDT, PSFM-DBT, and PSFM-TT. Experimental results on a widely used benchmark dataset PDB1075 and an independent dataset PDB186 showed that these four methods achieved state-of-the-art-performance, and PSFM-DBT outperformed other existing methods in this field. For practical applications, a user-friendly webserver of PSFM-DBT was established, which is available at http://bioinformatics.hitsz.edu.cn/PSFM-DBT/.

Keywords: PSFM-DBT; DNA binding protein; distance bigram transformation; PSFM

---

1. Introduction

DNA-binding proteins play crucial roles in various biological processes, such as DNA replication and repair, transcriptional regulation, the combination and separation of single-stranded DNA and other biological activities associated with DNA. Therefore, effective methods for identifying DNA-binding proteins are highly required.

There are some experimental recognition techniques for DNA-binding protein identification, such as filter binding assays, genetic analysis, chromatin immune precipitation on microarrays, and X-ray crystallography. However, these methods are both time consuming and expensive [1]. With the development of genomic and proteomic sequencing techniques, the number of protein sequences is growing rapidly. It is highly desired to develop fast and effective computational methods to identify the DNA binding proteins based on the protein sequences. In this regard, some computational methods based on machine learning algorithms have been proposed. These methods can be roughly divided into two groups: structure-based methods [2–8] and sequence-based methods. Stawiski et al. [7] analyzed the positive electrostatic patches in protein surface, and represented proteins with 12 features including the patch size, percent helix in patch, average surface area, hydrogen-bonding potential, three conserved special residues, and other features of the protein. These features were then inputted into a Neural Network (NN) for identifying DNA-binding proteins.

A webserver for the identification of DNA binding proteins (iDBPs) [9] recently was constructed for DNA binding protein identification, in which a random forest (RF) classifier was trained based on multiple structural features, such as electrostatic potential, cluster-based amino acid conservation patterns, secondary structure content of the patches, dipole moment and hydrogen-bonding potential. Song et al developed nDNA-Prot, which employed an imbalanced classifier [10]. Bhardwaj et al. [11] examined the sizes of positively charged patches on the surface of proteins, and used generated structural features to train a support vector machine (SVM) classifier. These structure-based methods achieved state-of-the-art performance. However, they require the structure information of proteins, which is not always available. In contrast, the sequence-based methods identify the DNA binding proteins only based on the sequence information of proteins, for example, Cai and Lin [12] proposed a method representing proteins employing pseudo amino acid composition (PseAAC) [13], in which amino acid composition, limited range correlation of hydrophobicity and solvent accessible surface area were taken into account. In method DNA-Prot [14], proteins was represented by various sequence properties, including frequency of amino acid, physical chemical properties, secondary structure, neutral amino acids, etc. Fang et al. [15] extracted protein features by using autocross-covariance (ACC) transform, pseudo amino acid composition, and dipeptide composition. Evolutionary profiles were introduced into this field by Kumar et al. [16]; they also developed a SVM-based predictor based on generated features. Recently, evolutionary profile was widely used in this field. Position specific score matrix distance transformation (PSSM-DT) [17] combined PSSM distance transformation with SVM. An improved DNA-binding protein prediction method (Local-DPP) [18] extracted local evolutionary information from some equally sized sub-PSSMs to represent proteins. Zhang et al. [19] proposed a new method in which feature vectors were extracted from PSSM, secondary structure, and physicochemical properties. They further improved the performance by using an improved Binary Firefly Algorithm (BFA) to filter noisy features and select optimal parameters for the classifier. Waris et al. [20] combined three different protein representations (dipeptide composition, split amino acid composition, and PSSM), and three machine learning algorithms (k Nearest Neighbor (KNN), SVM, and RF).

All these aforementioned methods have made great contributions to the development of this important field; the profile-based methods especially achieved state-of-the-art performance by incorporating evolutionary information into the predictors. Almost all of the machine-learning-based classifiers require fixed length feature vectors as inputs [21]. However, it is not an easy task to convert the profiles into feature vectors because a profile such as PSSM is a matrix with different dimensions. In this study, we employed four methods to extract the evolutionary information from Position Specific Frequency Matrix (PSFM), including Residue Probing Transformation (RPT) [22], Evolutionary Difference Transformation (EDT) [3], Distance-Bigram Transformation (DBT) [17,23,24], and Trigram Transformation (TT) [25]. The PSFMs were converted into fixed length feature vectors by these four methods, and then respectively combined with SVMs; four predictors for DNA binding protein identification were constructed, including PSFM-RPT, PSFM-EDT, PSFM-DBT and PSFM-TT. Experimental results on a widely used benchmark dataset and an independent dataset showed that these four methods achieved state-of-the-art-performance, and outperformed other existing methods in this field.

2. Result and Discussion

2.1. Impact of the Maximum Distance D

In order to evaluate the performance of the proposed methods, and select the optimized parameter, we explored the effect of the parameter D (see Equations (9) and (12)) in methods PSFM-EDT and PSFM-DBT. Taking into account the time cost, the predictive results were obtained by using 5-fold cross validation on benchmark dataset. The results of PSFM-EDT and PSFM-DBT with different values of D are shown in Figure 1a,b, respectively, from which we can see that PSFM-EDT and PSFM-DBT can achieve stable performance with different D values, and they achieved best performance when D = 7 and D = 4 respectively. Therefore, the parameter D of PSFM-EDT was set as 7 and the parameter D of PSFM-DBT was set as 4.

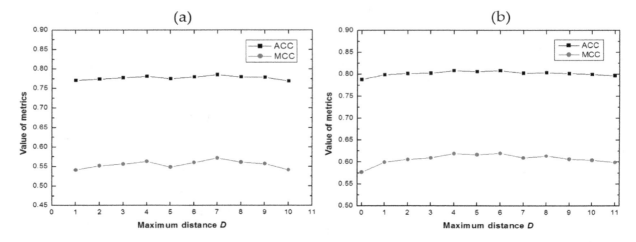

Figure 1. (**a**) The performance of Position Specific Frequency Matrix-Evolutionary Difference Transformation (PSFM-EDT) with different D on the benchmark dataset via five-cross validation. (**b**) The performance of Position Specific Frequency Matrix-Distance-Bigram Transformation (PSFM-DBT) with different D on the benchmark dataset via five-cross validation.

2.2. Comparison of the Four PSFM-Based Methods

The performance of the four proposed PSFM-based methods was shown in Table 1 by using jackknife test on benchmark dataset, and the corresponding ROC curves of these methods were shown in Figure 2a. From Table 1 and Figure 2a we can see that the PSFM-DBT is better than all the other methods. The reason is that PSFM-DBT incorporates more sequence-order effects by considering bigrams separated by different distances, which is more efficient than the other three approaches. Furthermore, a recent study showed that these sequence-order effects are critical for DNA binding protein identification [23].

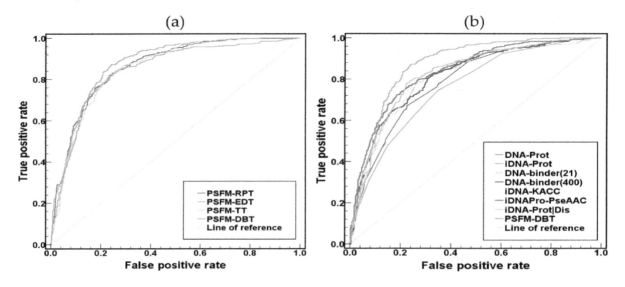

Figure 2. (**a**) The Receiver Operating Characteristic (ROC) curves of the four PSFM-based methods on the benchmark dataset using the jackknife tests. (**b**) The ROC curves of various methods on the benchmark dataset using the jackknife tests.

Table 1. The results of the four Position Specific Frequency Matrix (PSFM)-based methods on the benchmark dataset.

Method	ACC (%)	MCC	AUC (%)	SN (%)	SP (%)
PSFM-RPT [a]	78.88	0.5785	86.35	80.76	77.09
PSFM-EDT [b]	79.35	0.5868	84.49	78.86	**79.82**
PSFM-DBT [c]	**81.02**	**0.6224**	**87.12**	**84.19**	78.00
PSFM-TT [d]	79.16	0.5840	85.54	80.95	77.45

The results were obtained by jackknife test on benchmark dataset with SVM algorithm. The bold numbers represent the best values of the corresponding evaluation criteria in this table. [a] The parameters were: $c = 2^4$, $g = 2^6$; [b] The parameters were: $D = 7$, $c = 2^9$, $g = 2^{-2}$; [c] The parameters were: $D = 4$, $c = 2^3$, $g = 2^5$; [d] The parameters were: $c = 2^5$, $g = 2^{-9}$.

2.3. Comparison with Existing Methods

The performance of PSFM-DBT was compared with other existing methods on the benchmark dataset, including DNAbinder [16], DNA-Prot [14], iDNA-Prot [26], iDNA-KACC [27], PseDNA-Pro [17], iDNA-Prot|dis [23], iDNAPro-PseAAC [28], PSSM-DT [17] and Local-DPP [18]. Among these nine methods, DNAbinder, iDNAPro-PseAAC, PSSM-DT and Local-DPP are profile-based methods, and the other five methods are sequence-based methods. The performance of various methods was shown in Table 2 and Figure 2b, from which we can see that the profile-based methods achieved higher performance than other sequence-based methods, and PSFM-DBT obviously outperformed other methods, indicating that evolutionary information is critical for DNA binding protein identification, and PSFM-DBT is an efficient method. ACC represents the percentage of the samples which are correctly predicted among all samples; MCC explains the reliability of models; Sensitivity (SN) is an important measure, it presents the accuracy of predicting positive samples; Specificity (SP) denotes the percentage of true negative samples among negative samples; AUC is the area under ROC curve which gives a measure of the quality of binary classification methods, the larger AUC is, the better its predictive quality is.

Table 2. The performance of various methods on benchmark dataset.

Method	ACC (%)	MCC	AUC (%)	SN (%)	SP (%)
DNA-Prot	72.55	0.44	78.90	82.67	59.75
iDNA-Prot	75.40	0.50	76.10	83.81	64.73
DNAbinder (dimension 400)	73.58	0.47	81.50	66.47	80.36
DNAbinder (dimension 21)	73.95	0.48	81.40	68.57	79.09
PseDNA-Pro	76.55	0.53	N/A	79.61	73.63
iDNA-Prot|dis	77.30	0.54	82.60	79.40	75.27
iDNAPro-PseAAC	76.56	0.53	83.92	75.62	77.45
iDNA-KACC	75.16	0.50	83.00	77.52	72.90
PSSM-DT	79.96	0.62	86.50	78.00	**81.91**
Local-DPP	79.10	0.59	N/A	**84.80**	73.60
PSFM-DBT [a]	**81.02**	**0.62**	**87.12**	84.19	78.00

The results of all methods in the table were obtained by jackknife validation on benchmark dataset. The bold numbers represent the best values of the corresponding evaluation criteria in this table. [a] See the footnote of Table 1.

2.4. Independent Test

In this study, the four proposed PSFM-based methods were further evaluated on an independent dataset PDB186 constructed by Lou et al. [1]. It contains 93 DNA-binding proteins and 93 non-DNA-binding proteins selected from PDB. Because there are some proteins in benchmark dataset share more than 25% sequence identity with some proteins in independent dataset, this will lead to homology bias. In order to avoid this problem, the NCBI's BLASTCLUST [29] was employed to filter those proteins from the benchmark dataset which have more than 25% sequence identity to any protein in a same subset of the PDB186 dataset. Then we retrained the four proposed PSFM-based methods on such a reduced benchmark dataset, based on which the proteins in the independent dataset were predicted, and the results were shown in Table 3 and Figure 3a. PSFM-DBT achieved the top performance, which further demonstrates that it is a useful predictor for DNA binding protein identification.

Table 3. Performance of various methods on the independent dataset.

Method	ACC (%)	MCC	AUC (%)	SN (%)	SP (%)
DNA-Prot	61.80	0.240	N/A	69.90	53.80
iDNA-Prot	67.20	0.344	N/A	67.70	66.70
DNAbinder	60.80	0.216	60.70	57.00	64.50
DNABIND	67.70	0.355	69.40	66.70	68.80
DBPPred	76.90	0.538	79.10	79.60	74.20
iDNA-Prot|dis	72.00	0.445	78.60	79.50	64.50
iDNAPro-PseAAC-EL	71.50	0.442	77.80	82.80	60.2
iDNA-KACC-EL	79.03	0.611	81.40	**94.62**	63.44
PSSM-DT	80.00	**0.647**	87.40	87.09	72.83
Local-DPP	79.00	0.625	N/A	92.50	65.60
PSFM-TT	78.49	0.580	86.63	88.17	68.82
PSFM-RPT	79.57	0.594	85.67	84.95	**74.19**
PSFM-EDT	79.57	0.600	86.88	88.17	70.97
PSFM-DBT	**80.65**	0.624	**88.03**	90.32	70.97

The bold numbers represent the best values of the corresponding evaluation criteria in this table.

The number of DNA-binding proteins is much lower than that of the non DNA-binding proteins in the real world. In order to simulate real world applications, we evaluated the performance of PSFM-DBT on this independent dataset with different ratios of positive and negative samples, and the results were shown in Figure 3b, from which we can see that the ACC increases slightly as the ratio of positive samples increases, indicating that the PSFM-DBT can achieve stable performance and it is suitable for DNA binding protein prediction.

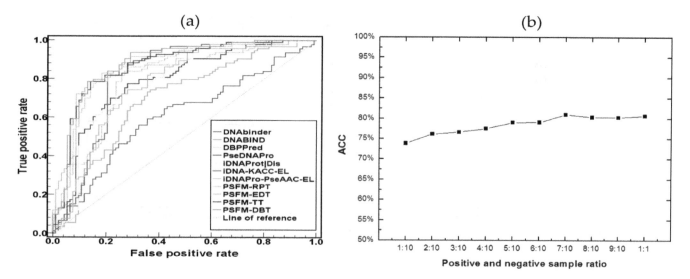

Figure 3. (a) The ROC curves of various methods on the independent dataset PDB186. (b) The performance of PSFM-DBT on the independent dataset with different ratios of positive samples.

2.5. Feature Analysis

To further investigate the importance of the features and to reveal the biological meaning of the features in proposed PSFM-DBT, we followed some previous studies [30,31] to calculate the discriminant weight vector in the feature space. The sequence-specific weight obtained from the SVM training process can be used to calculate the discriminant weight of each feature to measure the importance of the features. Given the weight vectors of the training set with N samples obtained from the kernel-based training $\mathbf{A} = [a_1, a_2, a_3, \ldots, a_N]$, the feature discriminant weight vector $\mathbf{W}$ in the feature space can be calculated by the following equation:

$$
\mathbf{W} = \mathbf{A} \cdot \mathbf{M} = \begin{bmatrix} a_1 \\ a_2 \\ \vdots \\ a_N \end{bmatrix}^{\mathbf{T}} \begin{bmatrix} m_{11} & m_{12} & \cdots & m_{1j} \\ m_{21} & m_{22} & \cdots & m_{2j} \\ \vdots & \vdots & \ddots & \vdots \\ m_{N1} & m_{N2} & \cdots & m_{Nj} \end{bmatrix} \tag{1}
$$

where $\mathbf{M}$ is the matrix of sequence representatives; $\mathbf{A}$ is the weight vectors of the training samples; N is the number of training samples; j is the dimension of the feature vector. The element in $\mathbf{W}$ represents the discriminative power of the corresponding feature.

In this study, the feature analysis was based on the predictor PSFM-DBT ($D = 4$). The discriminative weights of the 2000 features were calculated by Equation (1). Then we analyzed the features of amino acid composition and the features of amino acid bigrams respectively. The discriminant weights of the 400 features with $d = 0$ were visualized by a heatmap shown in Figure 4a. The 20 elements in the diagonal represent the 20 features of amino acids composition, from which we can see that the amino acid K (Lys) has the highest weight value among all the 20 features, indicating that amino acid K is critical for predicting the DNA binding proteins. For further exploration, all the discriminant weights of all the 20 features of amino acid composition were shown in Figure 4b. We can see that 10 amino acids show positive discriminative weights, while the other 10 amino acids show negative discriminative weights. The top five most discriminative amino acids are K (Lys), R (Arg), L (Leu), E (Glu) and T (Thr). It has been reported that the positively charged amino acids (such as Arg and Lys) and the polar amino acids (such as Thr and Ser) are important for a protein binding with a DNA sequence, and the acidic amino acids, such as D (Asp) and E (Glu), show low propensity for the interaction of protein and DNA [32,33]. However, amino acid Glu show positive discriminative weights in Figure 4b indicating that the bigram composition is more accurate than the amino acid composition.

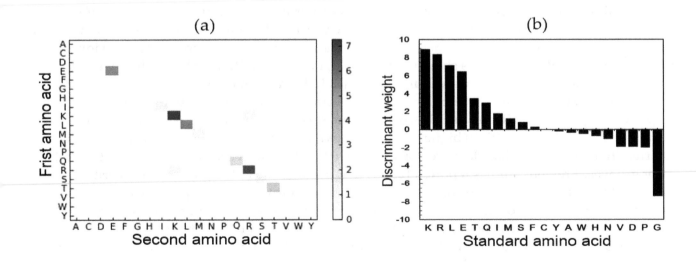

Figure 4. *Cont.*

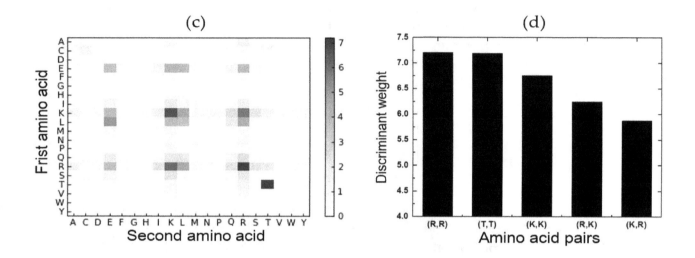

Figure 4. Feature analysis based on the features generated by PSFM-DBT. (**a**) The discriminant weights of the 400 features with $d = 0$. Each element in the figure represents the discriminant weight of the corresponding feature. The diagonal elements represent 20 features of amino acid composition. (**b**) The discriminant weights of the 20 amino acids according to amino acid composition. (**c**) The discriminant weights of the 400 standard amino acid pairs ($d = 1, 2, 3, 4$). Each element in the figure represents the sum of the discriminant weights of the corresponding bigrams, for example, the discriminant weight of bigrams (R, R) is $W_{(R, R)} = W_{(RR)} + W_{(R*R)} + W_{(R**R)} + W_{(R***R)}$, where * represents mismatch. The x-axis and y-axis represent the second amino acid and first amino acid in a bigram, respectively. (**d**) The discriminant weights of the top five most discriminant bigrams, including (R, R), (T, T), (K, K), (R, K) and (K, R).

Then we analyzed the rest of the 1600 features of amino acid bigrams obtained by PSFM-DBT with $d = 1, 2, 3, 4$. The weight values of the same kinds of bigrams with different d values were summed, and the results are shown in Figure 4c. We can see from this figure, the top five most discriminative amino acid bigrams are (R, R), (T, T), (K, K), (R, K) and (K, R), whose discriminant weights were shown in Figure 4d. These results further confirmed that the importance of amino acid R (Arg), T (Thr) and K (Lys). Interestingly, this conclusion is fully consistent with previous studies [32–35]. A specific DNA-binding protein 1IGN chain B was selected as an example to further explore the importance of the features in PSFM-DBT. 1IGNB is known as the yeast RAP1, a multifunctional protein binding with the telomeric DNA in the yeast *S. cerevisiae* via a sequence-specific manner, it is also involved in transcriptional regulation [36]. As shown in Figure 4d, bigrams (R, R) have the highest weight values among all the four bigrams. There are four kinds of (R, R) bigrams, including RR, R*R, R**R and R***R (* represents mismatch) with distance $d = 1, 2, 3, 4$ respectively. The distributions of these bigrams in the protein sequence 1IGNB and its 3D structure were shown in Figure 5a,c, respectively, from which we can see that most of the (R, R) bigrams were located in the DNA binding regions, except that two occurred in the structural disordered regions, and all (R, R) bigrams occurred in the area close to DNA major grooves. Previous studies reported [23,34] that the arginine rich region is indeed critical for the protein helix, and DNA major groove interaction by a mechanism known as 'phosphate bridging by an arginine-rich helix'. Moreover, we counted the numbers of these amino acid residues interacting with DNA in protein 1IGNB, the corresponding histogram is shown in Figure 5b, from which we can see that the positively charged amino acids (Arg, Lys and His) and the polar amino acids (Thr, Ser and Asn) are more likely to bind to DNA. This proved the correctness of the above conclusion, and explained the reason why the proposed PSFM-DBT predictor works well for DNA binding protein identification.

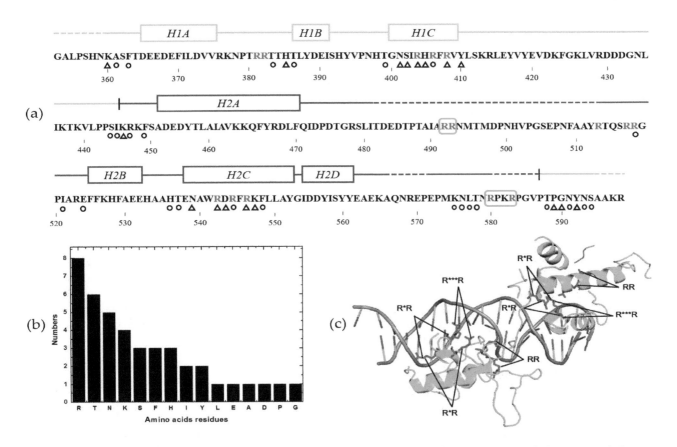

Figure 5. (**a**) The distributions of bigrams (R, R) in protein 1IGNB. The structural domains of this protein are color coded (orange represents domain 1, purple represents domain 2, and C-terminal tail is shown in blue). The open rectangles indicate the positions of helices, and broken lines mark regions of structural disorder. Residues interacting with DNA bases are indicated by triangles, and those contacting the phosphate backbone are indicated by circles. The two (R, R) bigrams shown in green rectangles are the two bigrams occurring in non-DNA-binding regions. (**b**) Histogram of the number of amino acid residues which binding with DNA in protein 1IGNB. (**c**) The distributions of bigrams (R, R) with different distances in the 3D structure of protein 1IGNB. The 3D structures of protein and DNA are shown in green and brown, respectively.

2.6. Web-Server Guide

We established an accessible web-server for the proposed PSFM-DBT predictor. Furthermore, for the convenience of the vast majority of experimental scientists, a step-by-step guide about how to use the web-server without the need to carefully understand the mathematical details was stated as follows.

Step 1. Open the web-server at http://bioinformatics.hitsz.edu.cn/PSFM-DBT/ and you will see the home page of PSFM-DBT, as shown in Figure 6. Click on the "ReadMe" button to see a brief introduction of the server and the caveat when using it.

Step 2. You can input the query sequences into the input box or directly upload your input data via the "Browse" button. The input sequence should be in the FASTA format. The examples of sequences in the FASTA format could be shown in the input box by clicking the Example button right above the input box.

Step 3. Click on the "Submit" button to execute the recognition program, then the predicted results will be shown in a new page. For example, if you use the four example protein sequences as the input, you will see on your computer screen that the first and second query sequences are DNA-binding proteins. The third and fourth are non-DNA-binding proteins.

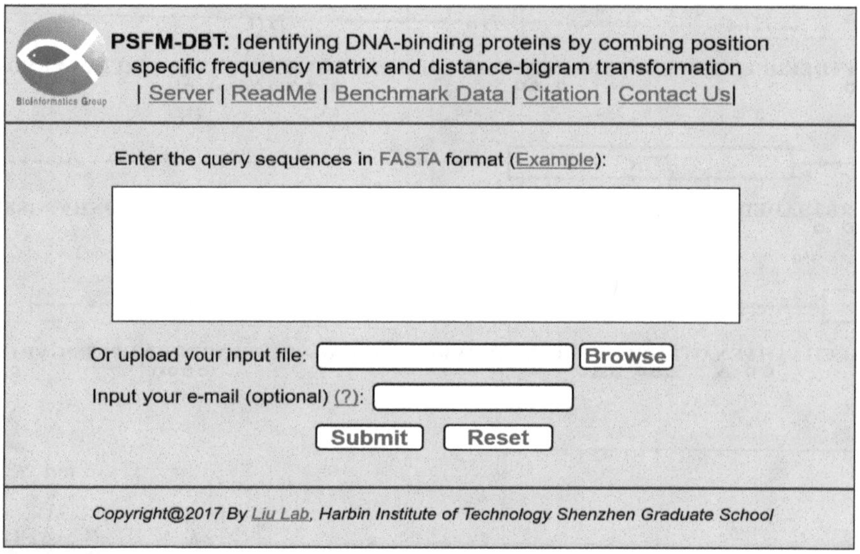

Figure 6. A semi-screenshot to show the home page of the web-server PSFM-DBT, which is available at http://bioinformatics.hitsz.edu.cn/PSFM-DBT/.

3. Methods and Materials

3.1. Dataset

The quality of the data set determines the quality of the research results. In the current study, we selected a widely used dataset PDB1075 [23] as the benchmark dataset. PDB1075 was constructed by Liu et al., which can be formulated as

$$\mathbb{S} = \mathbb{S}^+ \cup \mathbb{S}^- \tag{2}$$

where $\mathbb{S}^+$ is the subset of positive samples, $\mathbb{S}^-$ is the subset of negative samples and the symbol $\cup$ represents the "union" in the set theory. These proteins were all extracted from Protein Data Bank (PDB) released at December 2013, where DNA-binding proteins were obtained by searching the mmCIF keyword of 'DNA binding protein' through the advanced search interface and non-DNA-binding proteins were obtained by randomly extracting from PDB. To construct a high quality and non-redundant benchmark dataset, these proteins were filtered strictly according to the following criteria. (1) Remove all the sequences which have less than 50 amino acids or contain character of 'X'. (2) Using PISCES [37] to filter those sequences that have $\geq 25\%$ pairwise sequence similarity to any other in the same subset. Finally, the subset $\mathbb{S}^+$ consist of 525 DNA-binding proteins and the subset $\mathbb{S}^-$ consists of 550 non-DNA-binding proteins.

3.2. Protein Representation

One of the most challenging problems in machine learning-based methods for computational biology is how to effectively represent a biological sequence with a discrete model [38–40], because all the existing machine learning algorithms [41], such as NN, SVM, RF, and KNN can only handle vector rather than protein sequences with different lengths. To solve this problem, many researchers have proposed various methods. Previous experimental results showed that evolutionary information can obviously improve the performance of predictors for identifying DNA-binding proteins. In order to incorporate the evolutionary information into the predictors, we employed four feature extraction methods to extract the evolutionary information from the Position Specific Frequency Matrix (PSFM) [42]. PSFM and the four methods will be introduced in more detail in the following sections.

3.2.1. Position Specific Frequency Matrix

PSFM has been widely used in the field of predicting the structure and function of proteins [42,43]. Therefore, in this study, we employed the PSFM, which was generated by using PSI-BLAST [29] to search the target proteins against the non-redundant database NRDB90 [44] with default parameters, except the iteration and e-value were set as 10 and 0.001, respectively.

Given a protein sequence $\mathbf{P}$ with L amino acids, it can be formulated as:

$$\mathbf{P} = R_1R_2R_3R_4R_5 \cdots R_L \tag{3}$$

where R_1 represents the 1st residue, R_2 the 2nd residue, and so forth.

The PSFM profile can be represented as a matrix with dimensions of $20 \times L$ as follows:

$$\text{PSFM} = \begin{bmatrix} P_{1,1} & P_{1,2} & \cdots & P_{1,20} \\ P_{2,1} & P_{2,2} & \cdots & P_{2,20} \\ \vdots & \vdots & \ddots & \vdots \\ P_{L,1} & P_{L,2} & \cdots & P_{L,20} \end{bmatrix} \tag{4}$$

where 20 represents the number of standard amino acids, and L is the length of the query protein sequence. The element $P_{i,j}$ represents the occurrence probability of amino acid j at position i of the protein sequence, the rows of matrix represent the positions of the sequence, and the columns of the matrix represent the 20 standard amino acids. The sum of elements in each row is 1.

3.2.2. Residue Probing Transformation

RPT, first proposed by Jeong et al. [22], focuses on domains with similar conservation rates by grouping domain families based on their conservation scores in PSSM profiles. Because the idea is similar to the probe concept used in microarray technologies, it was called RPT. Each probe is a standard amino acid, and corresponds to a particular column in the PSFM profiles.

Given a PSFM (Equation (4)), it was divided into 20 groups according to 20 different standard amino acids, and for each group, we calculated the sum of the PSFM values in every column, leading to a feature vector of 20 dimension. Iteratively, for the 20 groups, the PSFM was translated into a Matrix $\mathbf{M}$ with 20×20 dimension, as follows:

$$\mathbf{M} = \begin{bmatrix} e_{1,1} & e_{1,2} & \cdots & e_{1,20} \\ e_{2,1} & e_{2,2} & \cdots & e_{2,20} \\ \vdots & \vdots & \ddots & \vdots \\ e_{20,1} & e_{20,2} & \cdots & e_{20,20} \end{bmatrix} \tag{5}$$

The $\mathbf{M}$ was then transferred into a feature vector of 400 dimension, as follows:

$$\mathbf{P} = [f(e_{1,1})\ f(e_{1,2})\ \cdots\ f(e_{i,j})\ \cdots\ f(e_{20,20})] \tag{6}$$

where $f(e_{i,j})$ was calculated by the following equation:

$$f(e_{i,j}) = \frac{e_{i,j}}{L}\ (i,j = 1, 2, \cdots, 20) \tag{7}$$

In this study, the amino acid composition of the 20 standard amino acids in PSFM was also incorporated into the RPT approach. As a result, the dimension of the corresponding feature vector is $400 + 20 = 420$.

3.2.3. Evolutionary Difference Transformation

EDT [3] is able to extract the information of the non-co-occurrence probability of two amino acids separated by a certain distance d in protein during the evolutionary process of the protein. The d is the distance between these two amino acids ($d = 1, 2, \ldots, L_{\min} - 1$, where $L_{\min}$ is the length of the shortest proteins in the benchmark dataset (Equation (2)). For example, $d = 1$ means the two amino acids are adjacent; $d = 2$ means there is one amino acid between the two amino acids; $d = 3$ means there are two amino acids between the two amino acids, and so forth.

For a given PSFM (Equation (4)), it can be transferred into a feature vector, as follows:

$$\mathbf{P} = [\psi_1\ \psi_2\ \cdots\ \psi_k\ \cdots\ \psi_\Omega] \tag{8}$$

where Ω is an integer reflecting the vector's dimension, its value is $D \times 400$; where D is the maximum value of d. The non-co-occurrence probability of two amino acids separated by distance d can be calculated by:

$$f(A_x, A_y|d) = \frac{1}{L-d} \sum_{i=1}^{L-d} \left(P_{i,x} - P_{i+d,y}\right)^2 \tag{9}$$

where $P_{i,x}$ ($P_{i+d,y}$) is the element in PSFM; A_x and A_y can be any of the 20 standard amino acids in the protein (Equation (3)).

Thus, each element in feature vector (Equation (8)) is obtained by

$$\begin{cases} \psi_1 = f(A_1, A_1|1) \\ \psi_2 = f(A_1, A_2|1) \\ \cdots \\ \psi_{400} = f(A_{20}, A_{20}|1) \\ \cdots \\ \psi_k = f(A_x, A_y|d) \\ \cdots \\ \psi_\Omega = f(A_{20}, A_{20}|D) \end{cases}, \ (1 \le d \le D) \tag{10}$$

3.2.4. Distance-Bigram Transformation

DBT [17,23,24] calculate the occurrence frequency of a combination of two amino acids separated by a certain distance along the protein sequence. The distance d is determined by the number of amino acids between the two amino acids of bigram. Some previous studies [17,23,24] have reported that the occurrence frequencies of amino acid pairs can well capture characteristics of proteins and they worked well for protein functionality annotation. To capture the characteristics of DNA-binding proteins, we represented proteins by combining PSFM with distance-bigram transformation, which can transform PSFM into fixed length feature vector.

For a given PSFM (Equation (4)), it can be transferred into a feature vector, as follows:

$$\mathbf{P} = [\psi_1\ \psi_2\ \cdots\ \psi_k\ \cdots\ \psi_\Omega] \tag{11}$$

where Ω is an integer to reflect the vector's dimension, its value is determined by D the maximum value of d. In order to incorporate the amino acid composition of the 20 standard amino acids in PSFM into the DBT approach, in this method, $d = 0$ was taken into account, therefore, $\Omega = 400 \times D + 400$.

The detail of DBT can be summarized mathematically as in the below equation.

$$f(A_x, A_y|d) = \frac{1}{L-d} \sum_{i=1}^{L-d} P_{i,x}P_{i+d,y} \tag{12}$$

where $P_{i,x}$ ($P_{i+d,y}$) is the element of the PSFM matrix; $f(A_x, A_y \mid d)$ represents the occurrence frequency of a bigram (standard amino acids A_x and A_y separated by a certain distance d) in evolutionary process.

Accordingly, each element in the feature vector (Equation (11)) is obtained by

$$
\left\{
\begin{array}{l}
\psi_1 = f(A_1, A_1 \mid 0) \\
\psi_2 = f(A_1, A_2 \mid 0) \\
\quad \cdots \\
\psi_{400} = f(A_{20}, A_{20} \mid 0) \\
\quad \cdots \\
\psi_k = f(A_x, A_y \mid d) \\
\quad \cdots \\
\psi_\Omega = f(A_{20}, A_{20} \mid D)
\end{array}
\right. \quad , \ (0 \leq d \leq D) \tag{13}
$$

3.2.5. Trigram Transformation

TT [25] is able to consider the local and global sequence-order effects by considering the trigrams along the protein sequences, the resulting feature vectors can be represented as:

$$
\mathbf{P} = [\psi_1 \, \psi_2 \, \cdots \, \psi_k \, \cdots \, \psi_{8000}] \tag{14}
$$

This technique can be summarized mathematically as shown in the below equation.

$$
f(A_x, A_y, A_z) = \sum_{i=1}^{L-2} P_{i,x} P_{i+1,y} P_{i+2,z} \tag{15}
$$

where $P_{i,x}$, $P_{i+1,y}$ and $P_{i+2,z}$ represent the corresponding elements in PSFM (Equation (4)); A_x, A_y and A_z can be any of the 20 standard amino acids in the protein (Equation (3)); $f(A_x, A_y, A_z)$ represents the occurrence frequency of trigram ($A_x A_y A_z$) in evolutionary process.

Accordingly, each element in the feature vector (Equation (14)) is obtained by

$$
\left\{
\begin{array}{l}
\psi_1 = f(A_1, A_1, A_1) \\
\psi_2 = f(A_1, A_1, A_2) \\
\quad \cdots \\
\psi_k = f(A_x, A_y, A_z) \\
\quad \cdots \\
\psi_{8000} = f(A_{20}, A_{20}, A_{20})
\end{array}
\right. \quad , \ (x, y, x = 1, 2, \cdots, 20) \tag{16}
$$

3.3. Support Vector Machine

SVM is a machine learning algorithm based on the structural-risk minimization principle of statistical learning theory. It was first presented by Vapnik [45] and has been widely used in bioinformatics. SVM is not only suitable for linear data, but also suitable for non-linear data. For linear data, SVM seek for an optimal hyper-plane to maximize the separation boundary between the positive instance and the negative instance, thereby separating the two instances. The nearest two points to the hyper-plane are called support vectors. For a non-linear model, SVM uses a non-linear transformation to map the input feature space to a high dimensional feature space where the samples can be well separated by an optimal hyper-plane. Kernel function is the most vital part for SVM; it determines the final performance of the SVM algorithm. There are some commonly used kernel functions for SVM, including Linear Function, Polynomial Function, Gaussian Function, Laplacian Function, Sigmoid Function and Radial Basis Function (RBF). SVM also can be used in the hierarchical classification [46]. Ensemble SVM may improve performance, too [47–49]. In the current study, an available SVM algorithm package called LIBSVM [50] was used to implement SVM algorithm, in which the RBF was chosen as the kernel function and the two parameters c and g were optimized by 5-fold cross validation on the benchmark.

3.4. Evaluation of Performance

In the current study, three commonly used methods were used to evaluate the performance of the proposed methods, including k-fold cross-validation, jackknife test and independent test. Moreover, sensitivity (SN), specificity (SP), accuracy (ACC), Matthew's correlation coefficient (MCC), the Receiver Operating Characteristic (ROC) curve [51] and the area under ROC curve (AUC) were selected as evaluation criteria. These criteria have been widely used in various studies for biological sequence annotation. They can be mathematically defined as follows:

$$\begin{cases} SN = \frac{TP}{TP+FN} \\ SP = \frac{TN}{TN+FP} \\ ACC = \frac{TP+TN}{TP+FP+TN+FN} \\ MCC = \frac{TP \times TN - FP \times FN}{\sqrt{(TP+FN) \times (TP+FP) \times (TN+FP) \times (TN+FN)}} \end{cases} \tag{17}$$

where TP is the number of true positive samples; TN is the number of true negative samples; FP is the number of false positive samples; and FN is the number of false negative samples. SN denote percentage of true positive samples among positive samples and SP denote percentage of true negative samples among negative samples. ACC represent the percentage of the samples which were correctly predicted among all samples. MCC explains the reliability of models, and its values range from -1 to 1, when MCC = -1 if all predictions are incorrect and when MCC = 1 if all predictions are correct. For MCC = 0, the prediction is no better than random. The ROC curve is a plot which is usually used to evaluate the performance of predictors. The AUC is the area under ROC curve which gives a measure of the quality of binary classification methods; the larger AUC, the better the predictive quality is.

4. Conclusions

To further improve the prediction accuracy and understand the binding regular patterns of DNA binding proteins, we explored and compared the performance of four feature extraction methods, including Residue Probing Transformation (RPT), Evolutionary Difference Transformation (EDT), Distance-Bigram Transformation (DBT), and Trigram Transformation (TT). Experimental results showed that PSFM-DBT achieved the best performance, and outperformed other existing methods in this field. This method was further evaluated on an independent dataset. Furthermore, some interesting patterns were discovered by analyzing the features generated PSFM-DBT, fully consistent with previous studies. Finally, a web server of the proposed PSFM-DBT predictor was established in order to help the users to use this method, which would be a useful tool for protein sequence analysis, especially for studying the structure and function of proteins. Future studies will focus on exploring advanced machine learning techniques to improve the performance of DNA binding protein prediction [52,53].

Acknowledgments: This work was supported by the National Natural Science Foundation of China (No. 61672184), the Natural Science Foundation of Guangdong Province (2014A030313695), Guangdong Natural Science Funds for Distinguished Young Scholars (2016A030306008), Scientific Research Foundation in Shenzhen (Grant No. JCYJ20150626110425228, JCYJ20170307152201596), and Guangdong Special Support Program of Technology Young talents (2016TQ03X618).

Author Contributions: Bin Liu conceived and designed the experiments; Jun Zhang performed the experiments; Bin Liu analyzed the data; Jun Zhang wrote the paper.

References

1. Lou, W.; Wang, X.; Chen, F.; Chen, Y.; Jiang, B.; Zhang, H. Sequence Based Prediction of DNA-Binding Proteins Based on Hybrid Feature Selection Using Random Forest and Gaussian Naive Bayes. *PLoS ONE* **2014**, *9*, e86703. [CrossRef] [PubMed]

2. Zhao, H.; Yang, Y.; Zhou, Y. Structure-based prediction of DNA-binding proteins by structural alignment and a volume-fraction corrected DFIRE-based energy function. *Bioinforma* **2010**, *26*, 1857–1863. [CrossRef] [PubMed]

3. Zhang, L.; Zhao, X.; Kong, L. Predict protein structural class for low-similarity sequences by evolutionary difference information into the general form of Chou's pseudo amino acid composition. *J. Theor. Biol.* **2014**, *355*, 105–110. [CrossRef] [PubMed]

4. Yu, X.; Cao, J.; Cai, Y.; Shi, T.; Li, Y. Predicting rRNA-, RNA-, and DNA-binding proteins from primary structure with support vector machines. *J. Theor. Biol.* **2006**, *240*, 175–184. [CrossRef] [PubMed]

5. Xia, J.; Zhao, X.; Huang, D. Predicting protein-protein interactions from protein sequences using meta predictor. *Amino Acids* **2010**, *39*, 1595–1599. [CrossRef] [PubMed]

6. Tjong, H.; Zhou, H. DISPLAR: An accurate method for predicting DNA-binding sites on protein surfaces. *Nucleic Acids Res.* **2007**, *35*, 1465–1477. [CrossRef] [PubMed]

7. Stawiski, E.W.; Gregoret, L.M.; Mandelgutfreund, Y. Annotating Nucleic Acid-Binding Function Based on Protein Structure. *J. Mol. Biol.* **2003**, *326*, 1065–1079. [CrossRef]

8. Shanahan, H.P.; Garcia, M.A.; Jones, S.; Thornton, J.M. Identifying DNA-binding proteins using structural motifs and the electrostatic potential. *Nucleic Acids Res.* **2004**, *32*, 4732–4741. [CrossRef] [PubMed]

9. Nimrod, G.; Schushan, M.; Szilagyi, A.; Leslie, C.; Bental, N. iDBPs: A web server for the identification of DNA binding proteins. *Bioinformatics* **2010**, *26*, 692–693. [CrossRef] [PubMed]

10. Song, L.; Li, D.; Zeng, X.; Wu, Y.; Guo, L.; Zou, Q. nDNA-prot: Identification of DNA-binding Proteins Based on Unbalanced Classification. *BMC Bioinform.* **2014**, *15*, 298. [CrossRef] [PubMed]

11. Bhardwaj, N.; Langlois, R.; Zhao, G.; Lu, H. Kernel-based machine learning protocol for predicting DNA-binding proteins. *Nucleic Acids Res.* **2005**, *33*, 6486–6493. [CrossRef]

12. Cai, Y.; Zhou, G.; Chou, K.-C. Support Vector Machines for Predicting Membrane Protein Types by Using Functional Domain Composition. *Biophys. J.* **2003**, *84*, 3257–3263. [CrossRef]

13. Chou, K.C. Prediction of protein cellular attributes using pseudo amino acid composition. *Proteins* **2001**, *43*, 246–255. [CrossRef] [PubMed]

14. Kumar, K.K.; Pugalenthi, G.; Suganthan, P.N. DNA-Prot: Identification of DNA binding proteins from protein sequence information using random forest. *J. Biomol. Struct. Dyn.* **2009**, *26*, 679–686. [CrossRef] [PubMed]

15. Fang, Y.; Guo, Y.; Feng, Y.; Li, M. Predicting DNA-binding proteins: Approached from Chou's pseudo amino acid composition and other specific sequence features. *Amino Acids* **2007**, *34*, 103–109. [CrossRef] [PubMed]

16. Kumar, M.; Gromiha, M.M.; Raghava, G.P. Identification of DNA-binding proteins using support vector machines and evolutionary profiles. *BMC Bioinform.* **2007**, *8*, 463. [CrossRef] [PubMed]

17. Liu, B.; Xu, J.; Fan, S.; Xu, R.; Zhou, J.; Wang, X. PseDNA-Pro: DNA-Binding Protein Identification by Combining Chou's PseAAC and Physicochemical Distance Transformation. *Mol. Inform.* **2015**, *34*, 8–17. [CrossRef]

18. Wei, L.; Tang, J.; Zou, Q. Local-DPP: An improved DNA-binding protein prediction method by exploring local evolutionary information. *Inf. Sci.* **2016**, *384*, 135–144. [CrossRef]

19. Zhang, J.; Gao, B.; Chai, H.; Ma, Z.; Yang, G. Identification of DNA-binding proteins using multi-features fusion and binary firefly optimization algorithm. *BMC Bioinform.* **2016**, *17*, 323. [CrossRef] [PubMed]

20. Waris, M.; Ahmad, K.; Kabir, M.; Hayat, M. Identification of DNA binding proteins using evolutionary profiles position specific scoring matrix. *Neurocomputing* **2016**, *199*, 154–162. [CrossRef]

21. Liu, S.; Wang, S.; Ding, H. Protein sub-nuclear location by fusing AAC and PSSM features based on sequence information. In Proceedings of the International Conference on Electronics Information and Emergency Communication, Beijing, China, 14 May 2015.

22. Jeong, J.C.; Lin, X.; Chen, X.-W. On Position-Specific Scoring Matrix for Protein Function Prediction. *IEEE/ACM Trans. Comput. Biol. Bioinform.* **2011**, *8*, 308–315. [CrossRef] [PubMed]

23. Liu, B.; Xu, J.; Lan, X.; Xu, R.; Zhou, J.; Wang, X.; Chou, K.-C. iDNA-Prot | dis: Identifying DNA-Binding Proteins by Incorporating Amino Acid Distance-Pairs and Reduced Alphabet Profile into the General Pseudo Amino Acid Composition. *PLoS ONE* **2014**, *9*, e106691. [CrossRef] [PubMed]

24. Saini, H.; Raicar, G.; Lal, S.P.; Dehzangi, A.; Imoto, S.; Sharma, A. Protein Fold Recognition Using Genetic Algorithm Optimized Voting Scheme and Profile Bigram. *J. Softw.* **2016**, *11*, 756–767. [CrossRef]

25. Paliwal, K.K.; Sharma, A.; Lyons, J.; Dehzangi, A. A tri-gram based feature extraction technique using linear probabilities of position specific scoring matrix for protein fold recognition. *IEEE Trans. Nanobiosci.* **2014**, *13*, 44–50. [CrossRef] [PubMed]

26. Lin, W.; Fang, J.; Xiao, X.; Chou, K.-C. iDNA-Prot: Identification of DNA binding proteins using random forest with grey model. *PLoS ONE* **2011**, *6*, e24756. [CrossRef] [PubMed]

27. Liu, B.; Wang, S.; Dong, Q.; Li, S.; Liu, X. Identification of DNA-binding proteins by combining auto-cross covariance transformation and ensemble learning. *IEEE Trans. NanoBiosci.* **2016**, *15*, 328–334. [CrossRef] [PubMed]

28. Liu, B.; Wang, S.; Wang, X. DNA binding protein identifcation by combining pseudo amino acid composition and profile-based protein representation. *Sci. Rep.* **2015**, *5*, 15497.

29. Altschul, S.F.; Madden, T.L.; Schaffer, A.A.; Zhang, J.; Zhang, Z.; Miller, W.; Lipman, D. J. Gapped BLAST and PSI-BLAST: A new generation of protein database search programs. *Nucleic Acids Res.* **1997**, *25*, 3389–3402. [CrossRef] [PubMed]

30. Liu, B.; Zhang, D.; Xu, R.; Xu, J.; Wang, X.; Chen, Q.; Dong, Q.; Chou, K.-C. Combining evolutionary information extracted from frequency profiles with sequence-based kernels for protein remote homology detection. *Bioinformatics* **2014**, *30*, 472–479. [CrossRef] [PubMed]

31. Liu, B.; Wang, X.; Lin, L.; Dong, Q.; Wang, X. A Discriminative Method for Protein Remote Homology Detection and Fold Recognition Combining Top-n-grams and Latent Semantic Analysis. *BMC Bioinform.* **2008**, *9*, 510. [CrossRef] [PubMed]

32. Mandelgutfreund, Y.; Schueler, O.; Margalit, H. Comprehensive Analysis of Hydrogen Bonds in Regulatory Protein DNA-Complexes: In Search of Common Principles. *J. Mol. Biol.* **1995**, *253*, 370–382. [CrossRef] [PubMed]

33. Jones, S.; Van Heyningen, P.; Berman, H.M.; Thornton, J.M. Protein-DNA interactions: A structural analysis. *J. Mol. Biol.* **1999**, *287*, 877–896. [CrossRef] [PubMed]

34. Tanaka, Y.; Nureki, O.; Kurumizaka, H.; Fukai, S.; Kawaguchi, S.; Ikuta, M.; Iwahara, J.; Okazaki, T.; Yokoyama, S. Crystal structure of the CENP-B protein–DNA complex: The DNA-binding domains of CENP-B induce kinks in the CENP-B box DNA. *EMBO J.* **2001**, *20*, 6612–6618. [CrossRef] [PubMed]

35. Szabóová, A.; Kuželka, O.; Železný, F.; Tolar, J. Prediction of DNA-binding propensity of proteins by the ball-histogram method using automatic template search. *BMC Bioinform.* **2012**, *13*, 1–11. [CrossRef] [PubMed]

36. Konig, P.; Giraldo, R.; Chapman, L.; Rhodes, D. The crystal structure of the DNA-binding domain of yeast RAP1 in complex with telomeric DNA. *Cell* **1996**, *85*, 125. [CrossRef]

37. Wang, G.; Dunbrack, R.L. PISCES: Recent improvements to a PDB sequence culling server. *Nucleic Acids Res.* **2005**, *33*, W94–W98. [CrossRef] [PubMed]

38. Liu, B.; Liu, F.; Fang, L.; Wang, X.; Chou, K.-C. repRNA: A web server for generating various feature vectors of RNA sequences. *Mol. Genet. Genom.* **2016**, *291*, 473–481. [CrossRef]

39. Zhu, L.; Deng, S.-P.; Huang, D.-S. A Two-Stage Geometric Method for Pruning Unreliable Links in Protein-Protein Networks. *IEEE Trans. Nanobiosci.* **2015**, *14*, 528–534.

40. Deng, S.-P.; Huang, D.-S. SFAPS: An R package for structure/function analysis of protein sequences based on informational spectrum method. *Methods* **2014**, *69*, 207–212. [CrossRef] [PubMed]

41. Zhao, Z.-Q.; Huang, D.-S.; Sun, B.-Y. Human face recognition based on multi-features using neural networks committee. *Pattern Recognit. Lett.* **2004**, *25*, 1351–1358. [CrossRef]

42. Liu, B.; Wang, X.; Chen, Q.; Dong, Q.; Lan, X. Using Amino Acid Physicochemical Distance Transformation for Fast Protein Remote Homology Detection. *PLoS ONE* **2012**, *7*, e46633. [CrossRef] [PubMed]

43. Wang, B.; Chen, P.; Huang, D.-S.; Li, J.-J.; Lok, T.-M.; Lyu, M.R. Predicting protein interaction sites from residue spatial sequence profile and evolution rate. *FEBS Lett.* **2006**, *580*, 380–384. [CrossRef]

44. Holm, L.; Sander, C. Removing near-neighbour redundancy from large protein sequence collections. *Bioinformatics* **1998**, *14*, 423–429. [CrossRef] [PubMed]

45. Cortes, C.; Vapnik, V. Support-Vector Networks. *Mach. Learn.* **1995**, *20*, 273–297. [CrossRef]

46. Li, D.; Ju, Y.; Zou, Q. Protein Folds Prediction with Hierarchical Structured SVM. *Curr. Proteom.* **2016**, *13*, 79–85. [CrossRef]

47. Chen, W.; Xing, P.; Zou, Q. Detecting N6-methyladenosine sites from RNA transcriptomes using ensemble Support Vector Machines. *Sci. Rep.* **2017**, *7*, 40242. [CrossRef] [PubMed]

48. Zou, Q.; Guo, J.; Ju, Y.; Wu, M.; Zeng, X.; Hong, Z. Improving tRNAscan-SE annotation results via ensemble classifiers. *Mol. Inform.* **2015**, *34*, 761–770. [CrossRef]

49. Zhu, L.; You, Z.-H.; Huang, D.-S. Increasing the reliability of protein–protein interaction networks via non-convex semantic embedding. *Neurocomputing* **2013**, *121*, 99–107. [CrossRef]

50. Chang, C.; Lin, C. LIBSVM: A library for support vector machines. *ACM Trans. Intell. Syst. Technol.* **2011**, *2*, 27. [CrossRef]

51. Sonego, P.; Kocsor, A.; Pongor, S. ROC analysis: Applications to the classification of biological sequences and 3D structures. *Brief. Bioinform.* **2008**, *9*, 198–209. [CrossRef] [PubMed]

52. Huang, D.-S. Radial basis probabilistic neural networks: Model and application. *Int. J. Pattern Recognit. Artif. Int.* **1999**, *13*, 1083–1101. [CrossRef]

53. Huang, D.S.; Du, J.-X. A constructive hybrid structure optimization methodology for radial basis probabilistic neural networks. *IEEE Trans. Neural Netw.* **2008**, *19*, 2099–2115. [CrossRef]

7

Protein Subcellular Localization with Gaussian Kernel Discriminant Analysis and its Kernel Parameter Selection

Shunfang Wang [1,*,†] **, Bing Nie** [1,†] **, Kun Yue** [1,*] **, Yu Fei** [2,*] **, Wenjia Li** [1] **and Dongshu Xu** [1]

[1] Department of Computer Science and Engineering, School of Information Science and Engineering, Yunnan University, Kunming 650504, China; bingn2017@gmail.com (B.N.); woshiwenzi666@gmail.com (W.L.); qq78316519@gmail.com (D.X.)

[2] School of Statistics and Mathematics, Yunnan University of Finance and Economics, Kunming 650221, China

* Correspondence: sfwang_66@ynu.edu.cn (S.W.); kyue@ynu.edu.cn (K.Y.); feiyukm@aliyun.com (Y.F.)

† These authors contributed equally to this work.

Abstract: Kernel discriminant analysis (KDA) is a dimension reduction and classification algorithm based on nonlinear kernel trick, which can be novelly used to treat high-dimensional and complex biological data before undergoing classification processes such as protein subcellular localization. Kernel parameters make a great impact on the performance of the KDA model. Specifically, for KDA with the popular Gaussian kernel, to select the scale parameter is still a challenging problem. Thus, this paper introduces the KDA method and proposes a new method for Gaussian kernel parameter selection depending on the fact that the differences between reconstruction errors of edge normal samples and those of interior normal samples should be maximized for certain suitable kernel parameters. Experiments with various standard data sets of protein subcellular localization show that the overall accuracy of protein classification prediction with KDA is much higher than that without KDA. Meanwhile, the kernel parameter of KDA has a great impact on the efficiency, and the proposed method can produce an optimum parameter, which makes the new algorithm not only perform as effectively as the traditional ones, but also reduce the computational time and thus improve efficiency.

Keywords: protein subcellular localization; kernel parameter selection; kernel discriminant analysis (KDA); Gaussian kernel function; dimension reduction

1. Introduction

Some proteins can only play the role in one specific place in the cell while others can play the role in several places in the cell [1]. Generally, a protein can function correctly only when it is localized to a correct subcellular location [2]. Therefore, protein subcellular localization prediction is an important research area of proteomics. It is helpful to predict protein function as well as to understand the interaction and regulation mechanism of proteins [3]. Now, many methods have been used to predict protein subcellular location, such as green fluorescent protein labeling [4], mass spectrometry [5], and so on. However, these traditional experimental methods usually have many technical limitations, resulting in high cost of time and money. Thus, prediction of protein subcellular location based on machine learning has become a focus research in bioinformatics [6–8].

When we use the methods of machine learning to predict protein subcellular location, we must extract features of protein sequences. We can get some vectors after feature extraction, and then we use the classifier to process these vectors. However, these vectors are usually complex due to their high dimensionality and nonlinear property. In order to improve the prediction accuracy of protein subcellular location, an appropriate nonlinear method for reducing data dimension should be used before classification. Kernel discriminant analysis (KDA) [9] is a nonlinear reductive dimension algorithm based on kernel trick that has been used in many fields such as facial recognition and fingerprint identification. The KDA method not only reduces data dimensionality but also makes use of the classification information. This paper newly introduces the KDA method to predict protein subcellular location. The algorithm of KDA first maps sample data to a high-dimensional feature space by a kernel function, and then executes linear discriminant analysis (LDA) in the high-dimensional feature space [10], which indicates that kernel parameter selection will significantly affect the algorithm performance.

There are some classical algorithms used to select the parameter of kernel function, such as genetic algorithm, grid searching algorithm, and so on. These methods have high calculation precision but large amounts of calculation. In an effort to reduce computational complexity, recently, Xiao et al. proposed a method based on reconstruction errors of samples and used it to select the parameters of Gaussian kernel principal component analysis (KPCA) for novelty detection [11]. Their methods are applied into the toy data sets and UCI (University of CaliforniaIrvine) benchmark data sets to demonstrate the correctness of the algorithm. However, their innovation in the KPCA method aims at dimensional reduction rather than discriminant analysis, which leads to unsatisfied classification prediction accuracy. Thus, it is necessary to improve the efficiency of the method in [11] especially for some complex data such as biological data.

In this paper, an improved algorithm of selecting parameters of Gaussian kernel in KDA is proposed to analyze complex protein data and predict subcellular location. By maximizing the differences of reconstruction errors between edge normal samples and interior normal samples, the proposed method not only shows the same effect as the traditional grid-searching method, but also reduces the computational time and improves efficiency.

2. Results and Discussion

In this section, the proposed method (in Section 3.4) and the grid-searching algorithm (in Section 4.4) are both applied to predict protein subcellular localization. We use two standard data sets as the experimental data. The two used feature expressions are generated from PSSM (position specific scoring matrix) [12], which are the PsePSSM (pseudo-position specific scoring matrix) [12] and the PSSM-S (AAO + PSSM-AAO + PSSM-SAC + PSSM-SD = PSSM-S) [13]. Here AAO means consensus sequence-based occurrence, PSSM-AAO means evolutionary-based occurrence or semi-occurrence of PSSM, PSSM-SD is segmented distribution of PSSM and PSSM-SAC is segmented auto covariance of PSSM. The k-nearest neighbors (KNN) is used as the classifier in which Euclidean distance is adopted for the distance between samples. The flow of experiments is as follows.

- First, for each standard data set, we use the PsePSSM algorithm and the PSSM-S algorithm to extract features, respectively. Then totally we obtain four sample sets, which are GN-1000 (Gram-negative with PsePSSM which contains 1000 features), GN-220 (Gram-negative with PSSM-S which contains 220 features), GP-1000 (Gram-positive with PsePSSM which contains 1000 features) and GP-220 (Gram-positive with PsePSSM which contains 220 features).
- Second, we use the proposed method to select the optimum kernel parameter for the Gaussian KDA model and then use KDA to reduce the dimension of sample sets. The same procedure is also carried out for the traditional grid-searching method to form a comparison with the proposed method.
- Finally, we use the KNN algorithm to classify the reduced dimensional sample sets and use some criterions to evaluate the results and give the comparison results.

Some detailed information in experiments is as follows. For every sample set, we choose the class that contains the most samples to form the training set [8]. Let $S = [0.1, 0.2, 0.3, 0.4, 1, 2, 3, 4]$ be a candidate set of the Gaussian kernel parameter, which is proposed at random. When we use the KDA algorithm to reduce dimension, the number of retained eigenvectors must be less than or equal to $C - 1$ (C is the number of classes). Therefore, for sample sets GN-1000 and GN-220, the number of retained eigenvectors, which is denoted as d, can be from 1 to 7. For the sample sets GP-1000 and GP-220, d can be 1, 2, and 3. As far as the parameter u is concerned, when it is 5–8% of the average number of samples, good classification can be achieved [14]. Besides, we demonstrate the robustness of the proposed method with the variation of u in Section 2.2. So here we simply pick a general value for u, say 8. To sum up, in the following experiments, when certain parameters need to be fixed, their default values are as follows. The value of d is 7 for sample sets GN-1000 and GN-220, and 3 for GP-1000 and GP-220; the value of u is 8 and the k value in KNN classifier is 20.

2.1. The Comparison Results of the Overall Accuracy

2.1.1. The Accuracy Comparison between the Proposed Method and the Grid-Searching Method

In this section, first, the proposed method and the grid-searching method are respectively used in the prediction of protein subcellular localization with different d values. The experimental results are presented in Figure 1.

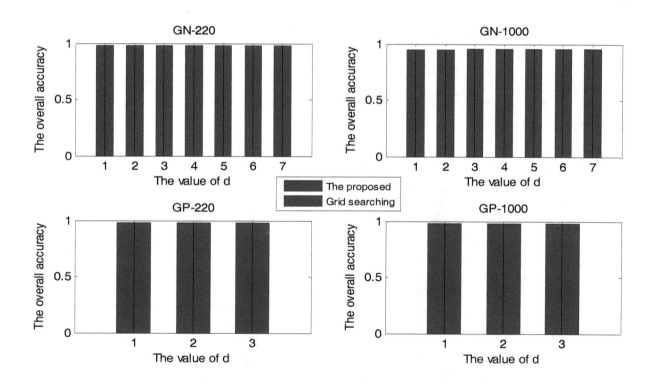

Figure 1. The overall accuracy versus d for four sample sets.

In Figure 1, all four sample sets suggest that when we use the KDA algorithm to reduce dimension, the larger the number of retained eigenvectors, the higher the accuracy. The overall accuracy of the proposed method is always the same as that of the grid-searching method, no matter which value of d. The proposed method is effective for selecting the optimal Gaussian kernel parameter.

Then, in the analyses and experiments, we find that superiority of the proposed method is the low runtime, which is demonstrated in Table 1 and Figure 2.

Table 1. The overall accuracy and the ratio of runtime for two methods.

Sample Sets	Overall Accuracy		Ratio (t_1/t_2)
GP-220 (PSSM-S)	The proposed method	0.9924	0.7087
	Grid searching method	0.9924	
GP-1000 (PsePSSM)	The proposed method	0.9924	0.7362
	Grid searching method	0.9924	
GN-220 (PSSM-S)	The proposed method	0.9801	0.7416
	Grid searching method	0.9801	
GN-1000 (PsePSSM)	The proposed method	0.9574	0.7687
	Grid searching method	0.9574	

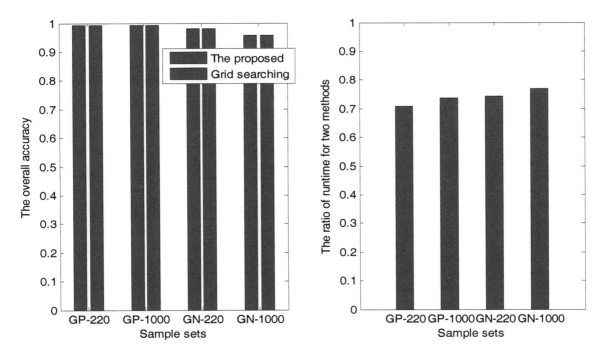

Figure 2. The overall accuracy and the ratio of runtime for two methods.

In Table 1, t_1 and t_2 are the runtimes of the proposed method and the grid-searching method, respectively. The overall accuracy and the ratio of t_1 and t_2 are presented in both Table 1 and Figure 2, from which we can see that for each sample set, the accuracy of the proposed method is always the same as that of the grid-searching method; meanwhile, the runtime of the former is about 70–80% of that of the latter, indicating that the proposed method has a higher efficiency than the grid-searching method.

2.1.2. The Comparison between Methods with and without KDA

In this experiment, we compare the overall accuracies between the cases of using KDA algorithm or not, with k values of the KNN classifier varying from 1 to 30. The experimental results are shown in Figure 3.

For each sample set, Figure 3 shows that the accuracy with KDA algorithm to reduce dimension is higher than that of without it. However, the kernel parameter has a great impact on the efficiency of the KDA algorithm, and the proposed method can be used to select the optimum parameter that makes the KDA perform perfect. Therefore, accuracy can be improved by using the proposed method to predict the protein subcellular localization.

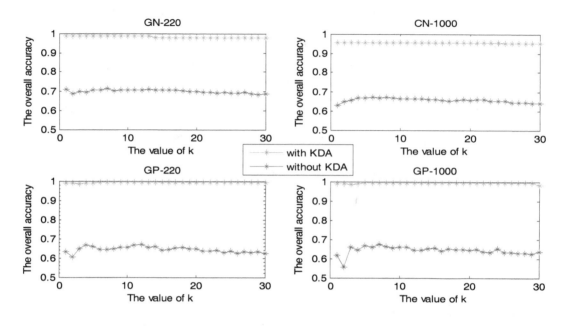

Figure 3. The overall accuracy versus k value with or without KDA algorithm.

2.2. The Robustness of the Proposed Method

In the proposed method, the value of u will have an impact on the radius value of neighborhood so that it can affect the number of the selected internal and edge samples. Figure 4 shows the experimental results when the value of u ranges from 6 to 10, in which the overall accuracies of the proposed method and the grid-searching method are given.

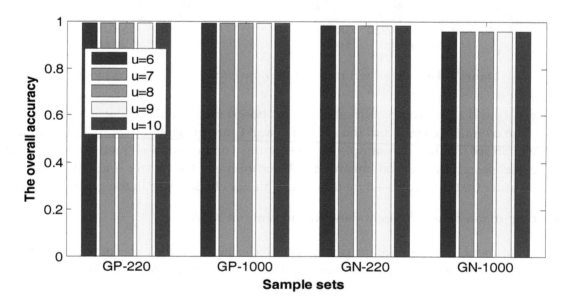

Figure 4. The overall accuracy for four sample sets with different u values.

It is easily seen from Figure 4 that the accuracy keeps invariable with different u values. The number of the selected internal and edge samples has little effect on the performance of the proposed method. Therefore, the method proposed in this paper has a good robustness.

2.3. Evaluating the Proposed Method with Some Regular Evaluation Criterions

In this subsection, we compute the values of some regular evaluation criterions with the proposed method for two standard data sets, which is show in Tables 2 and 3, respectively. In Table 3, "-" means an infinity value, corresponding to the cases when the denominator is 0 in MCC.

Table 2. The values of evaluation criterion with the proposed method for the Gram-positive.

Sample Set	Protein Subcellular Locations			
	Cell Membrane	Cell Wall	Cytoplasm	Extracell
	Sensitivity			
GP-220	1	0.9444	0.9904	0.9919
GP-1000	0.9943	0.9444	1	0.9837
	Specificity			
GP-220	0.9943	1	1	09950
GP-1000	0.9971	1	0.9937	0.9925
	Matthews coefficient correlation (MCC)			
GP-220	0.9914	0.9709	0.9920	0.9841
GP-1000	0.9914	0.9709	0.9921	0.9840
	Overall accuracy (Q)			
GP-220	0.9924			
GP-1000	0.9924			

Table 3. The values of evaluation criterion with the proposed method for the Gram-negative.

Sample Set	Protein Subcellular Locations							
	(1)	(2)	(3)	(4)	(5)	(6)	(7)	(8)
	Sensitivity							
GN-220	1	0.9699	1	0	0.9982	0	0.9677	1
GN-1000	1	0.9323	1	0	0.9659	0	0.9516	0.9556
	Specificity							
GN-220	0.9924	0.9902	1	1	0.9978	1	1	0.9953
GN-1000	0.9608	0.9872	1	1	0.9967	1	1	0.9992
	Matthews coefficient correlation (MCC)							
GN-220	0.9866	0.9324	1	-	0.9956	-	0.9823	0.9814
GN-1000	0.9346	0.8957	1	-	0.9681	-	0.9733	0.9712
	Overall accuracy (Q)							
GN-220	0.9801							
GN-1000	0.9574							

(1) Cytoplasm, (2) Extracell, (3) Fimbrium, (4) Flagellum, (5) Inner membrane, (6) Nucleoid, (7) Outer membrane, (8) Periplasm.

Tables 2 and 3 show that the values of the evaluation criterion are close to 1 for the proposed method. Then the selection of the kernel parameter using the proposed method will benefit the protein subcellular localization.

3. Methods

3.1. Protein Subcellular Localization Prediction Based on KDA

To improve the localization prediction accuracy, it is necessary to reduce dimension of high-dimensional protein data before subcellular classification. The flow of protein subcellular localization prediction is presented in Figure 5.

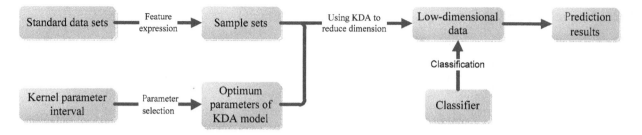

Figure 5. The flow of protein subcellular localization.

As shown in Figure 5, first, for a standard data set, some features of protein sequences such as PSSM-based expressions are extracted to form the sample sets. The specific feature expressions used in this paper are discussed in Section 4.2. Second, the kernel parameter is selected in an interval based on the sample sets to reach its optimal value in KDA model. Third, with this optimal value, we used the KDA to realize the dimension reduction of the sample sets. Lastly, the low dimensional data is treated by certain classifier to realize the classification and the final prediction.

In the whole process of Figure 5, dimension reduction with KDA is very important, in which the kernel selection is a key step and constructs the research focus of this paper. Kernel selection includes the choice of the type of kernel function and the choice of the kernel parameters. In this paper, Gaussian kernel function is adopted for KDA because of its good nature, learning performance, and catholicity. So, the emphasis of this study is to decide the scale parameter of the Gaussian kernel, which plays an important role in the process of dimensionality reduction and has a great influence on prediction results. We put forward a method for selecting the optimum Gaussian kernel parameter with the starting point of reconstruction error idea in [15].

3.2. Algorithm Principle

Kernel method constructs a subspace in the feature space by the kernel trick, which makes normal samples locate in or nearby this subspace, while novel samples are far from it. The reconstruction error is the distance of a sample from the feature space to the subspace [11], so the reconstruction errors of normal samples should be different from those of the novel samples. In this paper, we use the Gaussian KDA as the descending algorithms. Since the values of the reconstruction errors are influenced by the Gaussian kernel parameters, the reconstruction errors of normal samples should be differentiated from those of the novel samples by suitable parameters [11].

In the input space, we usually call the samples on the boundary as edge samples, and call those within the boundary as internal samples [16,17]. The edge samples are much closer to novel samples than the internal samples, while the internal samples are much closer to normal states than the edge samples [11]. We usually use the internal samples as the normal samples and use the edge samples as the novel samples, since there are no novel samples in data sets. Therefore, the principle is that the optimal kernel parameter makes the reconstruction errors have a reasonable difference between the internal samples and the edge samples.

3.3. Kernel Discriminant Analysis (KDA) and Its Reconstruction Error

KDA is an algorithm by applying kernel trick into linear discriminant analysis (LDA). LDA is an algorithm of linear dimensionality reduction together with classifying discrimination, which aims to find a direction that maximizes the between-class scatter while minimizing the within-class scatter [18]. In order to extend the LDA theory to the nonlinear data, Mika et al. proposed the KDA algorithm, which makes the nonlinear data linearly separable in a much higher dimensional feature space than before [9]. The principle of the KDA algorithm is shown as follows.

Suppose the N samples in X can be divided into C classes and the ith class contains N_i samples satisfying $N = \sum_{i=1}^{C} N_i$. The between-class scatter matrix S_b^{ϕ} and the within-class scatter matrix n_n^{ϕ} of X are defined in the following equations, respectively:

$$S_b^{\phi} = \sum_{i=1}^{C} N_i \left(m_i^{\phi} - m^{\phi} \right) \left(m_i^{\phi} - m^{\phi} \right)^T \qquad (1)$$

$$S_w^{\phi} = \sum_{i=1}^{C} \sum_{j=1}^{N_i} \left[\phi\left(x_j^i\right) - m_i^{\phi} \right] \left[\phi\left(x_j^i\right) - m_i^{\phi} \right]^T \qquad (2)$$

where $m_i^{\phi} = \frac{1}{N_i} \sum_{j=1}^{N_i} \phi\left(x_j^i\right)$ is the mean vector of the ith class, and $m^{\phi} = \frac{1}{N} \sum_{i=1}^{N} \phi(x_i)$ is the total mean of X. To find the optimal linear discriminant, we need to maximize J(W) as follows:

$$\max J(W) = \frac{W^T S_b^{\phi} W}{W^T S_w^{\phi} W} \qquad (3)$$

where $W = [w_1, w_2, \cdots, w_d]^T (1 \le d \le C-1)$ is a projection matrix, and $w_k (k = 1, 2, \cdots, d)$ is a column vector with N elements. Through certain algebra, it can be deduced that W is made up of the eigenvectors corresponding to the top d eigenvalues of $S_w^{\phi -1} S_b^{\phi}$. Also, the projection vector w_k can be represented by a linear combination of the samples in the feature space:

$$w_k = \sum_{j=1}^{N} a_j^k j(x_j) \qquad (4)$$

where a_j^k is a real coefficient. The projection of the sample X onto w_k is given by:

$$w_k^T \times \phi(x) = \sum_{i=1}^{N} a_i^k K(x, x_j) \qquad (5)$$

Let $a = [a^1, a^2, \cdots, a^d]^T$ be the coefficient matrix where $a^k = [a_1^k, a_2^k, \cdots, a_N^k]^T$ is the coefficient vector. Combining Equations (1)–(5), we can obtain the linear discriminant by maximizing the function J(a):

$$\max J(a) = \frac{a^T \widetilde{M} a}{a^T \widetilde{L} a} \qquad (6)$$

where $\widetilde{M} = \sum_{i=1}^{C} N_i (M_i - M)(M_i - M)^T$, $\widetilde{L} = \sum_{i=1}^{C} K_i \left(E - \frac{1}{N_i} I \right) K_i^T$, the kth component of the vector M_i is $(M_i)_k = \frac{1}{N_i} \sum_{j=1}^{N_i} K\left(x_k, x_j^i\right)$ $(k = 1, 2, \cdots, N)$, the kth component of the vector M is $(M)_k = \frac{1}{N} \sum_{j=1}^{N} K\left(x_k, x_j^i\right)$ $(k = 1, 2, \cdots, N)$, K_i is a $N \times N_i$ matrix with $(K_i)_{mn} = K(x_m, x_n^i)$, E is the $N_i \times N_i$ identity matrix, and $\frac{1}{N_i} I$ is the $N_i \times N_i$ matrix that all elements are $\frac{1}{N_i}$ [9]. Then, the projection matrix a is made up of the eigenvectors corresponding to the top d eigenvalues of $\widetilde{L}^{-1} \widetilde{M}$.

According to the KDA algorithm principle in (3) or (6), besides the Gaussian kernel parameter s, the number of retained eigenvectors d also affects the algorithm performance. Generally, in this paper, the proposed method is mainly used to screen an optimum S under a predetermined d value.

The Gaussian kernel function is defined as follows:

$$K(x_i, x_j) = \exp\left(-\frac{\|x_i - x_j\|^2}{\sigma^2}\right) \tag{7}$$

where σ is the scale parameter which is generally estimated by s. Note that $\|\phi(x)\|^2 = K(x, x) = 1$. The kernel-based reconstruction error is defined in the following equation:

$$\begin{aligned} RE(x) &= \|\phi(x) - W\, t(x)\|^2 = \|\phi(x)\|^2 - \|t(x)\|^2 \\ &= K(x, x) - \|t(x)\|^2 \end{aligned} \tag{8}$$

where $t(x)$ is the vector obtained by projecting $\phi(x)$ onto a projection matrix a.

3.4. The Proposed Method for Selecting the Optimum Gaussian Kernel Parameter

The method of kernel parameter selection relies on the reconstruction errors of the internal samples and the edge samples. Therefore, first we find a method to select the edge samples and the interior samples, then we propose the method for selection of the Gaussian kernel parameter.

3.4.1. The Method for Selecting Internal and Edge Samples

Li and Maguire present a border-edge pattern selection method (BEPS) to select the edge samples based on the local geometric information [16]. Xiao et al. [11] modified the BEPS algorithm so that it can select both the edge samples and internal samples. However, their algorithm has the risk of making all samples in the training set become the edge samples. For example, when all samples are distributed on a spherical surface in a three-dimensional space, every sample in the data set will be selected as the edge samples since its neighbors are all located on one side of its tangent plane. In order to solve this problem, this paper innovatively combines the ideas in [19,20] to select the internal and edge samples, respectively, which is not dependent on the local geometric information. The main principle is that the edge sample is usually surrounded by the samples belonging to other classes while the internal sample is usually surrounded by the samples belonging to its same class. Further, the edge samples are usually far from the centroid of this class, while the internal samples are usually close to the centroid. So, a sample will be selected as the edge sample if it is far from the centroid of this class and there are samples around it that belongs to other classes, otherwise it will be selected as the internal sample.

Specifically, suppose the ith class $X_i = \{x_1, x_2, \cdots, x_{N_i}\}$ in the sample set X is picked out as the training set. Denote c_i be the centroid of this class:

$$c_i = \frac{1}{N_i} \sum_{i=1}^{N_i} x_i \tag{9}$$

We use the median value m of the distances from all samples in a class to its centroid to measure the distance from a sample to the centroid of this class. A sample is conserved to be far from the centroid of this class if the distance from this sample to the centroid is greater than the median value. Otherwise, the sample is considered to be close to the centroid.

Denote $dist(x_i, x_j)$ as the distance between any two samples x_i and x_j, and $N_\varepsilon(x)$ as the ε-neighborhood of X:

$$N_\varepsilon(x) = \{y \,|\, dist(x, y) \le \varepsilon, y \in X\} \tag{10}$$

The value of neighborhood ε is given as follows. Let u be a given number which satisfies $0 < u < N_i$. $Density_u(X_i)$ is the mean radius of neighborhood of X_i for the given number u:

$$Density_u(X_i) = \frac{1}{N_i} \sum_{i=1}^{N_i} dist_u(x_i) \tag{11}$$

where $\text{dist}_u(x_i)$ is the distance from x_i to its uth nearest neighbor. So, $\text{Density}_u(X_i)$ is used as the value of ε for the training set X_i. The flow for the selection of the internal and edge samples is shown in Table 4.

Table 4. The Selection of Internal and Edge Samples.

Input: $X = \{X_1, X_2, \cdots, X_C\}$, the training set $X_i = \{x_1, x_2, \cdots, x_{N_i}\}$ $(1 \leq i \leq C)$.

1. Calculate the radius of neighborhood ε using Equation (11).
2. Calculate the centroid c_i of the ith class according to Equation (9).
3. Calculate the distances $\text{dist}_j (j = 1, 2, \cdots, N_i)$ from all samples in training set to c_i, respectively, and the median value m of them.
4. For each training sample x_j of the set X_i

- Calculate the $N_\varepsilon(x_j)$ according to Equation (10).
- If $\text{dist}_j > m$ and there are samples in $N_\varepsilon(x_j)$ belonging to other classes, x_j is selected as an edge sample.
- If $\text{dist}_j < m$ and no sample in $N_\varepsilon(x_j)$ belongs to other classes, x_j is selected as an internal sample.

Output: the selected internal sample set Ω_{in}, the selected edge sample set Ω_{ed}.

In Table 4, a sample X is considered to be the edge one when the distance from X to the centroid is larger than the median m and there are samples in $N_\varepsilon(x)$ belonging to other classes in this case. A sample X is considered to be the internal one when the distance from X to the centroid is less than m and in this case all samples of $N_\varepsilon(x)$ belong to this class.

3.4.2. The Proposed Method

In order to select the optimum kernel parameter, it is necessary to propose a criterion aiming to distinguish reconstruction errors of the edge samples from those of the internal samples. A suitable parameter not only maximizes the difference between reconstruction errors of the internal samples and those of the edge samples, but also minimizes the variance (or standard deviation) of reconstruction errors of the internal samples [11]. According to the rule, an improved objective function is proposed in this paper. The optimal Gaussian kernel parameter S is selected by maximizing this objective function.

$$s = \operatorname*{argmax}_s f(s) = \arg\max_s \frac{\|\text{RE}(\Omega_{\text{ed}})\|_\infty - \|\text{RE}(\Omega_{\text{in}})\|_\infty}{\text{std}\{\text{RE}(\Omega_{\text{in}})\}} \tag{12}$$

where $\|\cdot\|_\infty$ is the infinite norm which computes the maximum absolute component of a vector and $\text{std}(\cdot)$ is a function of the standard deviation. Note that in the objective function $f(s)$, our key improvement is to use the infinite norm to compute the size of reconstruction error vector since it can lead to a higher accuracy than many other measurements, which has been verified by a series of our experiments. The reason is probably that the maximum component is more reasonable to evaluate the size of a reconstruction error vector than others such as the 1-norm, p-norm $(1 < p < +\infty)$ and the minimum component of a reconstruction error vector in [11].

According to (8), when the number of retained eigenvectors is determined, we can select the optimum parameter s from a candidate set using the proposed method. The optimum parameter ensures that the Gaussian KDA algorithm performs well in dimensionality reduction, which improves the accuracy of protein subcellular location prediction. The proposed method for selecting the Gaussian kernel parameter can be presented in Table 5.

Table 5. The Method for Selecting the Gaussian KDA Parameter.

Input: A reasonable candidate set $S = \{s_1, s_2, \cdots, s_m\}$ for Gaussian kernel parameter, $X = \{X_1, X_2, \cdots, X_C\}$, the training set $X_i = \{x_1, x_2, \cdots, x_{N_i}\}$ $(1 \leq i \leq C)$, the number of retained eigenvectors d.

---

1. Get the internal sample set Ω_{in} and the edge sample set Ω_{ed} from the training set X_i using Algorithm 1.
2. For each parameter $s_i \in S$, $\quad i = 1, 2, \cdots, m$

- Calculate the kernel matrix K using Equation (7).
- Reduce dimension of the K using the Gaussian KDA algorithm.
- Calculate $RE(\Omega_{ed})$ and $RE(\Omega_{in})$ using Equation (8).
- Calculate the value of objective function $f(s_i)$ using Equation (12).

3. Select the optimum parameter $s = \underset{s_i \in S}{\operatorname{argmax}} f(s_i)$

Output: the optimum Gaussian kernel parameter S.

As the end of this section, we want to summarize the position of the proposed method in protein subcellular localization once more. First, two kinds of regularization forms of PSSM are used to extract the features in protein amino acid sequences. Then, the KDA method is performed on the extracted features for dimension reduction and discriminant analysis according to the KDA algorithm principle in Section 3.3 with formulas (1)–(6). During the procedure of KDA, the novelty of our work is to give a new method for selecting the Gaussian kernel parameter, which is summarized in Table 5. Finally, we choose the k-nearest neighbors (KNN) as the classifier to cluster the dimension-reduced data after KDA.

4. Materials

In this section, we introduce the other processes in Figure 5 except KDA model and its parameter selection, which are necessary materials for the whole experiment.

4.1. Standard Data Sets

In this paper, we use two standard datasets that have been widely used in the literature for Gram-positive and Gram-negative subcellular localizations [13], whose protein sequences all come from the Swiss-Prot database.

For the Gram-positive bacteria, the standard data set we found in the literature [13,14,21] is publicly available on http://www.csbio.sjtu.edu.cn/bioinf/Gpos-multi/Data.htm. There are 523 locative protein sequences in the data set that are distributed in four different subcellular locations. The number of proteins in each location is given in Table 6.

Table 6. The name and the size of each location for the Gram-positive data set.

No.	Subcellular Localization	Number of Proteins
1	cell membrane	174
2	cell wall	18
3	cytoplasm	208
4	extracell	123

For the Gram-negative bacteria, the standard data set of subcellular localizations is presented in the literature [13,22], which can be downloaded freely from http://www.csbio.sjtu.edu.cn/bioinf/Gneg-multi/Data.htm. The data set contains 1456 locative protein sequences located in eight different subcellular locations. The number of proteins in each location is shown in Table 7.

Table 7. The name and the size of each location for the Gram-negative data set.

No.	Subcellular Localization	Number of Proteins
1	cytoplasm	410
2	extracell	133
3	fimbrium	32
4	flagellum	12
5	inner membrane	557
6	nucleoid	8
7	outer membrane	124
8	periplasm	180

4.2. Feature Expressions and Sample Sets

In the prediction of protein subcellular localizations with machine learning methods, feature expressions are important information extracted from protein sequences, which have certain proper mathematical algorithms. There are many efficient algorithms used to extract features of protein sequences, in which two of them, PsePSSM [12] and PSSM-S [13], are used in this paper. The two methods rely on the position-specific scoring matrix (PSSM) for benchmarks which is obtained by using the PSI-BLAST algorithm to search the Swiss-Prot database with the parameter E-value of 0.01. The PSSM is defined as follows [12]:

$$
P_{PSSM} = \begin{bmatrix}
M_{1\to1} & M_{1\to2} & \cdots & M_{1\to20} \\
M_{2\to1} & M_{2\to2} & \cdots & M_{2\to20} \\
\vdots & \vdots & \vdots & \vdots \\
M_{i\to1} & M_{i\to2} & \cdots & M_{i\to20} \\
\vdots & \vdots & \vdots & \vdots \\
M_{L\to1} & M_{L\to2} & \cdots & M_{L\to20}
\end{bmatrix}
\tag{13}
$$

where $M_{i\to j}$ represents the score created in the case when the ith amino acid residue of the protein sequence is transformed to the amino acid type j during the evolutionary process [12].

Note that, usually, multiple alignment methods are used to calculate PSSM, whose chief drawback is being time-consuming. The reason why we select PSSM instead of simple multiple alignment in this paper to form the total normalized information content is as follows. First, since our focus is to demonstrate the effectiveness of dimensional reduction algorithm, we need to construct high-dimensional feature expressions such as PsePSSM and PSSM-S, whose dimensions are as high as 1000 and 220, respectively. Second, PSSM has many advantages, such as those described in [23]. As far as the information features are concerned, PSSM has produced the strongest discriminator feature between fold members of protein sequences. Multiple alignment methods are used to calculate PSSM, whose chief drawback is being time-consuming. However, in spite of the time-consuming nature of constructing a PSSM for the new sequence, the extracted feature vectors from PSSM are so informative that are worth the cost of their preparation [23]. Besides, for a new protein sequence, we only need to construct a PSSM for the first time, which could be used repeatedly in the future for producing new normalization forms such as PsePSSM and PSSM-S.

4.2.1. Pseudo Position-Specific Scoring Matrix (PsePSSM)

Let P be a protein sample, whose definition of PsePSSM is given as follows [12]:

$$
P_{Pse-PSSM}^{\xi} = \left[\overline{M}_1 \overline{M}_2 \cdots \overline{M}_{20} G_1^{\xi} G_2^{\xi} \cdots G_{20}^{\xi} \right]^{T} (\xi = 0, 1, 2, \cdots, 49)
\tag{14}
$$

$$\overline{M}_j = \frac{1}{L}\sum_{i=1}^{L} M_{i \to j}(j = 1, 2, \cdots, 20) \tag{15}$$

$$G_j^\xi = \frac{1}{L - \xi}\sum_{i=1}^{L-\xi}\left[M_{i \to j} - M_{(i+\xi) \to j}\right]^2 (j = 1, 2, \cdots, 20; \xi < L) \tag{16}$$

where L is the length of P, G_j^ξ is the correlation factor by coupling the ξ-most contiguous scores [22]. According to the definition of PsePSSM, a protein sequence can be represented by a 1000-dimensional vector.

4.2.2. PSSM-S

Dehzangi et al. [13] put forward a new feature extraction method, PSSM-S, which combines four components: AAO, PSSM-AAO, PSSM-SD, and PSSM-SAC. According to the definition of the PSSM-S, it can be represented a feature vector with 220 (20 + 20 + 80 + 100) elements.

4.2.3. Sample Sets

For the two benchmark data, PsePSSM and PSSM-S are used to extract features, respectively. Finally we get four experimental sample sets GN-1000, GN-220, GP-1000 and GP-220, shown in Table 8.

Table 8. Sample sets.

Sample Sets	Benchmarks for Subcellular Locations	Extraction Feature Method	The Number of Classes	The Dimension of Feature Vector	The Number of Samples
GN-1000	Gram-negative	PsePSSM	8	1000	1456
GN-220	Gram-negative	PSSM-S	8	220	1456
GP-1000	Gram-positive	PsePSSM	4	1000	523
GP-220	Gram-positive	PSSM-S	4	220	523

4.3. Evaluation Criterion

To evaluate the performance of the proposed method, we use Jackknife cross-validation, which has been widely used to predict protein subcellular localization [13]. The Jackknife test is the most objective and rigorous cross-validation procedure in examining the accuracy of a predictor, which has been used increasingly by investigators to test the power of various predictors [24,25]. In the Jackknife test (also known as leave-one-out cross-validation), every protein is removed one-by-one from the training dataset, and the predictor is trained by the remaining proteins. The isolated protein is then tested by the trained predictor [26]. Let x be a sample set with N samples. For each sample, it will be used as the test data, and the remaining N − 1 samples will be used to construct the training set [27]. In addition, we use some criterion to assess the experimental results, defined as follows [12]:

$$MCC(k) = \frac{TP_k \times TN_k - FN_k \times FP_k}{\sqrt{(TP_k + FN_k)(TP_k + FP_k)(TN_k + FP_k)(TN_k + FN_k)}} \times 100\% \tag{17}$$

$$Sen(k) = \frac{TP_k}{TP_k + FN_k} \times 100\% \tag{18}$$

$$Spe(k) = \frac{TN_k}{TP_k + FP_k} \times 100\% \tag{19}$$

$$Q = \frac{\sum_{k=1}^{C} TP_k}{N} \times 100\% \tag{20}$$

where TP is the number of true positive, TN is the number of true negative, FP is the number of false positive, and FN is the number of false negative [12]. The value of MCC (Matthews coefficient correlation) varies between −1 and 1, indicating when the classification effect goes from a bad to a good one. The values of Specificity (Spe), sensitivity (Sen), and the overall accuracy (Q) all vary between 0 and 1, and the classification effect is better when their values are closer to 1, while the classification effect is worse when their values are closer to 0 [13].

4.4. The Grid Searching Method Used as Contrast

In this section, we introduce a normal algorithm for searching S, the grid-searching algorithm, which is used as a contrast with the proposed algorithm in Section 3.4.

The grid-searching method is usually used to select the optimum parameter, whose steps are as follows for the candidate parameter set S [28].

- Compute the kernel matrix k for each parameter $s_i \in S$, $i = 1, 2, \cdots, m$.
- Use the Gaussian KDA to reduce the dimension of K.
- Use the KNN algorithm to classify the reduced dimensional samples.
- Calculate the classification accuracy.
- Repeat the above four steps until all parameters in S have been traversed. The parameter corresponding to the highest classification accuracy is selected as the optimum parameter.

5. Conclusions

Biological data is usually high-dimensional. As a result, it is necessary to reduce dimension to improve the accuracy of the protein subcellular localization prediction. The kernel discriminant analysis (KDA) based on Gaussian kernel function is a suitable algorithm for dimensional reduction in such applications. As is known to all, the selection of a kernel parameter affects the performance of KDA, and thus it is important to choose the proper parameter that makes this algorithm perform well. To handle this problem, we propose a method of the optimum kernel parameter selection, which relies on reconstruction error [15]. Firstly, we use a method to select the edge and internal samples of the training set. Secondly, we compute the reconstruction errors of the selected samples. Finally, we select the optimum kernel parameter that makes the objective function maximum.

The proposed method is applied to the prediction of protein subcellular locations for Gram-negative bacteria and Gram-positive bacteria. Compared with the grid-searching method, the proposed method gives higher efficiency and performance.

Since the performance of the proposed method largely depends on the selection of the internal and edge samples, in the future study, researchers may pay more attention to select more representative internal and edge samples from the biological data set to improve the prediction accuracy of protein subcellular localization. Besides this, it is also meaningful to research how to further improve the proposed method to make it suitable for selecting parameters of other kernels.

Acknowledgments: This research is supported by grants from National Natural Science Foundation of China (No. 11661081, No. 11561071 and No. 61472345) and Natural Science Foundation of Yunnan Province (2017FA032).

Author Contributions: Shunfang Wang and Bing Nie designed the research and the experiments. Wenjia Li, Dongshu Xu, and Bing Nie extracted the feature expressions from the standard data sets, and Bing Nie performed all the other numerical experiments. Shunfang Wang, Bing Nie, Kun Yue, and Yu Fei analyzed the experimental results. Shunfang Wang, Bing Nie, Kun Yue, and Yu Fei wrote this paper. All authors read and approved the final manuscript.

References

1. Chou, K.C. Some Remarks on Predicting Multi-Label Attributes in Molecular Biosystems. *Mol. Biosyst.* **2013**, *9*, 1092–1100. [CrossRef] [PubMed]
2. Zhang, S.; Huang, B.; Xia, X.F.; Sun, Z.R. Bioinformatics Research in Subcellular Localization of Protein. *Prog. Biochem. Biophys.* **2007**, *34*, 573–579.
3. Zhang, S.B.; Lai, J.H. Machine Learning-based Prediction of Subcellular Localization for Protein. *Comput. Sci.* **2009**, *36*, 29–33.

4. Huh, W.K.; Falvo, J.V.; Gerke, L.C.; Carroll, A.S.; Howson, R.W.; Weissman, J.S.; O'Shea, E.K. Global analysis of protein localization in budding yeast. *Nature* **2003**, *425*, 686–691. [CrossRef] [PubMed]

5. Dunkley, T.P.J.; Watson, R.; Griffin, J.L.; Dupree, P.; Lilley, K.S. Localization of organelle proteins by isotope tagging (LOPIT). *Mol. Cell. Proteom.* **2004**, *3*, 1128–1134. [CrossRef] [PubMed]

6. Hasan, M.A.; Ahmad, S.; Molla, M.K. Protein subcellular localization prediction using multiple kernel learning based support vector machine. *Mol. Biosyst.* **2017**, *13*, 785–795. [CrossRef] [PubMed]

7. Teso, S.; Passerini, A. Joint probabilistic-logical refinement of multiple protein feature predictors. *BMC Bioinform.* **2014**, *15*, 16. [CrossRef] [PubMed]

8. Wang, S.; Liu, S. Protein Sub-Nuclear Localization Based on Effective Fusion Representations and Dimension Reduction Algorithm LDA. *Int. J. Mol. Sci.* **2015**, *16*, 30343–30361. [CrossRef] [PubMed]

9. Baudat, G.; Anouar, F. Generalized Discriminant Analysis Using a Kernel Approach. *Neural Comput.* **2000**, *12*, 2385–2404. [CrossRef] [PubMed]

10. Zhang, G.N.; Wang, J.B.; Li, Y.; Miao, Z.; Zhang, Y.F.; Li, H. Person re-identification based on feature fusion and kernel local Fisher discriminant analysis. *J. Comput. Appl.* **2016**, *36*, 2597–2600.

11. Xiao, Y.C.; Wang, H.G.; Xu, W.L.; Miao, Z.; Zhang, Y.; Hang, L.I. Model selection of Gaussian kernel PCA for novelty detection. *Chemometr. Intell. Lab.* **2014**, *136*, 164–172. [CrossRef]

12. Chou, K.C.; Shen, H.B. MemType-2L: A Web server for predicting membrane proteins and their types by incorporating evolution information through Pse-PSSM. *Biochem. Biophys. Res. Commun.* **2007**, *360*, 339–345. [CrossRef] [PubMed]

13. Dehzangi, A.; Heffernan, R.; Sharma, A.; Lyons, J.; Paliwal, K.; Sattar, A. Gram-positive and Gram-negative protein subcellular localization by incorporating evolutionary-based descriptors into Chou's general PseAAC. *J. Theor. Biol.* **2015**, *364*, 284–294. [CrossRef] [PubMed]

14. Shen, H.B.; Chou, K.C. Gpos-PLoc: An ensemble classifier for predicting subcellular localization of Gram-positive bacterial proteins. *Protein Eng. Des. Sel.* **2007**, *20*, 39–46. [CrossRef] [PubMed]

15. Hoffmann, H. Kernel PCA for novelty detection. *Pattern Recogn.* **2007**, *40*, 863–874. [CrossRef]

16. Li, Y.; Maguire, L. Selecting Critical Patterns Based on Local Geometrical and Statistical Information. *IEEE Trans. Pattern Anal.* **2010**, *33*, 1189–1201.

17. Wilson, D.R.; Martinez, T.R. Reduction Techniques for Instance-Based Learning Algorithms. *Mach. Learn.* **2000**, *38*, 257–286. [CrossRef]

18. Saeidi, R.; Astudillo, R.; Kolossa, D. Uncertain LDA: Including observation uncertainties in discriminative transforms. *IEEE Trans. Pattern Anal.* **2016**, *38*, 1479–1488. [CrossRef] [PubMed]

19. Jain, A.K. Data clustering: 50 years beyond K-means. *Pattern Recogn. Lett.* **2010**, *31*, 651–666. [CrossRef]

20. Li, R.L.; Hu, Y.F. A Density-Based Method for Reducing the Amount of Training Data in kNN Text Classification. *J. Comput. Res. Dev.* **2004**, *41*, 539–545.

21. Chou, K.C.; Shen, H.B. Cell-PLoc 2.0: An improved package of web-servers for predicting subcellular localization of proteins in various organisms. *Nat. Sci.* **2010**, *2*, 1090–1103. [CrossRef]

22. Chou, K.C.; Shen, H.B. Large-Scale Predictions of Gram-Negative Bacterial Protein Subcellular Locations. *J. Proteome Res.* **2007**, *5*, 3420–3428. [CrossRef] [PubMed]

23. Kavousi, K.; Moshiri, B.; Sadeghi, M.; Araabi, B.N.; Moosavi-Movahedi, A.A. A protein fold classifier formed by fusing different modes of pseudo amino acid composition via PSSM. *Comput. Biol. Chem.* **2011**, *35*, 1–9. [CrossRef] [PubMed]

24. Shen, H.B.; Chou, K.C. Nuc-PLoc: A new web-server for predicting protein subnuclear localization by fusing PseAA composition and PsePSSM. *Protein Eng. Des. Sel.* **2007**, *20*, 561–567. [CrossRef] [PubMed]

25. Wang, T.; Yang, J. Using the nonlinear dimensionality reduction method for the prediction of subcellular localization of Gram-negative bacterial proteins. *Mol. Divers.* **2009**, *13*, 475.

26. Wei, L.Y.; Tang, J.J.; Zou, Q. Local-DPP: An improved DNA-binding protein prediction method by exploring local evolutionary information. *Inform. Sci.* **2017**, *384*, 135–144. [CrossRef]

27. Shen, H.B.; Chou, K.C. Gneg-mPLoc: A top-down strategy to enhance the quality of predicting subcellular localization of Gram-negative bacterial proteins. *J. Theor. Biol.* **2010**, *264*, 326–333. [CrossRef] [PubMed]

28. Bing, L.I.; Yao, Q.Z.; Luo, Z.M.; Tian, Y. Gird-pattern method for model selection of support vector machines. *Comput. Eng. Appl.* **2008**, *44*, 136–138.

8

Determination of Genes Related to Uveitis by Utilization of the Random Walk with Restart Algorithm on a Protein–Protein Interaction Network

Shiheng Lu [1], Yan Yan [1], Zhen Li [1], Lei Chen [2], Jing Yang [3], Yuhang Zhang [4], Shaopeng Wang [3] and Lin Liu [1,*]

[1] Department of Ophthalmology, Ren Ji Hospital, School of Medicine, Shanghai Jiao Tong University, Shanghai 200127, China; ludice@163.com (S.L.); hz2004yan@163.com (Y.Y.); lizhen1981_1@126.com (Z.L.)

[2] College of Information Engineering, Shanghai Maritime University, Shanghai 201306, China; chen_lei1@163.com

[3] School of Life Sciences, Shanghai University, Shanghai 200444, China; mercuryyangjing@sina.com (J.Y.); wsptfb@163.com (S.W.)

[4] Institute of Health Sciences, Shanghai Institutes for Biological Sciences, Chinese Academy of Sciences, Shanghai 200031, China; zhangyh825@163.com

* Correspondence: liulin20160929@hotmail.com

Academic Editor: Christo Z. Christov

Abstract: Uveitis, defined as inflammation of the uveal tract, may cause blindness in both young and middle-aged people. Approximately 10–15% of blindness in the West is caused by uveitis. Therefore, a comprehensive investigation to determine the disease pathogenesis is urgent, as it will thus be possible to design effective treatments. Identification of the disease genes that cause uveitis is an important requirement to achieve this goal. To begin to answer this question, in this study, a computational method was proposed to identify novel uveitis-related genes. This method was executed on a large protein–protein interaction network and employed a popular ranking algorithm, the Random Walk with Restart (RWR) algorithm. To improve the utility of the method, a permutation test and a procedure for selecting core genes were added, which helped to exclude false discoveries and select the most important candidate genes. The five-fold cross-validation was adopted to evaluate the method, yielding the average F1-measure of 0.189. In addition, we compared our method with a classic GBA-based method to further indicate its utility. Based on our method, 56 putative genes were chosen for further assessment. We have determined that several of these genes (e.g., *CCL4*, *Jun*, and *MMP9*) are likely to be important for the pathogenesis of uveitis.

Keywords: uveitis; protein–protein interaction; random walk with restart algorithm

1. Introduction

Uveitis is defined as an inflammation of the uveal tract, which is composed of the ciliary body, iris and choroid [1,2]. Uveitis is one of the leading causes of permanent and irreversible blindness in young and middle-aged people and accounts for 10–15% of blindness in the Western world [1–3]. Uveitis can be caused by infectious and non-infectious factors; the latter include Vogt–Koyanagi–Harada (VKH) syndrome, Behcet's disease (BD), acute anterior uveitis (AAU), birdshot chorioretinopathy (BCR) and some types of cancers. VKH is an autoimmune disease characterized by systemic disorders including poliosis, vitiligo, alopecia, auditory signs and disorders of the central nervous system [4,5]. BD is a chronic multi-systemic inflammatory disease characterized by nongranulomatous uveitis, oral ulcers and skin lesions [2,6]. AAU is the most common non-infectious cause of uveitis and is characterized by self-limiting and recurrent inflammation involving the ciliary and iris body [7]. BCR is a chronic, bilateral, and posterior uveitis that has an almost 100% genetic association with HLA-A29 [8]. Uveitis or uveitis masquerade syndrome could also be induced by some intraocular tumors, such as retinoblastoma and intraocular lymphoma, or their therapeutic approaches [9–15].

It has been reported that complex genetic mechanisms coupled with an aberrant immune response may be involved in the development of uveitis. In some cases, the pathogenesis of uveitis seemly has a different cause than those described above, such as sarcoidosis [16]. Mutations in different genes and gene families have been discovered in patients. In this study, we focused on the most important causes of uveitis and research for the putative genes involved in these processes. Human leukocyte antigens (HLAs) are the major molecules that are important for the development of uveitis, including uveitis associated with VKH (HLA-DR4, DRB1/DQA1), BD (HLA-B51), AAU (HLA-B27) and BCR (HLA-A29). In addition, genome-wide association studies revealed that abnormities of many non-HLA genes such as the interleukin (IL) family and the Signal transducer and activator of transcription 4 (STAT4) also participate in the progression of uveitis [17–19]. IL23R is associated with both VKH and AAU [20]. Furthermore, copy number variations (CNVs) of Toll-like receptors (TLRs), a family of cellular receptors that function in innate immune response, are associated with BD, VKH and AAU. These genes include TLRs 1–3, TLRs 5–7, and TLRs 9–10 [21]. SNPs of TLR4 were also shown to be involved in the development of BD [22]. In addition, it has been demonstrated that there is increased expression of T-bet and IFN-γ, two genes involved in the Th1 cell pathway, in uveitis patients [23]. The activator of STAT4 affects IL-17 production and is a shared risk factor for BD in different cohorts [17,24]. Finally, interleukins (notably IL-2, IL12B, IL18 and IL23R) are important cytokines that play a pathogenic role in the process of uveitis [2,17,25]. In this study, we mainly focused on the genes that play an important role in the immune system, transcription, or cell adhesion.

Using traditional methods, it is quite difficult to collect these large-scale data and analyze genes synthetically. The microarray is a widely used tool for the identification of novel genes. Microarray analysis has been used to determine a number of genes that are associated with uveitis, including the IL10 family and several other transcripts [16,26–29]. In recent years, computational analysis has been applied to identify virulence genes, but many of these genes were identified based on guilt by association (GBA) [30–32]. This approach assumes that the candidate genes, which are neighbors of disease genes, are more likely to be new virulence genes. Thus, the GBA-based methods only consider the neighbors of known disease genes to discover novel candidates. Therefore, these methods only examine part of the gene network. Random Walk with Restart (RWR) is another algorithm that identifies disease-related genes [33–35]. This algorithm utilizes a set of seed nodes that represent disease genes and performs random walking on the gene network. When the probabilities of all nodes are stable, the probability of a node gene correlating with disease is updated. The genes that correspond to nodes that have high probabilities may be potential novel candidate virulence genes. This method is useful for mining disease genes and to better explore the mechanism of disease. In addition, other studies have adopted the shortest path (SP) algorithm to identify novel disease genes [36–41]. By searching the shortest paths that connect any two validated disease genes, genes that are present in these paths could be extracted and considered as novel disease genes. An obvious advantage of the RWR or SP algorithms is that these algorithms utilize the entire gene network and consider more factors, therefore performing a more extensive and reliable analysis.

As discussed above, many genetic factors contribute to the pathogenesis of uveitis. In this study, we utilized computational analyses to build a genetic network based on previously known factors. A computational method was built to identify novel genes related to uveitis. First, a large network was constructed using human protein–protein interactions (PPIs). Next, the RWR algorithm was performed on the network using the validated uveitis-related genes as seed nodes, yielding several possible candidate genes. These candidate genes were filtered based on a set of criteria that were built by p-values and their associations with validated uveitis-related genes. To indicate the utility of the method, it was evaluated by five-fold cross-validation, resulting in the average F1-measure of 0.189. Furthermore, the proposed method was compared with a classic GBA-based method [30–32] to further prove its effectiveness for identification of uveitis-related genes. Through our method, 56 novel candidate genes were identified and extensively analyzed.

2. Results and Discussion

2.1. Results of Testing Random Walk with Restart (RWR)-Based Method

Before the RWR-based method was used to identify novel uveitis-related genes, five-fold cross-validation was adopted to evaluate its utility. For each part, the results yielded by the method on the rest four parts were counted as recall, precision and F1-measure, which are listed in Table 1. It can be observed that the average of recall, precision and F1-measure was 0.287, 0.141 and 0.189, respectively. Although these measurements are not very high, the RWR-based method is still acceptable due to the difficulties for identification of novel genes with given functions. Besides, the utility of the RWR-based method would be further proved by comparing it with other methods, which is described in Section 2.5.

Table 1. The performance of the Random Walk with Restrart (RWR)-based method yielded by five-fold cross-validation.

Index of Part	Recall	Precision	F1-Measure
1	0.172	0.089	0.118
2	0.172	0.088	0.116
3	0.379	0.177	0.242
4	0.310	0.141	0.194
5	0.400	0.211	0.276
Mean	0.287	0.141	0.189

2.2. RWR Genes

Based on the uveitis-related genes, the RWR algorithm yielded a probability for each gene in the PPI network, which indicated the likelihood of the gene being important for uveitis. Then, genes were selected that had probabilities larger than 10^{-5}. From our analysis, we obtained 3641 RWR genes, which are provided with their RWR probabilities in Supplementary Table S1.

2.3. Candidate Genes

According to the RWR-based method detailed in Section 3.3, RWR genes were filtered using a permutation test. For each RWR gene, a p-value was assigned to indicate whether the RWR gene is specific for uveitis. The p-value for each of the 3641 RWR genes is also provided in Supplementary Table S1. We found 1231 candidate genes that had a p-value < 0.05 (see the first 1231 genes in Supplementary Table S1).

The 1231 candidate genes were then further analyzed using the criteria outlined in Section 3.3. For each candidate gene, MIS (cf. Equation (3)) and MFS (cf. Equation (5)) were calculated, and the values for each gene are available in Supplementary Table S1. The threshold for MIS was set at 900, while 0.8 was used as the threshold for MFS. Finally, we obtained 56 Ensembl IDs (listed in Table 2) corresponding to core candidate genes. These genes were deemed to be highly related to uveitis and could be considered novel candidate genes. As intuitionistic evidence, a sub-network was plotted in Figure 1, which contains the putative and validated genes. Each putative gene had strong associations with validated genes, implying that they had functions similar to those of the validated genes and may be novel uveitis-related genes with high probabilities.

Table 2. Novel genes inferred by Random Walk with Restrart (RWR)-based method.

Ensembl ID	Gene Symbol	Description	Probability	p-Value	MIS	MFS
ENSP00000351671 [b]	CCL20	C-C motif chemokine ligand 20	1.65×10^{-4}	<0.001	999	0.841
ENSP00000250151 [b]	CCL4	C-C motif chemokine ligand 4	1.64×10^{-4}	<0.001	994	0.820
ENSP00000326432 [c]	CCR8	C-C motif chemokine receptor 8	8.90×10^{-5}	<0.001	951	0.816
ENSP00000313419 [b]	CD19	CD19 molecule	2.15×10^{-4}	<0.001	947	0.837
ENSP00000320084 [c]	CD276	CD276 molecule	1.91×10^{-4}	<0.001	955	0.823
ENSP00000359663 [b]	CD40LG	CD40 ligand	1.97×10^{-4}	<0.001	999	0.839
ENSP00000264246 [b]	CD80	CD80 molecule	2.18×10^{-4}	<0.001	999	0.820
ENSP00000283635 [c]	CD8A	CD8a molecule	1.91×10^{-4}	<0.001	990	0.815
ENSP00000296871 [c]	CSF2	Colony stimulating factor 2	2.71×10^{-4}	<0.001	992	0.875
ENSP00000225474 [c]	CSF3	Colony stimulating factor 3	1.55×10^{-4}	<0.001	916	0.829
ENSP00000379110 [b]	CXCL1	C-X-C motif chemokine ligand 1	1.69×10^{-4}	<0.001	973	0.827
ENSP00000306884 [b]	CXCL11	C-X-C motif chemokine ligand 11	1.28×10^{-4}	<0.001	999	0.818
ENSP00000286758 [b]	CXCL13	C-X-C motif chemokine ligand 13	1.49×10^{-4}	<0.001	986	0.806
ENSP00000293778 [b]	CXCL16	C-X-C motif chemokine ligand 16	1.02×10^{-4}	<0.001	952	0.800
ENSP00000296027 [b]	CXCL5	C-X-C motif chemokine ligand 5	1.11×10^{-4}	<0.001	958	0.811
ENSP00000354901 [b]	CXCL9	C-X-C motif chemokine ligand 9	2.13×10^{-4}	<0.001	999	0.883
ENSP00000295683 [c]	CXCR1	C-X-C motif chemokine receptor 1	8.67×10^{-5}	<0.001	999	0.833
ENSP00000319635 [b]	CXCR2	C-X-C motif chemokine receptor 2	1.02×10^{-4}	<0.001	999	0.851
ENSP00000229239 [c]	GAPDH	Glyceraldehyde-3-phosphate dehydrogenase	2.12×10^{-4}	<0.001	922	0.824
ENSP00000216341 [c]	GZMB	Granzyme B	2.46×10^{-4}	<0.001	991	0.829
ENSP00000364114 [c]	HLA-DRB5	Major histocompatibility complex, class II, DR β 5	2.27×10^{-4}	<0.001	948	0.822
ENSP00000304915 [a]	IL13	Interleukin 13	1.31×10^{-4}	<0.001	999	0.813
ENSP00000296545 [b]	IL15	Interleukin 15	1.85×10^{-4}	<0.001	946	0.806
ENSP00000263339 [b]	IL1A	Interleukin 1 α	1.82×10^{-4}	<0.001	996	0.820
ENSP00000263341 [b]	IL1B	Interleukin 1 β	3.58×10^{-4}	<0.001	999	0.873
ENSP00000259206 [a]	IL1RN	Interleukin 1 receptor antagonist	1.68×10^{-4}	<0.001	999	0.836
ENSP00000228534 [b]	IL23A	Interleukin 23 subunit A	2.87×10^{-4}	<0.001	998	0.844
ENSP00000369293 [b]	IL2RA	Interleukin 2 receptor subunit A	2.46×10^{-4}	<0.001	999	0.866
ENSP00000274520 [c]	IL9	Interleukin 9	1.27×10^{-4}	<0.001	965	0.806
ENSP00000360266 [b]	JUN	Jun proto-oncogene, AP-1 transcription factor subunit	3.22×10^{-4}	<0.001	999	0.831
ENSP00000361405 [b]	MMP9	Matrix metallopeptidase 9	1.70×10^{-4}	<0.001	971	0.833

Table 2. *Cont.*

ENSP00000379625 [a]	MYD88	Myeloid differentiation primary response 88	1.82×10^{-4}	<0.001	999	0.882
ENSP00000356346 [c]	PTPRC	Protein tyrosine phosphatase, receptor type C	2.18×10^{-4}	<0.001	994	0.826
ENSP00000331736 [c]	SELE	Selectin E	1.46×10^{-4}	<0.001	978	0.830
ENSP00000354394 [b]	STAT1	Signal transducer and activator of transcription 1	2.63×10^{-4}	<0.001	999	0.852
ENSP00000300134 [b]	STAT6	Signal transducer and activator of transcription 6	1.77×10^{-4}	<0.001	999	0.804
ENSP00000221930 [a]	TGFB1	Transforming growth factor β 1	2.90×10^{-4}	<0.001	997	0.832
ENSP00000416330 [c]	TGFBI	Transforming growth factor β induced	1.91×10^{-4}	<0.001	917	0.813
ENSP00000260010 [b]	TLR2	Toll like receptor 2	2.25×10^{-4}	<0.001	968	0.888
ENSP00000370034 [b]	TLR7	Toll like receptor 7	1.26×10^{-4}	<0.001	926	0.819
ENSP00000353874 [b]	TLR9	Toll like receptor 9	1.55×10^{-4}	<0.001	958	0.854
ENSP00000294728 [b]	VCAM1	Vascular cell adhesion molecule 1	2.23×10^{-4}	<0.001	968	0.882
ENSP00000292174 [c]	CXCR5	C-X-C motif chemokine receptor 5	1.14×10^{-4}	0.001	976	0.820
ENSP00000343204 [a]	JAK1	Janus kinase 1	1.21×10^{-4}	0.001	999	0.818
ENSP00000162749 [b]	TNFRSF1A	TNF Receptor superfamily member 1A	2.30×10^{-4}	0.001	999	0.826
ENSP00000304414 [c]	CXCR6	C-X-C motif chemokine receptor 6	9.27×10^{-5}	0.002	964	0.803
ENSP00000296795 [a]	TLR3	Toll like receptor 3	1.58×10^{-4}	0.002	966	0.858
ENSP00000231454 [c]	IL5	Interleukin 5	1.13×10^{-4}	0.004	991	0.803
ENSP00000222823 [a]	NOD1	Nucleotide binding oligomerization domain containing 1	7.72×10^{-5}	0.004	991	0.866
ENSP00000231449 [b]	IL4	Interleukin 4	2.55×10^{-4}	0.005	999	0.852
ENSP00000356438 [a]	PTGS2	Prostaglandin-endoperoxide synthase 2	1.92×10^{-4}	0.009	972	0.864
ENSP00000219244 [b]	CCL17	C-C motif chemokine ligand 17	1.20×10^{-4}	0.01	984	0.808
ENSP00000351273 [b]	CASP8	Caspase 8	9.66×10^{-5}	0.027	999	0.821
ENSP00000353483 [c]	MAPK8	Mitogen-activated protein kinase 8	1.03×10^{-4}	0.034	925	0.847
ENSP00000228280 [c]	KITLG	KIT ligand	9.60×10^{-5}	0.039	958	0.810
ENSP00000238682 [c]	TGFB3	Transforming growth factor β 3	5.37×10^{-5}	0.049	961	0.850

[a]: Genes with experiment evidence; [b]: Genes without experiment evidence but have significant relationship with uveitis; [c]: Genes without any evidence.

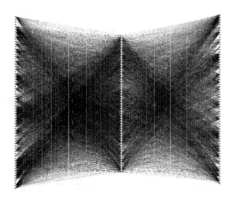

Figure 1. The sub-network of the large network containing Ensembl Identifications (IDs) of validated and putative uveitis-related genes. Blue nodes represent Ensembl IDs of validated uveitis-related genes. Green nodes represent Ensembl IDs of putative uveitis-related genes.

2.4. Analysis of Novel Genes

In this study, the RWR-based method yielded fifty-six genes that were deemed to have a significant correlation with uveitis. Detailed information for these genes is provided in Table 2.

2.4.1. Immune System Regulation Genes

CCL4 (C-C motif chemokine ligand 4) belongs to the cytokine family and is involved in immunoregulation and inflammation. It has been reported that *CCL4* is associated with BD immunopathogenesis [42]. In the majority of VKH cases, the expression of another family member *CCL17* was lower in cerebrospinal fluid than in serum, which indicated its potential function in VKH [43]. *CCL17* also could be inhibited by overexpression of *SOCS1* in the retina to regulate the recruitment of inflammatory cells [44]. The cytokine *CCL20* was considered to be a specific biomarker of *HLA-B27*-associated uveitis [45]. Our study revealed that *CCL4*, *CCL17* and *CCL20* likely play essential roles in uveitis.

CD40 ligand (also known as *CD154*) is a type II transmembrane glycoprotein that has structural homology to the proteins of the *TNF* (tumor necrosis factor) family [46–49]. The interaction between the *CD40* and *CD40 ligand* is important for both cellular and humoral immune responses [50]. The *CD40* and *CD40 ligand* interaction provides signals in T-cell priming and effecter functions [46,48,49,51–53], whereas monocyte and B-cell apoptosis could be inhibited by their interaction [54]. It has been demonstrated that the *CD40* ligand is associated with the immune-pathogenesis of several autoimmune diseases including AU (anterior uveitis) [54,55]. The *CD40* ligand is significantly expressed on T-cells in the peripheral blood of patients with AU [56]. The results of the RWR-based method revealed a MIS of 999 had a p-value < 0.001. Expression of *CD80* on dendritic cells (DCs) could be induced by activation of *NOD1* and *NOD2* and is involved in the pathogenesis of VKH syndrome [57]. In another report, it was found that BBR downregulated the expression of costimulatory molecules *CD40*, *CD80* and *CD86* on DCs [58]. The MIS and p-value of *CD80* were 999 and <0.001, respectively. We speculate that these molecules play key roles in uveitis, but their mechanism in uveitis must still be clarified.

CSF2 (colony stimulating factor 2) is a cytokine that functions as a hematological cell growth factor by stimulating stem cells to produce granulocytes and monocytes [59]. Three signaling pathways can be activated by *CSF2*: the JAK2/STAT pathway, the MAP pathway and the PI3K pathway [60–64]. *CSF2* is a valuable prognostic indicator and a therapeutic target in tumors [59]. *CSF2* expression in uveitis is reported as rare. However, in this study, the MIS of *CSF2* was 992 with a p-value < 0.001. We speculate that *CSF2* might be a key factor in the pathogenesis of uveitis.

Interleukins and their receptors are inflammatory cytokines that play an important role in immune system response. Many interleukins and their receptors are involved in uveitis, as discussed above. Our data showed that *IL13, IL15, IL1A, IL1B, IL1RN, IL4, IL5, IL9, IL23A* and *IL2RA* had MISs larger than 900 with p-values <0.05. It has been observed that the expression of *IL1A* is decreased in patients with clinically active BD, while the expression of *IL1B* is increased in patients with active, inactive or ocular BD [65]. IL1B has been associated with ocular Behcet's disease [66]. *IL-13* is a strong immunomodulatory cytokine which is a promising mode of treatment for uveitis [67–70]. *IL-15* and its receptor system is involved in the inflammatory process and pathogenesis of BD and the IL-15/Fc fusion protein has been shown to inhibit IRBP1-20 specific CD80+ T cell to decrease the severity of EAU [71,72]. An aberrantly high CNV of *IL23A* is a common risk factor for VKH and BD [73]. In mice, *IL-1RN* suppresses immune-mediated ocular inflammation and is considered a potential biomarker in the management of patients with uveitis [74]. Interleukin 2 receptor α (*IL2RA*) is a risk locus in various autoimmune diseases and a variant of this gene, *rs2104286*, was demonstrated to be strongly associated with intermediated uveitis [75]. An antibody against *IL2RA*, daclizumab is used to reduce intermediated uveitis [76]. However, *rs2104286* was not related to endogenous non-anterior uveitis [77]. EAU (experimental autoimmune uveoretinitis) disease severity was reduced in mice in which *IL-1B* expression was reduced in the retina through deletion of S100B, a Ca^{2+} binding protein [78]. In a Lewis rat model of EAU, *IL-2* and *IL-4* were produced in destructive foci in the retina and uveal tract. *IL-2* is thought to act as a cytotoxic effector, while *IL-4* is associated with a helper cell function [79]. In patients with BD, *IL-2* is more highly expressed, while *IL-4* is more lowly expressed [80]. Genetic findings suggest that more work should be done to evaluate both the molecular target and the inhibitor for personalized therapy.

TLR2, TLR3, TLR7 and *TLR9* belong to the Toll-like receptor (*TLR*) family, which are key factors in pathogen recognition and activation of innate immunity. *TLRs* are thought to be associated with infection and auto-inflammatory or autoimmune diseases, including uveitis [81,82]. Several autoimmune diseases, including BD, are associated with certain *TLR* gene polymorphisms [83,84]. A significant association has been found between polymorphism of TLR2 and ocular BD patients [85]. The expression of *TLR4* was significantly up-regulated in monocyte-derived macrophages from VKH patients [86]. The chitosan-mediated TLR3-siRNA transfection had a potential therapeutic effect on remitting uveitis [87]. In a Chinese Han population, a high copy number of TLR7 conferred risk for BD patients [88]. In the Japanese population, the homozygous genotypes and homozygous deplotype configuration of TLR9 SNPs was associated with the susceptibility to BD [89]. It has been reported that glucocorticoid could improve uveitis by downregulating *TLR7* and *TLR9* in peripheral blood of patients with uveitis [90]. In our analysis, *TLR2, TLR3, TLR7* and *TLR9* have MIS scores of 968, 966, 926 and 958, respectively. We argue that *TLR2, TLR3, TLR7* and *TLR8* play essential roles in uveitis and thus require more attention.

2.4.2. Transcription Associated Genes

Jun (also known as jun proto-oncogene) is a critical subunit of the transcription factor AP1, which is an important modulator of diverse biological processes such as cell proliferation, apoptosis and malignant transformation [91]. Jun is activated through phosphorylation at Ser 63 and Ser 73 by *JNK* [92,93]. A high level of *Jun* has been observed in various types of cancer including non-small cell lung cancer, oral squamous cell carcinoma, breast cancer and colorectal cancer [94–98]. Overexpression of *Jun* has led to aberrant tumor growth and progression and inhibited cell apoptosis [94]. The underlying mechanism of *Jun* as it relates to uveitis is still unclear. In a gene screen assay, it was found that expression of *Jun* showed a significantly higher index in experimental lens-induced uveitis rabbits [99]. In our analysis, *Jun* showed a significant index p-value and an MIS of 999; therefore, we propose that Jun may be an essential factor in uveitis.

STAT1 and *STAT6* encode transcription factors that belong to the *STAT* family, where phosphorylation is activated by receptor associated kinases. Atopic dermatitis associated uveitis may be driven by TH2-mediated inflammation [100]. IL-4 is a TH2 cytokine, and binding with its receptor can activate *STAT6* via the (Jak) Janus kinase/STAT signaling pathway to promote many immunomodulatory genes [101]. Furthermore, the Stat6 VT (STAT6 V547A/T548A) mouse model of atopic dermatitis exhibited uveitis symptoms [100]. In the STAT family, TH17 cells can be induced by *IL-2* and suppressed by *IL-27/STAT1* to contribute to uveitis [102]. In this study, *STAT1* and *STAT6* both had significant *p*-values and MIS of 999, and therefore we hypothesize that *STAT1/6* function has a putative role in uveitis.

2.4.3. Cell Adhesion and Signal Transduction Related Genes

Matrix metalloproteinase (*MMP*) are key factors for the degradation of extracellular matrix components and modification of cytokines, protease inhibitors, and cell surface signaling systems [103–106]. Polymorphisms on the *MMP-9* promoter can affect the development of visceral involvement in Korean people with BD [107]. In our RWR analysis, *MMP9* had an MIS of 971 and a *p*-value < 0.001, which suggests that *MMP-9* may be a novel susceptibility gene for uveitis.

VCAM1 (vascular cell adhesion molecule 1) belongs to the Ig superfamily and is a cell surface sialoglycoprotein expressed by cytokine-activated endothelium. This protein is mediated by leukocyte-endothelial cell adhesion and signal transduction [108,109]. *VCAM1* can be regulated by inflammatory cytokines such as *IL1B* [108]. Uveitis is closely associated with the immune system and immune-related proteins including the interleukin family. In our study, the MIS of *VCAM1* was 968 and the *p*-value was less than 0.05, which makes *VCAM1* a candidate gene for uveitis.

We detected 56 novel uveitis-related genes using the RWR-based method. These genes can be clustered into three categories, shown in Figure 2. Among these 56 potential genes, eight (8/56, 14.3%) genes, *IL13, IL1RN, JAK1, MYD88, NOD1, PTGS2, TGFB1* and *TLR3*, were considered as uveitis genes by experimental evidence [110–117], and 29 (29/56, 51.8%) genes (*CASP8, CCL17, CCL20, CCL4, CD19, CD40 LG, CD80, CXCL1, CXCL11, CXCL13, CXCL16, CXCL5, CXCL9, CXCR2, IL15, IL1A, IL1B, IL23A, IL2RA, IL4, JUN, MMP9, STAT1, STAT6, TLR2, TLR7, TLR9, TNFRSF1A* and *VCAM1*) had a correlation with uveitis. However, the pathogenesis is not clear. In our results, we found that these genes have a significant relationship with uveitis genes and therefore need more validation. There are few reports of the rest of the genes (19/56, 33.9%) (*CCR8, CD276, CD8A, CSF2, CSF3, CXCR1, CXCR5, CXCR6, GAPDH, GZMB, HLA-DB5, IL5, IL9, KITLG, MAPK8, PTPRC, SELE, TGFB3* and *TGFBI*) which participate in the process of uveitis. We considered that they might be novel uveitis genes and merit attention. We argue that some of these may be critical putative virulence genes for uveitis and could be interesting agents for the treatment of human uveitis.

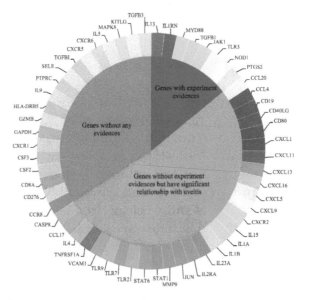

Figure 2. Clustering results of the 56 novel genes according to their evidences for being novel uveitis-related genes. Among 56 novel genes, eight have experiment evidence, 29 have significant relationship with uveitis but without experiment evidence, while no evidence can be found for the rest genes.

2.5. Comparison of Other Methods

The results listed in Sections 2.1–2.4 can partly prove the effectiveness of the RWR-based method. In this section, we compared our method with a classic GBA-based method [30–32], i.e., a method like the nearest neighbor algorithm (NNA). This method identified novel genes from neighbors of the uveitis-related genes in a network. For convenience, we directly used the PPI network that was adopted in the RWR-based method. In addition, we called a neighbor of a node is a nearer neighbor if the edge between them was assigned a higher weight due to the definition of the interaction score reported in STRING. The GBA-based method selected the k nearest neighbors of each uveitis-related genes and collected them together as the predicted genes of the method, where k is a predefined parameter.

The five-fold cross-validation method was also adopted to test the GBA-based method, which used the same partition in testing the RWR-based method. Because we do not know the best value of k, we tried the following values: 1, 2, 3, 4, 5, 6, 7, 8, 9, 10, 20, 30, 40, 50, 60, 70, 80, 90, and 100. The testing results are provided in Supplementary Table S2. The best performance of the GBA-based method with different parameter k on each part is shown in Table 3. Compared with the testing results of RWR-based method, also listed in Table 3 for convenience, we can see that GBA-based method provides higher recalls sometimes, however, it always provides lower precisions, indicating the GBA-based method can yield more false positive genes. If only considering the F1-measure, we can conclude that F1-measures of the RWR-based method are always higher than those of the GBA-based method. It is indicated that the RWR-based method is superior to GBA-based method for identification of uveitis-related genes.

Table 3. Comparison of the RWR-based method and GBA-based method.

Index of Part	RWR-Based Method			GBA-Based Method			
	Recall	Precision	F1-Measure	Best Value of k	Recall	Precision	F1-Measure
1	0.172	0.089	0.118	1	0.207	0.061	0.094
2	0.172	0.088	0.116	1	0.207	0.059	0.092
3	0.379	0.177	0.242	3	0.345	0.039	0.069
4	0.310	0.141	0.194	1	0.172	0.052	0.079
5	0.400	0.211	0.276	3	0.500	0.061	0.109

3. Materials and Methods

3.1. Materials

Uveitis-related genes were collected from literatures indexed by PubMed (http://www.ncbi.nlm.nih.gov/pubmed/). The keywords "uveitis" and "genes" were used to search the literature in PubMed, which resulted in the collection of 744 papers. Among them, 98 review papers that generally summarized uveitis-related genes were manually reviewed. From those 98 papers, 121 genes were chosen from 96 reviews reporting the functional genes that may be important for uveitis or for specific uveitis symptoms. These genes are provided in Supplementary Table S3. In total, 146 Ensembl IDs for these genes were also determined and are provided in Supplementary Table S3.

3.2. Protein-protein Interaction (PPI) Network

PPIs are useful for the investigation of genetic disorders because they play an essential role in intracellular and intercellular biochemical processes. Many computational methods have been developed using this information, such as the prediction engines for the identification of protein functions [118–120] and methods for identification of novel disease genes [36–38]. Several methods were built based on the hypothesis that two proteins in a PPI are more likely to share similar functions. Thus, we can infer novel genes related to uveitis using PPI information and the uveitis-related genes mentioned in Section 3.1.

In this study, we used the PPI information retrieved from STRING (Search Tool for the Retrieval of Interacting Genes/Proteins, Version 9.1, http://string-db.org/) [121] to construct the PPI network that the RWR algorithm can be applied. To access the PPI information in STRING, we downloaded the file "protein.links.v9.1 txt.gz". Because "9606" is the organism code for the human interactome in STRING, lines in this file that started with "9606" were extracted, obtaining 2,425,314 human PPIs involving 20,770 proteins. According to STRING, these PPIs were derived from the following four sources: (1) genomic context; (2) high-throughput experiments; (3) (conserved) co-expression; and (4) previous knowledge. Thus, the information in STRING contained both the direct (physical) and the indirect (functional) association between proteins, therefore STRING could widely measure the associations between proteins. Each PPI contained two Ensembl IDs and one score that ranged between 150 and 999, which indicated the strength of the interaction. An interaction with a high score meant this interaction has a high probability of occurring. For each interaction containing proteins p_a and p_b, the score was denoted by $S(p_a,p_b)$. The PPI network defined the 20,770 proteins as the nodes, and two nodes were adjacent if and only if their corresponding proteins can form a PPI. Additionally, each edge in the network represented a PPI; thus, we assigned a weight to each edge, which was defined as the score of its corresponding PPI. From our analysis, a PPI network containing 20,770 nodes and 2,425,314 edges was obtained.

3.3. RWR-Based Method

The RWR algorithm was executed on the PPI network using validated genes as seed nodes to search possible genes. Then, a permutation test was executed to exclude false discoveries found by RWR. The remaining candidate genes with strong associations to validated genes were selected for further analysis. The pseudo-codes of the RWR-based method are listed in Table 4.

Table 4. The pseudo-code of the RWR-based method.

RWR-Based Method
Input: Ensembl IDs of uveitis-related genes, a PPI network **Output:** A number of putative uveitis-related genes

1.	Execute the RWR algorithm on the PPI network using the Ensembl IDs of uveitis-related genes as seed nodes, yielding a probability for each gene in the network; genes with probabilities higher than 10^{-5} were selected and called RWR genes;
2.	Execute a permutation test, producing the p-value for each RWR gene; select RWR genes with p-values less than 0.05; the remaining genes were called candidate genes;
3.	For each candidate gene, calculate its MIS (cf. Equation (3)) and MFS (cf. Equation (5)); select candidate genes with MISs no less than 900 and MFSs larger than 0.8;
4.	Output the remaining candidate genes as the putative uveitis-related genes.

3.3.1. Searching Possible Genes Using the RWR Algorithm

RWR is a type of ranking algorithm [33]. Based on a seed node or a set of seed nodes, it simulates a walker that starts from the nodes and randomly walks in a network. Here, 146 Ensembl IDs listed in Supplementary Table S3 were deemed as seed nodes. Starting from these nodes, we attempted to discover novel nodes (genes) related to uveitis. In the beginning of the RWR algorithm, a 20,770-D vector P_0 was constructed, in which each composition represented the probability that a node in the network was a uveitis-related gene. Because the 146 Ensembl IDs represented validated uveitis-related genes, their compositions in P_0 were set to 1/146, while others were set to zero. Then, the RWR algorithm repeatedly updated this probability vector until it became stable. We designated P_i to represent the probability vector after the i-th step was executed. The probability vector was updated according to the following equation:

$$P_{i+1} = (1-r)A^T P_i + r P_0 \qquad (1)$$

where A represented the column-wise normalized adjacency matrix of the PPI network and r was set to 0.8. When $\|P_{t+1} - P_t\|_{L_1} < 10^{-6}$, the update procedure was stopped, and P_{t+1} was the output of the RWR algorithm.

According to the probability vector yielded by the RWR algorithm, some nodes received high probabilities. It was apparent that their corresponding genes are more likely to be uveitis-related genes. To avoid missing possible uveitis-related genes, we set a probability threshold of 10^{-5}. The corresponding genes of these nodes were designated as RWR genes.

In this study, we used the RWR program on the heterogeneous network that was implemented in Matlab and proposed by Li and Patra [122]. The code can be downloaded at http://www3.ntu.edu. sg/home/aspatra/research/Yongjin_BI2010.zip. By setting the special values of some parameters, this program could be used to execute the RWR algorithm on a single network.

3.3.2. Excluding False Discoveries Using the Permutation Test

Based on the validated uveitis-related genes and RWR algorithm, new RWR genes were accessed. However, this result was influenced by the structure of the constructed PPI network, i.e., some RWR genes were selected due to the structure of the network and they were not necessarily unique to uveitis. Furthermore, if we randomly selected some nodes in the network as seed nodes of the RWR algorithm, these genes were still selected for and were therefore deemed as likely to be false positive. To control for these genes, a permutation test was executed. We randomly constructed 1000 Ensembl ID sets, denoted by $E_1, E_2, \ldots, E_{1000}$, consisting of 146 Ensembl IDs. For each set, the Ensembl IDs were deemed seed nodes of the RWR algorithm. Each RWR gene was given a probability. Thus, there were 1000 probabilities for 1000 sets and one probability for 146 Ensembl IDs of the uveitis-related genes for each RWR gene. Then, a measurement, called the p-value, was counted for each RWR gene g, which was defined as:

$$p - \text{value}(g) = \frac{\Theta}{1000} \tag{2}$$

where Θ represented the number of randomly constructed sets where the probability assigned to g was larger than that for the 146 Ensembl IDs of uveitis-related genes. Clearly, an RWR gene with a high p-value indicated that the gene was not specific for uveitis and should be discarded. RWR genes with p-values less than 0.05 were selected for further analysis as potential candidate genes for uveitis.

3.3.3. Selection of Core Genes by Associations with Validated Genes

We hypothesized that, of the candidate genes, some may have a strong correlation with uveitis. To further select core candidate genes, two criteria were designed. Candidate genes satisfying both criteria were selected for additional analysis. Candidate genes that had the strongest associations with uveitis-related genes were more likely to be novel uveitis-related genes. Thus, for each candidate gene g, we calculated the maximum interaction score (MIS) as follows:

$$MIS(g) = \max\{S(g, g') : g' \text{ is a uveitis} - \text{related gene}\} \tag{3}$$

A high MIS suggested that the candidate gene was closely related to at least one uveitis-related gene, indicating that it was a novel uveitis-related gene with a high probability. According to STRING, a score of 900 was the cut-off for the highest confidence level. Therefore, candidate genes with MISs larger than 900 were selected.

Validated uveitis-related genes have strong associations with specific gene ontology (GO) terms and Kyoto Encyclopedia of Genes and Genomes (KEGG) pathways. Therefore, candidate genes that had similar associations with uveitis GO terms and KEGG pathways were more likely to be novel uveitis-related genes. We performed GO term (KEGG pathway) [123–126] enrichment analysis for candidate genes and uveitis-related genes. The representation of a gene g on all GO terms and KEGG pathways was encoded into a vector ES (g) using this theory. This vector can be obtained by an in-house program using the R function phyper. The R code used was "score <− −log10 (phyper (numWdrawn− 1, numW, numB, numDrawn, lower.tail = FALSE))," where numW, numB, and numDrawn are the number of genes annotated to the GO term or KEGG pathway, the number of genes not annotated to the GO term or KEGG pathway, and the number of neighbors of gene g and numWdrawn is the number of neighbors of gene g that are also annotated to the GO term or KEGG pathway. The relativity of the two genes g and g' on GO terms and KEGG pathways was measured by

$$\Gamma(g, g') = \frac{ES(g) \cdot ES(g')}{\|ES(g)\| \cdot \|ES(g')\|} \tag{4}$$

A high outcome of Equation (4) indicated that g and g' have a similar relationship in terms of GO terms and KEGG pathways. For any candidate gene g, we calculated the maximum function score (MFS) using the following equation:

$$MFS(g) = \max\{\Gamma(g, g') : g' \text{is a uveitis} - \text{related gene}\} \tag{5}$$

Candidate genes with high MFSs were selected. In this equation, we set 0.8 as the threshold of MFS to select essential candidate genes.

3.4. Methods for Testing RWR-Based Method

In this study, we designed the RWR-based method to identify novel uveitis-related genes. However, it is necessary to test its effectiveness in advance. Here, the five-fold cross-validation [127] was employed. In detail, 146 Ensembl IDs of uveitis-related genes were randomly and equally divided into five parts. Then, Ensembl IDs in each part were singled out in turn and other Ensembl IDs in the rest four parts were used as the seed nodes in the RWR-based method. For each part, the results yielded by a good identification method on the rest four parts should satisfy the following conditions: (I) the results can recover a high proportion of the Ensembl IDs in the part; and (II) the results cannot contain several Ensembl IDs that are not in the part. Thus, recall and precision were employed to evaluate the results yielded by the RWR-based method, which can be calculated by

$$\begin{cases} \text{recall} = \frac{TP}{TP+FN} \\ \text{precision} = \frac{TP}{TP+FP} \end{cases} \tag{6}$$

where TP represented the number of Ensembl IDs in the part that can be recovered by the method, FN represented the number of Ensembl IDs in the part that cannot be recovered by the method and FP represented the number of Ensembl IDs that were yielded by the method and not in the part. In addition, to evaluate the predicted results on the whole, the F1-measure was also adopted, which can be computed by

$$\text{F1} - \text{measure} = \frac{2 \times \text{recall} \times \text{precision}}{\text{recall} + \text{precision}} \tag{7}$$

It is clear that a high F1-measure means the good performance of the method.

4. Conclusions

This study presented a computational method to determine novel uveitis-related genes. Using the RWR algorithm and certain screening criteria, 56 putative genes were accessed. Extensive analysis of the obtained genes confirmed that several genes are associated with the pathogenesis of uveitis. We hope that the identified novel genes may be used as material to study uveitis and that the proposed method can be extended to other diseases.

Acknowledgments: This study was supported by the National Natural Science Foundation of China (31371335), Natural Science Foundation of Shanghai (17ZR1412500).

Author Contributions: Lin Liu conceived and designed the experiments; Shiheng Lu, Yan Yan performed the experiments; Shiheng Lu, Zhen Li, Lei Chen and Jing Yang analyzed the data; Shiheng Lu, Yuhang Zhang, Shaopeng Wang contributed reagents/materials/analysis tools; Shiheng Lu wrote the paper.

References

1. Miserocchi, E.; Foqliato, G.; Modorati, G.; Bandello, F. Review on the worldwide epidemiology of uveitis. *Eur. J. Ophthalmol.* **2013**, *23*, 705–717. [CrossRef] [PubMed]

2. Fraga, N.A.; Oliveira, M.F.; Follador, I.; Rocha, B.O.; Rêgo, V.R. Psoriasis and uveitis: A literature review. *An. Bras. Dermatol.* **2012**, *87*, 877–883. [CrossRef] [PubMed]

3. Kulkarni, P. Review: Uveitis and immunosuppressive drugs. *J. Ocul. Pharmacol. Ther.* **2001**, *17*, 181–187. [CrossRef] [PubMed]

4. Murakami, S.; Inaba, Y.; Mochizuki, M.; Nakajima, A.; Urayama, A. A nation-wide survey on the occurrence of Vogt-Koyanagi-Harada disease in Japan. *Jpn. J. Ophthalmol.* **1994**, *98*, 389–392.

5. Moorthy, R.S.; Inomata, H.; Rao, N.A. Vogt-Koyanagi-Harada syndrome. *Surv. Ophthalmol.* **1995**, *39*, 265–292. [CrossRef]

6. Evereklioglu, C. Current concepts in the etiology and treatment of Behcet disease. *Surv. Ophthalmol.* **2005**, *50*, 297–350. [CrossRef] [PubMed]

7. Chang, J.H.; Wakefield, D. Uveitis: A global perspective. *Ocul. Immunol. Inflamm.* **2002**, *10*, 263–279. [CrossRef] [PubMed]

8. Priem, H.A.; Kijlstra, A.; Noens, L.; Baarsma, G.S.; de Laey, J.J.; Oosterhuis, J.A. HLA typing in birdshot chorioretinopathy. *Am. J. Ophthalmol.* **1988**, *105*, 182–185. [CrossRef]

9. Chaput, F.; Amer, R.; Baglivo, E.; Touitou, V.; Kozyreff, A.; Bron, D.; Bodaghi, B.; LeHoang, P.; Bergstrom, C.; Grossniklaus, H.E.; et al. Intraocular T-cell Lymphoma: Clinical Presentation, Diagnosis, Treatment, and Outcome. *Ocul. Immunol. Inflamm.* **2016**, *22*, 1–10. [CrossRef] [PubMed]

10. Kitazawa, K.; Nagata, K.; Yamanaka, Y.; Kuwahara, Y.; Lehara, T.; Kinoshita, S.; Sotozono, C. Diffuse Anterior Retinoblastoma with Sarcoidosis-Like Nodule. *Case Rep. Ophthalmol.* **2015**, *6*, 443–447. [CrossRef] [PubMed]

11. Catala-Mora, J.; Parareda-Salles, A.; Vicuña-Muñoz, C.G.; Medina-Zurinaga, M.; Prat-Bartomeu, J. [Uveitis masquerade syndrome as a presenting form of diffuse retinoblastoma]. *Arch. Soc. Esp. Oftalmol.* **2009**, *84*, 477–480. [PubMed]

12. All-Ericsson, C.; Economou, M.A.; Landau, I.; Seregard, S.; Träisk, F. Uveitis masquerade syndromes: Diffuse retinoblastoma in an older child. *Acta Ophthalmol. Scand.* **2007**, *85*, 569–570. [CrossRef] [PubMed]

13. Jovanovic, S.; Jovanović, Z.; Paović, J.; Teperković, V.S.; Pesić, S.; Marković, V. Two cases of uveitis masquerade syndrome caused by bilateral intraocular large B-cell lymphoma. *Vojnosanit. Pregl.* **2013**, *70*, 1151–1154. [CrossRef] [PubMed]

14. Shen, K.; Smith, S.V.; Lee, A.G. Acute myelogenous leukemia presenting with uveitis, optic disc edema, and granuloma annulare: Case report. *Can. J. Ophthalmol.* **2016**, *51*, e153–e155. [CrossRef] [PubMed]

15. Miserocchi, E.; Cimminiello, C.; Mazzola, M.; Russo, V.; Modorati, G.M. New-onset uveitis during CTLA-4 blockade therapy with ipilimumab in metastatic melanoma patient. *Can. J. Ophthalmol.* **2015**, *50*, e2–e4. [CrossRef] [PubMed]

16. Rosenbaum, J.T.; Pasadhika, S.; Crouser, E.D.; Choi, D.; Harrington, C.A.; Lewis, J.A.; Austin, C.R.; Diebel, T.N.; Vance, E.E.; Braziel, R.M.; et al. Hypothesis: Sarcoidosis is a STAT1-mediated disease. *Clin. Immunol.* **2009**, *132*, 174–183. [CrossRef] [PubMed]

17. Hou, S.; Yang, Z.; Du, L.; Jiang, Z.; Shu, Q.; Chen, Y.; Li, F.; Zhou, Q.; Ohno, S.; Chen, R.; et al. Identification of a susceptibility locus in STAT4 for Behcet's disease in Han Chinese in a genome-wide association study. *Arthritis. Rheum.* **2012**, *64*, 4104–4113. [CrossRef] [PubMed]

18. Remmers, E.F.; Cosan, F.; Kirino, Y.; Ombrello, M.J.; Abaci, N.; Satorius, C.; Le, J.M.; Yang, B.; Korman, B.D.; Cakiris, A.; et al. Genome-wide association study identifies variants in the MHC class I, IL10, and IL23R-IL12RB2 regions associated with Behcet's disease. *Nat. Genet.* **2010**, *42*, 698–702. [CrossRef] [PubMed]

19. Mizuki, N.; Meguro, A.; Ota, M.; Ohno, S.; Shiota, T.; Kawagoe, T.; Ito, N.; Kera, J.; Okada, E.; Yatsu, K.; et al. Genome-wide association studies identify IL23R-IL12RB2 and IL10 as Behcet's disease susceptibility loci. *Nat. Genet.* **2010**, *42*, 703–706. [CrossRef] [PubMed]

20. Robinson, P.C.; Claushuis, T.A.; Cortes, A.; Martin, T.M.; Evans, D.M.; Leo, P.; Mukhopadhyay, P.; Bradbury, L.A.; Cremin, K.; Harris, J.; et al. Genetic dissection of acute anterior uveitis reveals similarities and differences in associations observed with ankylosing spondylitis. *Arthritis Rheumatol.* **2015**, *67*, 140–151. [CrossRef] [PubMed]

21. Fang, J.; Chen, L.; Tang, J.; Hou, S.; Liao, D.; Ye, Z.; Wang, C.; Cao, Q.; Kijlstra, A.; Yang, P. Association Between Copy Number Variations of TLR7 and Ocular Behcet's Disease in a Chinese Han Population. *Investig. Ophthalmol. Vis. Sci.* **2015**, *56*, 1517–1523. [CrossRef] [PubMed]

22. Kirino, Y.; Zhou, Q.; Ishigatsubo, Y.; Mizuki, N.; Tugal-Tutkun, I.; Seyahi, E.; Özyazgan, Y.; Ugurlu, S.; Erer, B.; Abaci, N. Targeted resequencing implicates the familial Mediterranean fever gene *MEFV* and the toll-like receptor 4 gene *TLR4* in Behcet disease. *Proc. Natl. Acad. Sci. USA* **2013**, *110*, 8134–8139. [CrossRef] [PubMed]

23. Li, B.; Yang, P.; Chu, L.; Zhou, H.; Huang, X.; Zhu, L.; Kijlstra, A. T-bet expression in the iris and spleen parallels disease expression during endotoxin-induced uveitis. Graefe's Arch. *Clin. Exp. Ophthalmol.* **2007**, *245*, 407–413. [CrossRef] [PubMed]

24. Kirino, Y.; Bertsias, G.; Ishigatsubo, Y.; Mizuki, N.; Tugal-Tutkun, I.; Seyahi, E.; Ozyazgan, Y.; Sacli, F.S.; Erer, B.; Inoko, H.; et al. Genome-wide association analysis identifies new susceptibility loci for Behcet's disease and epistasis between HLA-B*51 and ERAP1. *Nat. Genet.* **2013**, *45*, 202–207. [CrossRef] [PubMed]

25. Jiang, Z.; Yang, P.; Hou, S.; Du, L.; Xie, L.; Zhou, H.; Kijlstra, A. IL-23R gene confers susceptibility to Behcet's disease in a Chinese Han population. *Ann. Rheum. Dis.* **2010**, *69*, 1325–1328. [CrossRef] [PubMed]

26. Smith, J.R.; Choi, D.; Chipps, T.J.; Pan, Y.; Zamora, D.O.; Davies, M.H.; Babra, B.; Powers, M.R.; Planck, S.R.; Rosenbaum, J.T. Unique gene expression profiles of donor-matched human retinal and choroidal vascular endothelial cells. *Investig. Ophthalmol. Vis. Sci.* **2007**, *48*, 2676–2684. [CrossRef] [PubMed]

27. Li, Z.; Liu, B.; Maminishkis, A.; Mahesh, S.P.; Yeh, S.; Lew, J.; Lim, W.K.; Sen, H.N.; Clarke, G.; Buggage, R. Gene expression profiling in autoimmune noninfectious uveitis disease. *J. Immunol.* **2008**, *181*, 5147–5157. [CrossRef] [PubMed]

28. Ohta, K.; Kikuchi, T.; Miyahara, T.; Yoshimura, N. DNA microarray analysis of gene expression in iris and ciliary body of rat eyes with endotoxin-induced uveitis. *Exp. Eye Res.* **2005**, *80*, 401–412. [CrossRef] [PubMed]

29. Li, Z.; Mzhesh, S.P.; Liu, B.; Yeh, S.; Lew, J.; Lim, W.; Levy Clarke, G.; Buggage, R.; Nussenblatt, R.B. Gene Expression Profiling of Non-infectious Uveitis Patients Using Pathway Specific cDNA Microarray Analysis. *Investig. Ophthalmol. Vis. Sci.* **2007**, *48*, 1505. [CrossRef]

30. Oliver, S. Guilt-by-association goes global. *Nature* **2000**, *403*, 601–603. [CrossRef] [PubMed]

31. Oti, M.; Snel, B.; Huynen, M.A.; Brunner, H.G. Predicting disease genes using protein-protein interactions. *J. Méd. Genet.* **2006**, *43*, 691–698. [CrossRef] [PubMed]

32. Krauthammer, M.; Kaufmann, C.A.; Conrad Gilliam, T.; Rzhetsky, A. Molecular triangulation: Bridging linkage and molecular-network information for identifying candidate genes in Alzheimer's disease. *Proc. Natl. Acad. Sci. USA* **2004**, *101*, 15148–15153. [CrossRef] [PubMed]

33. Kohler, S.; Bauer, S.; Horn, D.; Robinson, P.N. Walking the interactome for prioritization of candidate disease genes. *Am. J. Hum. Genet.* **2008**, *82*, 949–958. [CrossRef] [PubMed]

34. Li, Y.; Li, J. Disease gene identification by random walk on multigraphs merging heterogeneous genomic and phenotype data. *BMC Genom.* **2012**, *13*, S27. [CrossRef] [PubMed]

35. Jiang, R.; Gan, M.X.; He, P. Constructing a gene semantic similarity network for the inference of disease genes. *BMC Syst. Biol.* **2011**, *5*, S2. [CrossRef] [PubMed]

36. Chen, L.; Hao, X.Z.; Huang, T.; Shu, Y.; Huang, G.H.; Li, H.P. Application of the shortest path algorithm for the discovery of breast cancer related genes. *Curr. Bioinform.* **2016**, *11*, 51–58. [CrossRef]

37. Zhang, J.; Yang, J.; Yang, T.; Huang, T.; Shu, Y.; Chen, L. Identification of novel proliferative diabetic retinopathy related genes on protein-protein interaction network. *Neurocomputing* **2016**, *217*, 63–72. [CrossRef]

38. Gui, T.; Dong, X.; Li, R.; Li, Y.; Wang, Z. Identification of Hepatocellular Carcinoma–Related Genes with a Machine Learning and Network Analysis. *J. Comput. Biol.* **2015**, *22*, 63–71. [CrossRef] [PubMed]

39. Chen, L.; Yang, J.; Huang, T.; Kong, X.; Lu, L.; Cai, Y.D. Mining for novel tumor suppressor genes using a shortest path approach. *J. Biomol. Struct. Dyn.* **2016**, *34*, 664–675. [CrossRef] [PubMed]

40. Chen, L.; Huang, T.; Zhang, Y.H.; Jiang, Y.; Zheng, M.; Cai, Y.D. Identification of novel candidate drivers connecting different dysfunctional levels for lung adenocarcinoma using protein-protein interactions and a shortest path approach. *Sci. Rep.* **2016**, *6*, 29849. [CrossRef] [PubMed]

41. Chen, L.; Wang, B.; Wang, S.; Yang, J.; Hu, J.; Xie, Z.; Wang, Y.; Huang, T.; Cai, Y.D. OPMSP: A computational method integrating protein interaction and sequence information for the identification of novel putative oncogenes. *Protein Pept. Lett.* **2016**, *23*, 1081–1094. [CrossRef] [PubMed]

42. Oguz, A.K.; Yılmaz, S.T.; Oygür, Ç.Ş.; Çandar, T.; Sayın, I.; Kılıçoğlu, S.S.; Ergün, İ.; Ateş, A.; Özdağ, H.; Akar, N. Behcet's: A Disease or a Syndrome? Answer from an Expression Profiling Study. *PLoS ONE* **2016**, *11*, e0149052. [CrossRef] [PubMed]

43. Miyazawa, I.; Abe, T.; Narikawa, K.; Feng, J.; Misu, T.; Nakashima, I.; Fujimori, J.; Tamai, M.; Fujihara, K.; Itoyama, Y. Chemokine profile in the cerebrospinal fluid and serum of Vogt-Koyanagi-Harada disease. *J. Neuroimmunol.* **2005**, *158*, 240–244. [CrossRef] [PubMed]

44. Yu, C.R.; Mahdi, R.R.; Oh, H.M.; Amadi-Obi, A.; Levy-Clarke, G.; Burton, J.; Eseonu, A.; Lee, Y.; Chan, C.C.; Egwuagu, C.E. Suppressor of cytokine signaling-1 (SOCS1) inhibits lymphocyte recruitment into the retina and protects SOCS1 transgenic rats and mice from ocular inflammation. *Investig. Ophthalmol. Vis. Sci.* **2011**, *52*, 6978–6986. [CrossRef] [PubMed]

45. Abu El-Asrar, A.M.; Berghmans, N.; Al-Obeidan, S.A.; Mousa, A.; Opdenakker, G.; van Damme, J.; Struyf, S. The Cytokine Interleukin-6 and the Chemokines CCL20 and CXCL13 Are Novel Biomarkers of Specific Endogenous Uveitic Entities. *Investig. Ophthalmol. Vis. Sci.* **2016**, *57*, 4606–4613. [CrossRef] [PubMed]

46. Hollenbaugh, D.; Grosmaire, L.S.; Kullas, C.D.; Chalupny, N.J.; Braesch-Andersen, S.; Noelle, R.J.; Stamenkovic, I.; Ledbetter, J.A.; Aruffo, A. The human T cell antigen gp39, a member of the TNF gene family, is a ligand for the CD40 receptor: Expression of a soluble form of gp39 with B cell co-stimulatory activity. *EMBO J.* **1992**, *11*, 4313–4321. [PubMed]

47. Lane, P.; Traunecker, A.; Hubele, S.; Inui, S.; Lanzavecchia, A.; Gray, D. Activated human T cells express a ligand for the human B cell-associated antigen CD40 which participates in T cell-dependent activation of B lymphocytes. *Eur. J. Immunol.* **1992**, *22*, 2573–2578. [CrossRef] [PubMed]

48. Noelle, R.J.; Ledbetter, J.A.; Aruffo, A. CD40 and its ligand, an essential ligand-receptor pair for thymus-dependent B-cell activation. *Immunol. Today* **1992**, *13*, 431–433. [CrossRef]

49. Fanslow, W.C.; Srinivasan, S.; Paxton, R.; Gibson, M.G.; Spriggs, M.K.; Armitage, R.J. Structural characteristics of CD40 ligand that determine biological function. *Semin. Immunol.* **1994**, *6*, 267–278. [CrossRef] [PubMed]

50. Howard, L.M.; Miga, A.J.; Vanderlugt, C.L.; Dal Canto, M.C.; Laman, J.D.; Noelle, R.J.; Miller, S.D. Mechanisms of immunotherapeutic intervention by anti-CD40L (CD154) antibody in an animal model of multiple sclerosis. *J. Clin. Investig.* **1999**, *103*, 281–290. [CrossRef] [PubMed]

51. Casamayor-Palleja, M.; Khan, M.; MacLennan, I.C. A subset of CD4+ memory T cells contains preformed CD40 ligand that is rapidly but transiently expressed on their surface after activation through the T cell receptor complex. *J. Exp. Med.* **1995**, *181*, 1293–1301. [CrossRef] [PubMed]

52. Stuber, E.; Strober, W.; Neurath, M. Blocking the CD40L-CD40 interaction in vivo specifically prevents the priming of T helper 1 cells through the inhibition of interleukin 12 secretion. *J. Exp. Med.* **1996**, *183*, 693–698. [CrossRef] [PubMed]

53. Grewal, I.S.; Flavell, R.A. CD40 and CD154 in cell-mediated immunity. *Annu. Rev. Immunol.* **1998**, *16*, 111–135. [CrossRef] [PubMed]

54. Ogard, C.; Sorensen, T.L.; Krogh, E. Increased CD40 ligand in patients with acute anterior uveitis. *Acta Ophthalmol. Scand.* **2005**, *83*, 370–373. [CrossRef] [PubMed]

55. Balashov, K.E.; Smith, D.R.; Khoury, S.J.; Hafler, D.A.; Weiner, H.L. Increased interleukin 12 production in progressive multiple sclerosis: Induction by activated CD4+ T cells via CD40 ligand. *Proc. Natl. Acad. Sci. USA* **1997**, *94*, 599–603. [CrossRef] [PubMed]

56. Ang, M.; Cheung, G.; Vania, M.; Chen, J.; Yang, H.; Li, J.; Chee, S.P. Aqueous cytokine and chemokine analysis in uveitis associated with tuberculosis. *Mol. Vis.* **2012**, *18*, 565–573. [PubMed]

57. Deng, B.; Ye, Z.; Li, L.; Zhang, D.; Zhu, Y.; He, Y.; Wang, C.; Wu, L.; Kijlstra, A.; Yang, P. Higher Expression of NOD1 and NOD2 is Associated with Vogt-Koyanagi-Harada (VKH) Syndrome But Not Behcet's Disease (BD). *Curr. Mol. Med.* **2016**, *16*, 424–435. [CrossRef] [PubMed]

58. Yang, Y.; Qi, J.; Wang, Q.; Du, L.; Zhou, Y.; Yu, H.; Kijlstra, A.; Yang, P. Berberine suppresses Th17 and dendritic cell responses. *Investig. Ophthalmol. Vis. Sci.* **2013**, *54*, 2516–2522. [CrossRef] [PubMed]

59. Lee, Y.Y.; Wu, W.J.; Huang, C.N.; Li, C.C.; Li, W.M.; Yeh, B.W.; Liang, P.I.; Wu, T.F.; Li, C.F. CSF2 Overexpression Is Associated with STAT5 Phosphorylation and Poor Prognosis in Patients with Urothelial Carcinoma. *J. Cancer* **2016**, *7*, 711–721. [CrossRef] [PubMed]

60. Jucker, M.; Feldman, R.A. Identification of a new adapter protein that may link the common β subunit of the receptor for granulocyte/macrophage colony-stimulating factor, interleukin (IL)-3, and IL-5 to phosphatidylinositol 3-kinase. *J. Biol. Chem.* **1995**, *270*, 27817–27822. [PubMed]

61. Bittorf, T.; Jaster, R.; Brock, J. Rapid activation of the MAP kinase pathway in hematopoietic cells by erythropoietin, granulocyte-macrophage colony-stimulating factor and interleukin-3. *Cell Signal.* **1994**, *6*, 305–311. [CrossRef]

62. Kimura, A.; Rieger, M.A.; Simone, J.M.; Chen, W.; Wickre, M.C.; Zhu, B.M.; Hoppe, P.S.; O'Shea, J.J.; Schroeder, T.; Hennighausen, L. The transcription factors STAT5A/B regulate GM-CSF-mediated granulopoiesis. *Blood* **2009**, *114*, 4721–4728. [CrossRef] [PubMed]

63. Mui, A.L.; Wakao, H.; Harada, N.; O'Farrell, A.M.; Miyajima, A. Interleukin-3, granulocyte-macrophage colony-stimulating factor, and interleukin-5 transduce signals through two forms of STAT5. *J. Leukoc. Biol.* **1995**, *57*, 799–803. [PubMed]

64. Feldman, G.M.; Rosenthal, L.A.; Liu, X.; Hayes, M.P.; Wynshaw-Boris, A.; Leonard, W.J.; Hennighausen, L.; Finbloom, D.S. STAT5A-deficient mice demonstrate a defect in granulocyte-macrophage colony-stimulating factor-induced proliferation and gene expression. *Blood* **1997**, *90*, 1768–1776. [PubMed]

65. Taheri, S.; Borlu, M.; Evereklioglu, C.; Ozdemir, S.Y.; Ozkul, Y. mRNA Expression Level of Interleukin Genes in the Determining Phases of Behcet's Disease. *Ann. Dermatol.* **2015**, *27*, 291–297. [CrossRef] [PubMed]

66. Liang, L.; Tan, X.; Zhou, Q.; Zhu, Y.; Tian, Y.; Yu, H.; Kijlstra, A.; Yang, P. IL-1β triggered by peptidoglycan and lipopolysaccharide through TLR2/4 and ROS-NLRP3 inflammasome-dependent pathways is involved in ocular Behcet's disease. *Investig. Ophthalmol. Vis. Sci.* **2013**, *54*, 402–414. [CrossRef] [PubMed]

67. Roberge, F.G.; de Smet, M.D.; Benichou, J.; Kriete, M.F.; Raber, J.; Hakimi, J. Treatment of uveitis with recombinant human interleukin-13. *Br. J. Ophthalmol.* **1998**, *82*, 1195–1198. [CrossRef] [PubMed]

68. Marie, O.; Thillaye-Goldenberg, B.; Naud, M.C.; de Kozak, Y. Inhibition of endotoxin-induced uveitis and potentiation of local TNF-α and interleukin-6 mRNA expression by interleukin-13. *Investig. Ophthalmol. Vis. Sci.* **1999**, *40*, 2275–2282.

69. Lemaitre, C.; Thillaye-Goldenberg, B.; Naud, M.C.; de Kozak, Y. The effects of intraocular injection of interleukin-13 on endotoxin-induced uveitis in rats. *Investig. Ophthalmol. Vis. Sci.* **2001**, *42*, 2022–2030.

70. De Kozak, Y.; Omri, B.; Smith, J.R.; Naud, M.C.; Thillaye-Goldenberg, B.; Crisanti, P. Protein kinase C ζ (PKC ζ) regulates ocular inflammation and apoptosis in endotoxin-induced uveitis (EIU)—Signaling molecules involved in EIU resolution by PKC ζ inhibitor and interleukin-13. *Am. J. Pathol.* **2007**, *170*, 1241–1257. [CrossRef] [PubMed]

71. Xia, Z.J.; Kong, X.L.; Zhang, P. [In vivo effect of recombined IL-15/Fc fusion protein on EAU]. *Sichuan Da Xue Xue Bao Yi Xue Ban* **2008**, *39*, 944–949. [PubMed]

72. Choe, J.Y.; Lee, H.; Kim, S.G.; Kim, M.J.; Park, S.H.; Kim, S.K. The distinct expressions of interleukin-15 and interleukin-15 receptor α in Behcet's disease. *Rheumatol. Int.* **2013**, *33*, 2109–2115. [CrossRef] [PubMed]

73. Hou, S.; Liao, D.; Zhang, J.; Fang, J.; Chen, L.; Qi, J.; Zhang, Q.; Liu, Y.; Bai, L.; Zhou, Y.; et al. Genetic variations of IL17F and IL23A show associations with Behcet's disease and Vogt-Koyanagi-Harada syndrome. *Ophthalmology* **2015**, *122*, 518–523. [CrossRef] [PubMed]

74. Lim, W.K.; Fujimoto, C.; Ursea, R.; Mahesh, S.P.; Silver, P.; Chan, C.C.; Gery, I.; Nussenblatt, R.B. Suppression of immune-mediated ocular inflammation in mice by interleukin 1 receptor antagonist administration. *Arch. Ophthalmol.* **2005**, *123*, 957–963. [CrossRef] [PubMed]

75. Lindner, E.; Weger, M.; Steinwender, G.; Griesbacher, A.; Posch, U.; Ulrich, S.; Wegscheider, B.; Ardjomand, N.; El-Shabrawi, Y. *IL2RA* gene polymorphism rs2104286 A>G seen in multiple sclerosis is associated with intermediate uveitis: Possible parallel pathways? *Investig. Ophthalmol. Vis. Sci.* **2011**, *52*, 8295–8299. [CrossRef] [PubMed]

76. Nussenblatt, R.B.; Fortin, E.; Schiffman, R.; Rizzo, L.; Smith, J.; van Veldhuisen, P.; Sran, P.; Yaffe, A.; Goldman, C.K.; Waldmann, T.A.; et al. Treatment of noninfectious intermediate and posterior uveitis with the humanized anti-Tac mAb: A phase I/II clinical trial. *Proc. Natl. Acad. Sci. USA* **1999**, *96*, 7462–7466. [CrossRef] [PubMed]

77. Cenit, M.C.; Marquez, A.; Cordero-Coma, M.; Fonollosa, A.; Adan, A.; Martinez-Berriotxoa, A.; Llorenc, V.; Diaz Valle, D.; Blanco, R.; Canal, J.; et al. Evaluation of the IL2/IL21, IL2RA and IL2RB genetic variants influence on the endogenous non-anterior uveitis genetic predisposition. *BMC Med. Genet.* **2013**, *14*, 52. [CrossRef] [PubMed]

78. Niven, J.; Hoare, J.; McGowan, D.; Devarajan, G.; Itohara, S.; Gannage, M.; Teismann, P.; Crane, I. S100B Up-Regulates Macrophage Production of IL1β and CCL22 and Influences Severity of Retinal Inflammation. *PLoS ONE* **2015**, *10*, e0132688. [CrossRef] [PubMed]

79. Charteris, D.G.; Lightman, S.L. Comparison of the expression of interferon gamma, IL2, IL4, and lymphotoxin mRNA in experimental autoimmune uveoretinitis. *Br. J. Ophthalmol.* **1994**, *78*, 786–790. [CrossRef] [PubMed]

80. Shahram, F.; Nikoopour, E.; Rezaei, N.; Saeedfar, K.; Ziaei, N.; Davatchi, F.; Amirzargar, A. Association of interleukin-2, interleukin-4 and transforming growth factor-β gene polymorphisms with Behcet's disease. *Clin. Exp. Rheumatol.* **2011**, *29*, S28–S31. [PubMed]

81. Chang, J.H.; McCluskey, P.; Wakefield, D. Expression of toll-like receptor 4 and its associated lipopolysaccharide receptor complex by resident antigen-presenting cells in the human uvea. *Investig. Ophthalmol. Vis. Sci.* **2004**, *45*, 1871–1878. [CrossRef]

82. Chang, J.H.; McCluskey, P.J.; Wakefield, D. Toll-like receptors in ocular immunity and the immunopathogenesis of inflammatory eye disease. *Br. J. Ophthalmol.* **2006**, *90*, 103–108. [CrossRef] [PubMed]

83. Meguro, A.; Ota, M.; Katsuyama, Y.; Oka, A.; Ohno, S.; Inoko, H.; Mizuki, N. Association of the toll-like receptor 4 gene polymorphisms with Behcet's disease. *Ann. Rheum. Dis.* **2008**, *67*, 725–727. [CrossRef] [PubMed]

84. Song, G.G.; Choi, S.J.; Ji, J.D.; Lee, Y.H. Toll-like receptor polymorphisms and vasculitis susceptibility: Meta-analysis and systematic review. *Mol. Biol. Rep.* **2013**, *40*, 1315–1323. [CrossRef] [PubMed]

85. Fang, J.; Hu, R.; Hou, S.; Ye, Z.; Xiang, Q.; Qi, J.; Zhou, Y.; Kijlstra, A.; Yang, P. Association of *TLR2* gene polymorphisms with ocular Behcet's disease in a Chinese Han population. *Investig. Ophthalmol. Vis. Sci.* **2013**, *54*, 8384–8392. [CrossRef] [PubMed]

86. Liang, L.; Tan, X.; Zhou, Q.; Tian, Y.; Kijlstra, A.; Yang, P. TLR3 and TLR4 But not TLR2 are Involved in Vogt-Koyanagi-Harada Disease by Triggering Proinflammatory Cytokines Production Through Promoting the Production of Mitochondrial Reactive Oxygen Species. *Curr. Mol. Med.* **2015**, *15*, 529–542. [CrossRef] [PubMed]

87. Chen, S.; Yan, H.; Sun, B.; Zuo, A.; Liang, D. Subretinal transfection of chitosan-loaded TLR3-siRNA for the treatment of experimental autoimmune uveitis. *Eur. J. Pharm. Biopharm.* **2013**, *85*, 726–735. [CrossRef] [PubMed]

88. Fang, J.; Chen, L.; Tang, J.H.; Hou, S.P.; Liao, D.; Ye, Z.; Wang, C.K.; Cao, Q.F.; Kijlstra, A.; Yang, P.Z. Association Between Copy Number Variations of TLR7 and Ocular Behcet's Disease in a Chinese Han Population. *Investig. Ophthalmol. Vis. Sci.* **2015**, *56*, 1517–1523. [CrossRef] [PubMed]

89. Sakamoto, N.; Sekine, H.; Kobayashi, H.; Sato, Y.; Ohira, H. Association of the toll-like receptor 9 gene polymorphisms with Behcet's disease in a Japanese population. *Fukushima J. Med. Sci.* **2012**, *58*, 127–135. [CrossRef] [PubMed]

90. Cui, H.P.; Pei, Y.X.; Li, G.F.; Lou, Y.R. Effect of glucocorticoid on cytokines TLR9 and TLR7 in peripheral blood for patients with uveitis. *Exp. Ther. Med.* **2016**, *12*, 3893–3896. [CrossRef] [PubMed]

91. Shaulian, E. AP-1—The Jun proteins: Oncogenes or tumor suppressors in disguise? *Cell Signal.* **2010**, *22*, 894–899. [CrossRef] [PubMed]

92. Smeal, T.; Binetruy, B.; Mercola, D.; Grover-Bardwick, A.; Heidecker, G.; Rapp, U.R.; Karin, M. Oncoprotein-mediated signalling cascade stimulates c-Jun activity by phosphorylation of serines 63 and 73. *Mol. Cell Biol.* **1992**, *12*, 3507–3513. [CrossRef] [PubMed]

93. Pulverer, B.J.; Kyriakis, J.M.; Avruch, J.; Nikolakaki, E.; Woodgett, J.R. Phosphorylation of c-jun mediated by MAP kinases. *Nature* **1991**, *353*, 670–674. [CrossRef] [PubMed]

94. Qing, H.; Gong, W.; Che, Y.; Wang, X.; Peng, L.; Liang, Y.; Wang, W.; Deng, Q.; Zhang, H.; Jiang, B. PAK1-dependent MAPK pathway activation is required for colorectal cancer cell proliferation. *Tumour Biol.* **2012**, *33*, 985–994. [CrossRef] [PubMed]

95. Wang, C.Y.; Chen, C.L.; Tseng, Y.L.; Fang, Y.T.; Lin, Y.S.; Su, W.C.; Chen, C.C.; Chang, K.C.; Wang, Y.C.; Lin, C.F. Annexin A2 silencing induces G2 arrest of non-small cell lung cancer cells through p53-dependent and -independent mechanisms. *J. Biol. Chem.* **2012**, *287*, 32512–32524. [CrossRef] [PubMed]

96.   Gonzalez-Villasana, V.; Gutierrez-Puente, Y.; Tari, A.M. Cyclooxygenase-2 utilizes Jun N-terminal kinases to induce invasion, but not tamoxifen resistance, in MCF-7 breast cancer cells. *Oncol. Rep.* **2013**, *30*, 1506–1510. [PubMed]

97.   Gao, L.; Huang, S.; Ren, W.; Zhao, L.; Li, J.; Zhi, K.; Zhang, Y.; Qi, H.; Huang, C. Jun activation domain-binding protein 1 expression in oral squamous cell carcinomas inversely correlates with the cell cycle inhibitor p27. *Med. Oncol.* **2012**, *29*, 2499–2504. [CrossRef] [PubMed]

98.   Song, X.; Tao, Y.G.; Deng, X.Y.; Jin, X.; Tan, Y.N.; Tang, M.; Wu, Q.; Lee, L.M.; Cao, Y. Heterodimer formation between c-Jun and Jun B proteins mediated by Epstein-Barr virus encoded latent membrane protein 1. *Cell Signal.* **2004**, *16*, 1153–1162. [CrossRef] [PubMed]

99.   Rocha, G.; Duclos, A.; Chalifour, L.E.; Baines, M.G.; Antecka, E.; Deschenes, J. Analysis of gene expression during experimental uveitis in the rabbit. *Can. J. Ophthalmol.* **1996**, *31*, 228–233. [PubMed]

100. Turner, M.J.; DaSilva-Arnold, S.; Luo, N.; Hu, X.; West, C.C.; Sun, L.; Hall, C.; Bradish, J.; Kaplan, M.H.; Travers, J.B.; et al. STAT6-mediated keratitis and blepharitis: A novel murine model of ocular atopic dermatitis. *Investig. Ophthalmol. Vis. Sci.* **2014**, *55*, 3803–3808. [CrossRef] [PubMed]

101. Tepper, R.I.; Levinson, D.A.; Stanger, B.Z.; Campos-Torres, J.; Abbas, A.K.; Leder, P. IL-4 induces allergic-like inflammatory disease and alters T cell development in transgenic mice. *Cell* **1990**, *62*, 457–467. [CrossRef]

102. Amadi-Obi, A.; Yu, C.R.; Liu, X.; Mahdi, R.M.; Clarke, G.L.; Nussenblatt, R.B.; Gery, I.; Lee, Y.S.; Egwuagu, C.E. TH17 cells contribute to uveitis and scleritis and are expanded by IL-2 and inhibited by IL-27/STAT1. *Nat. Med.* **2007**, *13*, 711–718. [CrossRef] [PubMed]

103. Malla, N.; Sjoli, S.; Winberg, J.O.; Hadler-Olsen, E.; Uhlin-Hansen, L. Biological and pathobiological functions of gelatinase dimers and complexes. *Connect. Tissue Res.* **2008**, *49*, 180–184. [CrossRef] [PubMed]

104. Murphy, G.; Nagase, H. Progress in matrix metalloproteinase research. *Mol. Asp. Med.* **2008**, *29*, 290–308. [CrossRef] [PubMed]

105. Sivak, J.M.; Fini, M.E. Mmps in the eye: Emerging roles for matrix metalloproteinases in ocular physiology. *Prog. Retin. Eye Res.* **2002**, *21*, 1–14. [CrossRef]

106. Nagase, H.; Visse, R.; Murphy, G. Structure and function of matrix metalloproteinases and TIMPs. *Cardiovasc. Res.* **2006**, *69*, 562–573. [CrossRef] [PubMed]

107. Lee, Y.J.; Kang, S.W.; Baek, H.J.; Choi, H.J.; Bae, Y.D.; Kang, E.H.; Lee, E.Y.; Lee, E.B.; Song, Y.W. Association between matrix metalloproteinase 9 promoter polymorphisms and Behcet's disease. *Hum. Immunol.* **2010**, *71*, 717–722. [CrossRef] [PubMed]

108. Quillard, T.; Coupel, S.; Coulon, F.; Fitau, J.; Chatelais, M.; Cuturi, M.C.; Chiffoleau, E.; Charreau, B. Impaired Notch4 activity elicits endothelial cell activation and apoptosis: Implication for transplant arteriosclerosis. Arterioscler. *Thromb. Vasc. Biol.* **2008**, *28*, 2258–2265. [CrossRef] [PubMed]

109. Verginelli, F.; Adesso, L.; Limon, I.; Alisi, A.; Gueguen, M.; Panera, N.; Giorda, E.; Raimondi, L.; Ciarapica, R.; Campese, A.F.; et al. Activation of an endothelial Notch1-Jagged1 circuit induces VCAM1 expression, an effect amplified by interleukin-1β. *Oncotarget* **2015**, *6*, 43216–43229. [PubMed]

110. Crosson, J.N.; Laird, P.W.; Debiec, M.; Bergstrom, C.S.; Lawson, D.H.; Yeh, S. Vogt-Koyanagi-Harada-like syndrome after CTLA-4 inhibition with ipilimumab for metastatic melanoma. *J. Immunother.* **2015**, *38*, 80–84. [CrossRef] [PubMed]

111. Yu, C.R.; Kim, S.H.; Mahdi, R.M.; Egwuagu, C.E. SOCS3 deletion in T lymphocytes suppresses development of chronic ocular inflammation via upregulation of CTLA-4 and expansion of regulatory T cells. *J. Immunol.* **2013**, *191*, 5036–5043. [CrossRef] [PubMed]

112. Shimizu, J.; Izumi, T.; Arimitsu, N.; Fujiwara, N.; Ueda, Y.; Wakisaka, S.; Yoshikawa, H.; Kaneko, F.; Suzuki, T.; Takai, K.; et al. Skewed TGFβ/Smad signalling pathway in T cells in patients with Behcet's disease. *Clin. Exp. Rheumatol.* **2012**, *30*, S35–S39. [PubMed]

113. Li, Q.; Sun, B.; Dastgheib, K.; Chan, C.C. Suppressive effect of transforming growth factor β1 on the recurrence of experimental melanin protein-induced uveitis: Upregulation of ocular interleukin-10. *Clin. Immunol. Immunopathol.* **1996**, *81*, 55–61. [CrossRef] [PubMed]

114. Sharma, R.K.; Gupta, A.; Kamal, S.; Bansal, R.; Singh, N.; Sharma, K.; Virk, S.; Sachdeva, N. Role of Regulatory T Cells in Tubercular Uveitis. *Ocul. Immunol. Inflamm.* **2016**, 1–10. [CrossRef] [PubMed]

115. Fabiani, C.; Vitale, A.; Lopalco, G.; Iannone, F.; Frediani, B.; Cantarini, L. Different roles of TNF inhibitors in acute anterior uveitis associated with ankylosing spondylitis: State of the art. *Clin. Rheumatol.* **2016**, *35*, 2631–2638. [CrossRef] [PubMed]

116. Hatemi, I.; Hatemi, G.; Pamuk, O.N.; Erzin, Y.; Celik, A.F. TNF-α antagonists and thalidomide for the management of gastrointestinal Behcet's syndrome refractory to the conventional treatment modalities: A case series and review of the literature. *Clin. Exp. Rheumatol.* **2015**, *33*, S129–S137. [PubMed]

117. Bharadwaj, A.S.; Schewitz-Bowers, L.P.; Wei, L.; Lee, R.W.; Smith, J.R. Intercellular adhesion molecule 1 mediates migration of Th1 and Th17 cells across human retinal vascular endothelium. *Investig. Ophthalmol. Vis. Sci.* **2013**, *54*, 6917–6925. [CrossRef] [PubMed]

118. Hu, L.; Huang, T.; Shi, X.; Lu, W.C.; Cai, Y.D.; Chou, K.C. Predicting functions of proteins in mouse based on weighted protein-protein interaction network and protein hybrid properties. *PLoS ONE* **2011**, *6*, e14556. [CrossRef] [PubMed]

119. Hu, L.L.; Huang, T.; Cai, Y.D.; Chou, K.C. Prediction of Body Fluids where Proteins are Secreted into Based on Protein Interaction Network. *PLoS ONE* **2011**, *6*, e22989. [CrossRef] [PubMed]

120. Chen, L.; Zhang, Y.H.; Huang, T.; Cai, Y.D. Identifying novel protein phenotype annotations by hybridizing protein-protein interactions and protein sequence similarities. *Mol. Genet. Genom.* **2016**, *291*, 913–934. [CrossRef] [PubMed]

121. Jensen, L.J.; Kuhn, M.; Stark, M.; Chaffron, S.; Creevey, C.; Muller, J.; Doerks, T.; Julien, P.; Roth, A.; Simonovic, M. STRING 8-a global view on proteins and their functional interactions in 630 organisms. *Nucleic Acids Res.* **2009**, *37*, D412–D416. [CrossRef] [PubMed]

122. Li, Y.; Patra, J.C. Genome-wide inferring genephenotype relationship by walking on the heterogeneous network. *Bioinformatics* **2010**, *26*, 1219–1224. [CrossRef] [PubMed]

123. Yang, J.; Chen, L.; Kong, X.; Huang, T.; Cai, Y.-D. Analysis of Tumor Suppressor Genes Based on Gene Ontology and the KEGG Pathway. *PLoS ONE* **2014**, *9*, e107202. [CrossRef] [PubMed]

124. Huang, T.; Zhang, J.; Xu, Z.P.; Hu, L.L.; Chen, L.; Shao, J.L.; Zhang, L.; Kong, X.Y.; Cai, Y.D.; Chou, K.C. Deciphering the effects of gene deletion on yeast longevity using network and machine learning approaches. *Biochimie* **2012**, *94*, 1017–1025. [CrossRef] [PubMed]

125. Zhang, J.; Xing, Z.; Ma, M.; Wang, N.; Cai, Y.-D.; Chen, L.; Xu, X. Gene Ontology and KEGG Enrichment Analyses of Genes Related to Age-Related Macular Degeneration. *BioMed Res. Int.* **2014**, *2014*, 450386. [CrossRef] [PubMed]

126. Chen, L.; Zhang, Y.-H.; Zheng, M.; Huang, T.; Cai, Y.-D. Identification of compound-protein interactions through the analysis of gene ontology, KEGG enrichment for proteins and molecular fragments of compounds. *Mol. Genet. Genom.* **2016**, *291*, 2065–2079. [CrossRef] [PubMed]

127. Kohavi, R. A study of cross-validation and bootstrap for accuracy estimation and model selection. In Proceedings of the 14th International joint Conference on artificial intelligence, Montreal, QC, Canada, 20–25 August 1995.

9

Prediction of Protein Hotspots from Whole Protein Sequences by a Random Projection Ensemble System

Jinjian Jiang [1,2], **Nian Wang** [1], **Peng Chen** [3,*], **Chunhou Zheng** [4] **and Bing Wang** [5,*]

[1]   School of Electronics and Information Engineering, Anhui University, Hefei 230601, China;
     jiangjj@aqnu.edu.cn (J.J.); wn_xlb@ahu.edu.cn (N.W.)
[2]   School of Computer and Information, Anqing Normal University, Anqing 246133, China
[3]   Institute of Health Sciences, Anhui University, Hefei 230601, China
[4]   School of Electronic Engineering & Automation, Anhui University, Hefei 230601, China; zhengch99@126.com
[5]   School of Electrical and Information Engineering, Anhui University of Technology, Ma'anshan 243032, China
*   Correspondence: pchen@ahu.edu.cn (P.C.); wangbing@ustc.edu (B.W.)

Abstract: Hotspot residues are important in the determination of protein-protein interactions, and they always perform specific functions in biological processes. The determination of hotspot residues is by the commonly-used method of alanine scanning mutagenesis experiments, which is always costly and time consuming. To address this issue, computational methods have been developed. Most of them are structure based, i.e., using the information of solved protein structures. However, the number of solved protein structures is extremely less than that of sequences. Moreover, almost all of the predictors identified hotspots from the interfaces of protein complexes, seldom from the whole protein sequences. Therefore, determining hotspots from whole protein sequences by sequence information alone is urgent. To address the issue of hotspot predictions from the whole sequences of proteins, we proposed an ensemble system with random projections using statistical physicochemical properties of amino acids. First, an encoding scheme involving sequence profiles of residues and physicochemical properties from the AAindex1 dataset is developed. Then, the random projection technique was adopted to project the encoding instances into a reduced space. Then, several better random projections were obtained by training an IBk classifier based on the training dataset, which were thus applied to the test dataset. The ensemble of random projection classifiers is therefore obtained. Experimental results showed that although the performance of our method is not good enough for real applications of hotspots, it is very promising in the determination of hotspot residues from whole sequences.

Keywords: random projection; hot spots; IBk; ensemble system

1. Introduction

Hotspot residues contribute a large portion of the binding energy of one protein in complex with another protein [1,2], which are always surrounded by residues contributing less binding energy. These are not uniformly distributed for the binding energy of proteins over their interaction surfaces [1]. Hotspots are important in the binding and the stability of protein-protein interactions and thus key to perform specific functions in the protein [3,4]. Actually, hotspots are difficult to determine. A common determination method is the method of alanine scanning mutagenesis experiments, which identify a hotspot if a change in its binding free energy is larger than a predefined threshold when the residue is mutated to alanine. However, this method is costly and time consuming.

Several databases stored experimental and computational hotspot residues and the details of hotspots' properties. The first database for storing experimental hotspots was the Alanine Scanning Energetics Database (ASEdb) by the use of alanine scanning energetics experiments [5]. Another database is the Binding Interface Database (BID) developed by Fischer et al., which mined the primary scientific literature for detailed data about protein interfaces [6]. These databases are commonly used in previous works on hotspot identification. The Protein-protein Interactions Thermodynamic Database (PINT) is another database that mainly accumulates the thermodynamic data of interacting proteins upon binding along with all of the experimentally-measured thermodynamic data (Kd, Ka, ΔG, ΔH and ΔCp) for wild-type and mutant proteins [7]. It contains 1513 entries in 129 protein–protein complexes from 72 original research articles, where only 33 entries have complete 3D structures deposited in PDB (Protein Data Bank), in the first release of PINT. Recently, Moal et al. built the SKEMPI (Structural Kinetic and Energetic database of Mutant Protein Interactions) that has collected 3047 binding free energy changes from 85 protein-protein complexes from the literature [8].

Although some databases stored hotspot residues, few of the protein complexes were solved. Computational approaches were proposed to identify hotspot residues, and they were complementary to the experimental methods. Some methods predicted hotspots by energy function-based physical models [3,9–11], molecular dynamics simulation-based approaches [12–14], evolutionary conservation-based methods [4,15,16] and docking-based methods [17,18]. Some methods adopted machine learning methods for the hotspot prediction, such as graph-based approaches [19], neural network [20], decision tree [3,21], SVM (Support Vector Machine) [22], random forest [23] and the consensus of different machine learning methods [24], combining features of solvent accessibility, conservation, sequence profiles and pairing potential [20,23,25–29].

All of the previous methods were developed to identify hotspots from a part of residues in the interface regions. They always worked on selected datasets containing almost the same numbers of hotspots and non-hotspots. The ratio of the number of hotspots to that of residues in whole datasets is around 20~50%, for example: BID contains 54 hotspots and 58 non-hotspots; 58 hotspots and 91 non-hotspots are in the ASEdb dataset; and SKEMPI contains 196 hotspots and 777 non-hotspots [29]. However, no more than 2% of the residues in whole protein sequences are hotspots. The issue of identifying hotspots from whole protein sequences in our study is more difficult than others, but more interesting.

Most hotspot prediction methods are structure-based, which cannot be applied to protein complexes without the information of protein structures [3,22,23]. Therefore, identifying hotspots from the protein sequence only is important. Moreover, few works identified hotspot residues from the whole protein sequences. To address these issues, here, we propose a method that predicts hotspots from the whole protein sequences using physicochemical characteristics extracted from amino acid sequences. A random projection ensemble classifier system is developed for the hotspot predictions. The system involves an encoding scheme integrating sequence profiles of residues and the statistical physicochemical properties of amino acids from the AAindex1 (Amino Acid index1 database) dataset. Then, the random projection technique was adopted to obtain a reduced input space, but to retain the structure of the original space. Several better classifiers with the IBk algorithm are obtained after the use of random projections. The ensemble of good classifiers is therefore constructed. Experimental results showed that our method performs well in hotspot predictions for the whole protein sequences.

2. Results

2.1. Performance of the Hotspot Prediction

In the running of the random projection-based classifier, different random projections in Equation (1) construct different classifiers. After running the classifier 100 times, 100 classifiers with random projections R are formed and trained on the training subset D_{tr}^k. As a result, 100 predictions are obtained. All of the classifiers are ranked in terms of the prediction measure $F1$. The ensemble of several top N classifiers is then tested on the test subset D_{ts}^k. In this work, the ASEdb0 is regarded as the training dataset, and the test dataset is BID0; while the predictions on the ASEdb0 dataset are also tested by training on the BID0 dataset.

Table 1 shows the performance of the top individual classifiers trained by the ASEdb0 dataset and the prediction performance on the BID0 dataset. The individual classifiers are ranked in terms of the *F1* measure in the training process. The top classifiers yield good predictions on the BID0 dataset. It achieves an F1 of 0.109, as well as a precision of 0.069 at a sensitivity of 0.259 in the training process and, therefore, yields an F1 of 0.315, as well as a precision of 0.220 at a sensitivity of 0.558 in the test process. Here, the dimensionality of the original data is reduced from 7072 to only five.

Table 2 shows the performance comparison of the ensembles of the top *N* classifiers. In the classifier ensemble, the majority vote technique was applied to the ensemble, i.e., one residue will be identified as the hotspot if half of the *N* classifiers predict it to be the hotspot. Here, seven ensembles of the number of top classifiers are listed, i.e., 2, 3, 5, 10, 15, 25 and 50. From Table 2, it can be seen that the ensemble of the top three classifiers with the majority vote yields a good performance compared with other classifier ensembles. It yields an *MCC* (Matthews Correlation Coefficient) of 0.428, as well as a precision of 0.245 at a sensitivity of 0.793, for testing on the ASEdb0 dataset by training on the BID0 dataset; and it yields an *MCC* of 0.601, as well as a precision of 0.440 at a sensitivity of 0.846, for testing on the BID0 dataset by training on the ASEdb0 dataset. The ensemble of the top three classifiers resulted in a dramatic improvement, compared with the top three individual classifiers. The reason for the improvement is most likely in that a suitable random projection makes the classifier more diverse, where the detailed results are not shown here. Previous methods also showed that the ensemble of more diverse classifiers yielded more efficient predictions [30].

It seems that the more top classifiers the ensemble contains, the worse the ensemble performs. The ensemble with the top 50 classifiers performs the worst both for testing on the ASEdb0 and the BID0 dataset. Therefore, a suitable number of top classifiers can improve the predictions of hotspot residues. Moreover, our method on the BID0 dataset performs better than that on the ASEdb0 dataset, maybe because of the larger ratio of hotspots to the total residues in BID0 (1.831%) than that in ASEdb0 (1.445%).

Table 1. Prediction performance of individual classifiers with the reduced dimension of 5 on the Binding Interface Database 0 (BID0) test dataset training by Alanine Scanning Energetics Database 0 (ASEdb) dataset. There are 50 top individual classifiers listed here for a simple comparison between classifiers. Here measures of *"Sen"*, *"Prec"*, *"F1"* and *"MCC"* denote Sensitivity, Precision, F-Measure, and Matthews Correlation Coefficient, respectively.

No.	Training				Test			
	Sen	MCC	Prec	F1	Sen	MCC	Prec	F1
1	0.259	0.110	0.069	0.109	0.558	0.332	0.220	0.315
2	0.069	0.125	0.250	0.108	0.558	0.357	0.250	0.345
3	0.138	0.080	0.070	0.093	0.212	0.141	0.122	0.155
4	0.069	0.085	0.129	0.090	0.500	0.274	0.173	0.257
5	0.121	0.075	0.071	0.089	0.308	0.194	0.150	0.201
6	0.069	0.083	0.125	0.089	0.096	0.040	0.044	0.060
7	0.069	0.076	0.108	0.084	0.269	0.136	0.096	0.141
8	0.069	0.076	0.108	0.084	0.269	0.129	0.090	0.135
9	0.138	0.071	0.061	0.084	0.558	0.364	0.259	0.354
10	0.138	0.069	0.058	0.082	0.346	0.226	0.173	0.231
11	0.069	0.071	0.098	0.081	0.135	0.038	0.037	0.058
12	0.086	0.066	0.075	0.080	0.615	0.337	0.205	0.308
13	0.052	0.080	0.150	0.077	0.577	0.317	0.196	0.293
14	0.052	0.076	0.136	0.075	0.404	0.227	0.153	0.222
15	0.069	0.064	0.083	0.075	0.135	0.082	0.080	0.100
16	0.052	0.074	0.130	0.074	0.577	0.323	0.203	0.300
17	0.052	0.074	0.130	0.074	0.596	0.279	0.153	0.243
18	0.069	0.062	0.080	0.074	0.404	0.225	0.151	0.220

Table 1. Cont.

No.	Training				Test			
	Sen	*MCC*	*Prec*	*F1*	*Sen*	*MCC*	*Prec*	*F1*
19	0.069	0.062	0.080	0.074	0.308	0.152	0.102	0.153
20	0.052	0.072	0.125	0.073	0.115	0.030	0.033	0.052
21	0.121	0.058	0.052	0.073	0.192	0.135	0.123	0.150
22	0.052	0.067	0.111	0.071	0.288	0.150	0.105	0.154
23	0.190	0.064	0.044	0.071	0.577	0.281	0.159	0.249
24	0.069	0.056	0.070	0.070	0.269	0.145	0.105	0.151
25	0.086	0.054	0.057	0.069	0.423	0.171	0.095	0.155
26	0.086	0.053	0.057	0.068	0.212	0.079	0.057	0.090
27	0.086	0.051	0.054	0.066	0.365	0.218	0.156	0.218
28	0.052	0.058	0.091	0.066	0.250	0.091	0.060	0.097
29	0.052	0.057	0.088	0.065	0.481	0.237	0.141	0.218
30	0.034	0.095	0.286	0.062	0.519	0.241	0.136	0.215
31	0.034	0.095	0.286	0.062	0.346	0.204	0.146	0.206
32	0.052	0.050	0.073	0.061	0.173	0.095	0.081	0.110
33	0.138	0.048	0.039	0.061	0.442	0.271	0.190	0.266
34	0.052	0.049	0.071	0.060	0.231	0.115	0.085	0.124
35	0.224	0.055	0.035	0.060	0.346	0.186	0.127	0.186
36	0.034	0.078	0.200	0.059	0.250	0.161	0.131	0.172
37	0.207	0.052	0.034	0.059	0.519	0.273	0.167	0.252
38	0.034	0.074	0.182	0.058	0.365	0.238	0.181	0.242
39	0.034	0.064	0.143	0.056	0.192	0.083	0.064	0.096
40	0.052	0.044	0.061	0.056	0.231	0.146	0.120	0.158
41	0.052	0.042	0.059	0.055	0.135	0.070	0.065	0.088
42	0.103	0.038	0.036	0.054	0.327	0.145	0.091	0.143
43	0.103	0.037	0.036	0.053	0.192	0.111	0.093	0.125
44	0.034	0.049	0.095	0.051	0.077	0.013	0.025	0.037
45	0.069	0.035	0.040	0.051	0.154	0.054	0.046	0.071
46	0.121	0.034	0.031	0.050	0.423	0.231	0.151	0.222
47	0.224	0.041	0.028	0.050	0.288	0.172	0.129	0.179
48	0.241	0.037	0.026	0.046	0.308	0.152	0.102	0.153
49	0.052	0.030	0.040	0.045	0.442	0.210	0.125	0.195
50	0.155	0.031	0.026	0.045	0.462	0.252	0.162	0.240

Table 2. Prediction performance of the ensemble of the top N classifiers with reduced instance dimension of 5 on the two datasets.

Test Set	No. Dimension	*Sen*	*MCC*	*Prec*	*F1*
ASEdb0	2	0.224	0.322	0.481	0.306
	3	0.793	0.428	0.245	0.374
	5	0.897	0.383	0.177	0.295
	10	1.000	0.299	0.103	0.186
	15	1.000	0.219	0.062	0.116
	25	1.000	0.149	0.036	0.070
	50	1.000	0.081	0.021	0.041
BID0	2	0.385	0.260	0.200	0.263
	3	0.846	0.601	0.440	0.579
	5	1.000	0.461	0.226	0.369
	10	1.000	0.283	0.096	0.175
	15	1.000	0.222	0.066	0.124
	25	1.000	0.145	0.038	0.074
	50	1.000	0.078	0.024	0.046

Furthermore, the performance comparison of ensembles with different numbers of reduced instance dimensions by the random projection technique was investigated. The ensembles of random projections with seven reduced dimensions were built, i.e., the dimensions of 1, 2, 5, 10, 20, 50 and 100. The ensemble with the reduced dimension of five performs the best among the seven ensembles, while the ensemble of the top three classifiers with instance dimension of one also performs well in the hotspot predictions for the whole sequences of proteins, which yields an *MCC* of 0.475, as well as a precision of 0.704 at a sensitivity of 0.328. Table 3 shows the performance comparison of the classifier ensemble with different numbers of reduced dimensions on the BID0 test dataset.

Table 3. Prediction performance of the ensemble of the top 3 classifiers with different reduced instance dimensions on the BID0 test dataset.

No. Dimension	*Sen*	*MCC*	*Prec*	*F1*
1	0.328	0.475	0.704	0.447
2	0.328	0.352	0.396	0.358
5	0.846	0.601	0.440	0.579
10	0.846	0.499	0.310	0.454
20	0.481	0.240	0.144	0.221
50	0.500	0.274	0.173	0.257
100	0.538	0.252	0.141	0.224

This study adopted the window length technique to encode input instances of classifiers; however, the sliding window technique makes the performance of the classifier varied. To show which window length makes the classifiers better for a specific type of dataset, several windows with different lengths were investigated. Figure 1 shows the prediction performance on different sliding windows on the BID0 dataset. Among the seven sliding windows, the window with length 13 performs the best, which yields an *F1* of 0.579. It should be mentioned here that classifier ensembles with a suitable window length perform better than those with a smaller or bigger length.

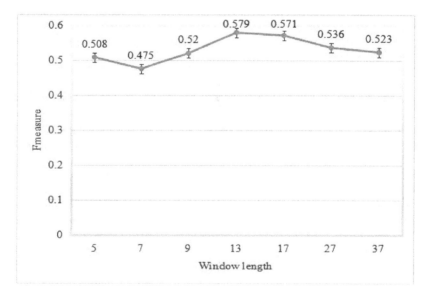

Figure 1. Prediction performance for different sliding windows in instance encoding on the BID0 dataset training by the ASEdb0 dataset. The symbol "I" for each window denotes the calculation error of prediction performance in *F1*.

2.2. Comparison with Other Methods

So far, few works identified hotspots from the whole protein sequences by sequence information alone. Some top hotspot predictors did the predictions based on protein structures. Most of hotspot prediction methods predicted hotspots from protein-protein interfaces or from some benchmark datasets, such as ASEdb0 and BID0, which contained approximately the same hotspots and non-hotspots. Therefore, the random predictor is used to compare with our method. The random predictor was run 100 times, and the average performance was calculated. Furthermore, for prediction comparison, the tool of ISIS (Interaction Sites Identified from Sequence) [20] on the PredictProtein server [31] was adopted, which has been applied in hotspot predictions on the dataset of interface residues [20]. ISIS is a machine learning-based method that identified interacting residues from the sequence alone. Similar to our method, although the method was developed using transient protein-protein interfaces from complexes of experimentally-known 3D structures, it only used the sequence and predicted 2D information. In PredictProtein, it predicted a residue as a hotspot if the prediction score of the residue was bigger than 21, otherwise being non-hotspot residues. Since PredictProtein currently cannot process short input sequences less than 17 residues, protein sequences in PDB names "1DDMB" and "2NMBB" were removed from the BID0 test set. We tested all of the sequences of more than 17 residues on the BID0 dataset, and the performance of hotspot predictions on the dataset was obtained. The predictions of ISIS method can be referred to the Supplementary Materials.

Table 4 lists the hotspot prediction comparison in detail. Our method developed a random projection ensemble system yielding a final precision of 0.440 at a sensitivity of 0.846 by the use of sequence information only. Results showed that our method outperforms the random predictor. Furthermore, our method outperformed the ISIS method. Actually, ISIS was developed to identify protein-protein interactions. The power of ISIS for the identification of hotspot residues was poor. It can predict nine of 47 real hotspots correctly; however, 2920 non-hotspots were predicted to be hotspots in the BID0 dataset.

Table 4. Performance comparison of the three methods on the BID0 dataset by training on the ASEdb0 dataset.

Method	Type	Sen	MCC	Prec	F1
Our Method	Random Projection	0.846	0.601	0.440	0.579
ISIS	Neural Networks	0.191	0.030	0.026	0.046
	Random Predictor	0.983	0.000	0.018	0.035

We also show the performance of classifier ensemble in several measures based on the measure of sensitivity. Figure 2 illustrates the performance of the ensemble classifier with the majority vote for the test set BID0. Although it is very difficult to identify hotspots from the whole protein sequences, our method yields a good result based on sequence information only.

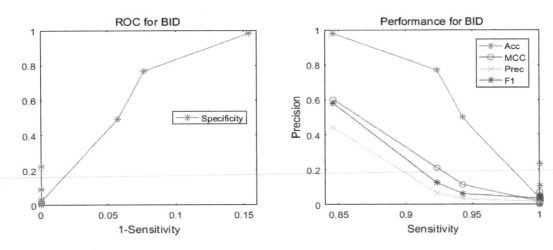

Figure 2. The performance of our method for testing on the BID0 dataset by training on the ASEdb0 dataset. The left graph illustrates the ROC (receiver operating characteristic) curve, and the right one shows the four measure curves with respect to sensitivity.

2.3. Case Study of Hotspot Predictions

To show the performance of our method on a single protein chain, hotspot predictions for chain "A" of protein PDB:1DDM are illustrated in Figure 3. Protein 1DDM is an in vivo complex containing a phosphotyrosine-binding (PTB) domain (chain "A") of the cell fate determinant Numb, which can bind a diverse array of peptide sequences in vitro, and a peptide containing an amino acid sequence "NMSF" derived from the Numb-associated kinase (Nak) (chain "B"). The Numb PTB domain is in complex with the Nak peptide. The chain "A" contains 135 amino acid residues, where only residues E144, I145, C150 and C198 are real hotspot residues in complex with the chain "B" of the protein (which contains 11 amino acid residues; see Figure 3c). Our method correctly predicted the first three true hotspots, and hotspot residue 198 was predicted as a non-hotspot, while residues 69, 112, 130 and 160 were wrongly predicted as hotspot residues. All of them are located at the surface of the protein structure. The results of ISIS are also illustrated in Figure 3b. The ISIS method cannot identify the four true hotspot residues, although most of the hotspot predictions are located at the surface of the protein.

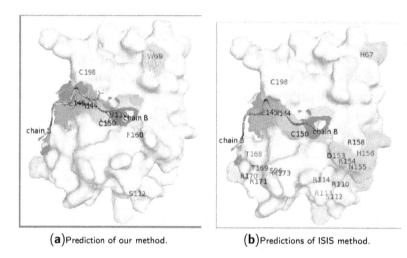

(a) Prediction of our method.   (b) Predictions of ISIS method.

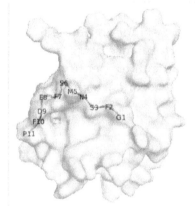

(c) Complex structure of protein PDB:1DDM.

Figure 3. Case study for the complex of protein PDB:1DDM. The subgraphs (**a,b**) are shown for the prediction comparison of our method and the ISIS method, respectively, where the chain B of protein 1DDM is colored in wheat. The subgraph (**c**) illustrates the cartoon structure of the protein complex, where the chain B of protein 1DDM is colored in green. Here, red residues are the hotspots that are predicted correctly; green residues are non-hotspots that are predicted to be hotspots; while yellow ones are real hotspots that are predicted to be non-hotspot residues. All other residues are correctly predicted as non-hotspots.

3. Materials and Methods

3.1. Hot Spot Definitions

As we know, a residue is defined as a hotspot by the change of the binding free energy ($\Delta\Delta G$) higher than a threshold, if mutated to alanine. Several thresholds were adopted in previous works. Many works defined residues as hotspots when their $\Delta\Delta G$s are higher than 2.0 kcal/mol, and other residues with $\Delta\Delta G$ from 0–2.0 kcal/mol were defined as non-hotspots [21–23]. Ofran et al. used another definition that defined residues with $\Delta\Delta G$ above 2.5 kcal/mol as hotspots and those with $\Delta\Delta G = 0$ kcal/mol (i.e., no change in binding energy) as non-hotspots [20]. Moreover, Tuncbag and colleagues defined hotspots as those with $\Delta\Delta G$ higher than 2.0 kcal/mol and non-hotspots as those with $\Delta\Delta G$ from 0–0.4 kcal/mol [24]. Previous works also investigated several definitions of hotspots [26,29]. They concluded that different definitions of hotspots and non-hotspots yield different ratios of the number of hotspots to that of non-hotspots and, therefore, change the performances of hotspot prediction methods [26,29]. In this paper, residues higher than 2.0 kcal/mol are defined as hotspots and all other residues in the whole protein sequences as non-hotspots, no matter if their position is in interfaces, surfaces or any other regions.

3.2. Datasets

Since this work addresses the issue of hotspot residue predictions for the whole sequences of proteins, the definitions of hotspot residues are the same as those of the ASEdb and BID datasets, while all of the other residues in the protein sequences are considered as non-hotspot residues.

Two commonly-used benchmark datasets are used in this work. The first dataset is ASEdb [5]. To clean the proteins in ASEdb, protein sequences in the dataset were removed when the sequence identity between any two sequences was higher than 35%. Based on the hotspot definition in this study, we constructed a new ASEdb0 dataset consisting of 58 hotspots from the ASEdb dataset and 3957 non-hotspots of the other residues in whole protein sequences, totaling 4015 residues in our new ASEdb0 dataset.

The BID dataset [6] is the other one used in this work. The dataset was filtered in the same manner as the ASEdb dataset. As a result, we constructed a new BID0 dataset consisting of 54 hotspots from the BID dataset and 2895 non-hotspots from the rest of the residues in the whole protein sequences, totaling 2949 residues in our new BID0 dataset. The data in the two datasets came from different complexes and were mutually exclusive. Table 5 lists the composition of hotspots and non-hotspots.

Table 5. The details of the hotspot datasets.

Dataset	Hot Spots	Non-Hotspots	Total Residues	Ratio [§]
BID0	54	2895	2949	1.831%
ASEdb0	58	3957	4015	1.445%
BID	54	58	112	48.214%
ASEdb	58	91	149	38.926%

[§] The ratio of the number of hotspots to that of total residues in the dataset.

3.3. Feature Encoding Scheme

The AAindex1 database [32] contained 544 numerical indices representing various physicochemical and biochemical properties of amino acids. It collected published indices with a set of 20 numerical values representing different properties of amino acids. It also contained the results of cluster analysis using the correlation coefficient as the distance between two indices. All data were derived from published literature.

The protein sequence profile of one amino acid is a set of 20 numerical values representing the evolution of the amino acid residue, where each value represents the frequency by which residue was mutated into another amino acid residue. It can be used to recognizing remote homologs and plays an important role in protein sequence database search, protein structure/function prediction and phylogenetic analysis. Protein sequence profiles are always obtained by a BLAST (Basic Local Alignment Search Tool) program, such as the commonly-used program of PSI-BLAST (Position-Specific Iterative Basic Local Alignment Search Tool) [33]. Therefore, for the residue R_i of one protein sequence, the multiplication MSK_i^j of the sequence profile SP_i of residue R_i and one physicochemical amino acid property AAP^j can represent the statistical evolution of the amino acid property [34–36], i.e., $MSK_i^j = SP_i \times AAP^j$, where SP_i and AAP^j are both vectors of 1×20. The multiplication for residue R_i results in a set of 20 numerical vectors MSK_i^j. The standard deviation STD_i^j of the multiplication is then obtained. For residue R_i, the 544 amino acid AAindex1 properties yield a set of 544 standard deviations $STD_i = STD_i^j, j = 1\ 544$, which will be used for encoding residue R_i. Our previous work has shown that the standard deviations of the multiplications can reflect the evolutionary variance of the residue R_i along with the amino acid property AAP^j [29,34,35].

To encode the residue R_i in one protein sequence, a sliding window involving residues centered at the residue R_i is considered, i.e., several neighboring residues are used to represent the center residue R_i. Therefore, a set of $winLen \times 544 = 7072$ numerical values represents the residue R_i, where $winLen = 13$ is the sliding window length in this work. A similar vector representation can be found in our previous work [29,34,35]. For the residue R_i, it is represented by a 1×7072 vector V_i, whose corresponding target value T_i is 1 or 0, denoting whether the residue is a hotspot or not. Therefore, our method is developed to learn the relationship between input vectors V and the corresponding target array T and tries to make its output $Y = f(V)$ as close to the target T as possible.

3.4. IBk Classifier Ensemble by the Random Mapping Technique

The random projection technique can be traced back to the work done by Ritter and Kohonen [37], which reduced the dimensionality of the representations of the word contexts by replacing each dimension of the original space by a random direction in a smaller-dimensional space. From the literature [37,38], it seems surprising that random mapping can reduce the dimensionality of the data in a manner that preserves enough structure of the original dataset to be useful. Kaski used both analytical and empirical evidence to explain the reason why the random mapping method worked well in high-dimensional spaces [39].

Given the original data, $X \in \Re^{N \times L1}$, let the linear random projection be the multiplication of the original instances by a random matrix $R \in \Re^{L1 \times L2}$, where the element in the matrix ranges from 0–1. The matrix R is composed of random elements, and each column has been normalized to unity. The projection:

$$X^R = XR = \sum_i (x_i \times r_i) \tag{1}$$

yields a dimensionality-reduced instance $X^R \in \Re^{N \times L2}$ from dimension $L1$ to $L2$, where x_i is the i-th sample of the original data, r_i is the i-th column of the random matrix and $L2 \ll L1$. In Equation (1), each original instance with dimension $L1$ has been replaced by a random, non-orthogonal direction $L2$ in the reduced-dimensional space [39]. Therefore, the dimensionality of the original instance is reduced from 7072 to a rather small value.

The dimension-reduced instances are then input into the classifier with the IBk algorithm. The IBk algorithm, implementing the k-nearest neighbor algorithm, is a type of instance-based learning, where the function is only approximated locally, and all computations are deferred until classification. The simplest of the IBk algorithms among machine learning algorithms was adopted since we want to ensemble diverse classifiers and expect to yield good results. Previous results showed that the generalization error caused by one classifier can be compensated by other classifiers; therefore, the ensemble of some diverse classifiers can yield significant improvement [40].

In the hotspot prediction, the multiplication of the k-th random projection R_k on the original instances (X, Y) forms a set of instances $D^k = \{(X_i^{R_k}, Y_i)\}$, $i = 1, ..., N$, where N and K denote the number of training instances and that of random projections, respectively. For the k-th random projection, the instances D^k are generated from the original instances (X, Y) as an input to an IBk classifier, and thus, it forms a classifier $IBk_k(x)$, where x is a training instance. To train the classifier $IBk_k(x)$, the instance set D^k is divided into training dataset D^k_{tr} and test dataset D^k_{ts} by 10-fold cross-validation. For training the classifier, the training dataset D^k_{tr} is divided into training subset D^{k-tr}_{tr} and test subset D^{k-ts}_{tr} again. The training process retains the top classifiers on some random projections, and in the test process, they are applied to test the test dataset D^k_{ts}.

After running random projection 100 times, top classifiers in the *F1* measure are retained for testing the test dataset D^k_{ts}. The ensemble of top classifiers yields the final predictions. The mjority vote technique was always used in classifier ensemble and often made a dramatic improvement [41]. Here, a residue is predicted as a hotspot if half of the classifiers identified it as positive Class 1, otherwise it is a non-hotspot residue.

Moreover, since the hotspot dataset is extremely imbalanced, containing only 1.4% of hotspots, balancing the dataset is necessary to avoid the overfitting of the classifier. Therefor, the training dataset D^{k-tr}_{tr} is resampled and then consists of positive instances and negative instances with roughly the same number. The ensemble system can be seen i Figure 4.

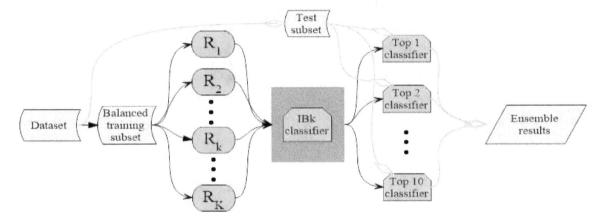

Figure 4. The flowchart of the ensemble system for the hotspot prediction. Here, R_k means the k-th random projection. The IBk implements k-Nearest Neighbors (KNN) algorithm. Here the black arrows denote the flow of the training subset, while the blue ones are that of the test subset.

3.5. Hot Spot Prediction Evaluation

To evaluate hotspot predictions, in this work, we adopted four evaluation measures to show the ability of our model objectively. They are the criteria of sensitivity (*Sen*), precision (*Prec*), F-measure (*F1*) and Matthews correlation coefficient (*MCC*) [34,42] and shown below:

$$
Sen = \frac{TP}{TP + FN}, \quad Prec = \frac{TP}{TP + FP}
$$
$$
F1 = 2 \times \frac{Prec \times Sen}{Prec + Sen}
$$
$$
MCC = \frac{TP \times TN - FP \times FN}{\sqrt{(TP + FN)(TP + FP)(TN + FP)(TN + FN)}},
$$

(2)

where *TP* (true positive) is the number of correctly-predicted hotspot residues; *FP* (false positive) is the number of false positives (incorrectly over-predicted non-hotspot residues); *TN* (true negative) is the number of correctly-predicted non-hotspot residues; and *FN* (false negative) is false negative, i.e., incorrectly under-predicted hotspot residues.

4. Conclusions

This paper proposes an ensemble method based on the random projection technique that predicts hotspots from the whole sequences of proteins, using physicochemical characteristics of amino acids. The classifier system involves an encoding scheme integrating sequence profiles of residues and statistical physicochemical properties of amino acids from the AAindex1 dataset. Then, the random projection technique was adopted to obtain a reduced input space for the original input instances, but retaining the structure of the original space. Several top classifiers are obtained after the use of random projections. The ensemble of the top classifiers is therefore constructed. The classifier with random projection ran 50 times, and 50 classifiers were sorted in the *F1* measure in the training step. Applying the 50 classifiers to the test dataset yielded the final hotspot predictions. Results showed that the ensemble of the top three classifiers yields better performance in hotspot predictions. Moreover, random projections with different reduced dimensions were investigated, and the projection with the dimension of five performs the best. To select the most effective sliding window, several sliding windows were investigated for encoding instances, and a window with a length of 13 was chosen finally, which performed the best among the eight windows. It is suggested that our method is promising in computational hotspot prediction for the whole protein sequence.

Acknowledgments: This work was supported by the National Natural Science Foundation of China (Nos. 61672035, 61300058, 61472282 and 61271098) and the Project Foundation of Natural Science Research in Universities of Anhui Province in China (No. KJ2017A355).

Author Contributions: Jinjian Jiang and Peng Chen conceived and designed the experiments; Jinjian Jiang and Peng Chen performed the experiments; Jinjian Jiang and Nian Wang analyzed the data; Nian Wang and Bing Wang contributed reagents/materials/analysis tools; Jinjian Jiang and Peng Chen wrote the paper. All authors proved the final manuscript.

Abbreviations

kNN	k-Nearest Neighbor
Sen	Sensitivity
Prec	Precision
F1	F-Measure
MCC	Matthews Correlation Coefficient
ASEdb	Alanine Scanning Energetics Database
BID	Binding Interface Database
SKEMPI	Structural Kinetic and Energetic Database of Mutant Protein Interactions

References

1. Clackson, T.; Wells, J.A. A hot spot of binding energy in a hormone-receptor interface. *Science* **1995**, *267*, 383–386.
2. Bogan, A.A.; Thorn, K.S. Anatomy of hot spots in protein interfaces. *J. Mol. Biol.* **1998**, *280*, 1–9.
3. Kortemme, T.; Baker, D. A simple physical model for binding energy hot spots in protein-protein complexes. *Proc. Natl. Acad. Sci. USA* **2002**, *99*, 14116–14121.
4. Keskin, O.; Ma, B.; Nussinov, R. Hot regions in protein-protein interactions: The organization and contribution of structurally conserved hot spot residues. *J. Mol. Biol.* **2005**, *345*, 1281–1294.
5. Thorn, K.S.; Bogan, A.A. ASEdb: A database of alanine mutations and their effects on the free energy of binding in protein interactions. *Bioinformatics* **2001**, *17*, 284–285.
6. Fischer, T.B.; Arunachalam, K.V.; Bailey, D.; Mangual, V.; Bakhru, S.; Russo, R.; Huang, D.; Paczkowski, M.; Lalchandani, V.; Ramachandra, C.; et al. The binding interface database (BID): A compilation of amino acid hot spots in protein interfaces. *Bioinformatics* **2003**, *19*, 1453–1454.
7. Kumar, M.D.S.; Gromiha, M.M. PINT: Protein-protein interactions thermodynamic database. *Nucleic Acids Res.* **2006**, *34*, D195–D198.
8. Moal, I.H.; Fernández-Recio, J. SKEMPI: A structural kinetic and energetic database of mutant protein interactions and its use in empirical models. *Bioinformatics* **2012**, *28*, 2600–2607.

9.   Guerois, R.; Nielsen, J.E.; Serrano, L. Predicting changes in the stability of proteins and protein complexes: A study of more than 1000 mutations. *J. Mol. Biol.* **2002**, *320*, 369–387.

10.   Gao, Y.; Wang, R.; Lai, L. Structure-based method for analyzing protein-protein interfaces. *J. Mol. Model.* **2004**, *10*, 44–54.

11.   Schymkowitz, J.; Borg, J.; Stricher, F.; Nys, R.; Rousseau, F.; Serrano, L. The FoldX web server: An online force field. *Nucleic Acids Res.* **2005**, *33*, W382–W388.

12.   Huo, S.; Massova, I.; Kollman, P.A. Computational alanine scanning of the 1:1 human growth hormone-receptor complex. *J. Comput. Chem.* **2002**, *23*, 15–27.

13.   Rajamani, D.; Thiel, S.; Vajda, S.; Camacho, C.J. Anchor residues in protein-protein interactions. *Proc. Natl. Acad. Sci. USA* **2004**, *101*, 11287–11292.

14.   Gonzalez-Ruiz, D.; Gohlke, H. Targeting protein-protein interactions with small molecules: Challenges and perspectives for computational binding epitope detection and ligand finding. *Curr. Med. Chem.* **2006**, *13*, 2607–2625.

15.   Ma, B.; Elkayam, T.; Wolfson, H.; Nussinov, R. Protein-protein interactions: Structurally conserved residues distinguish between binding sites and exposed protein surfaces. *Proc. Natl. Acad. Sci. USA* **2003**, *100*, 5772–5777.

16.   Del Sol, A.; O'Meara, P. Small-world network approach to identify key residues in protein-protein interaction. *Proteins* **2005**, *58*, 672–682.

17.   Guharoy, M.; Chakrabarti, P. Conservation and relative importance of residues across protein-protein interfaces. *Proc. Natl. Acad. Sci. USA* **2005**, *102*, 15447–15452.

18.   Grosdidier, S.; Fernandez-Recio, J. Identification of hot-spot residues in protein-protein interactions by computational docking. *BMC Bioinform.* **2008**, *9*, 447.

19.   Brinda, K.V.; Kannan, N.; Vishveshwara, S. Analysis of homodimeric protein interfaces by graph-spectral methods. *Protein Eng.* **2002**, *15*, 265–277.

20.   Ofran, Y.; Rost, B. Protein-protein interaction hotspots carved into sequences. *PLoS Comput. Biol.* **2007**, *3*, e119.

21.   Darnell, S.J.; Page, D.; Mitchell, J.C. An automated decision-tree approach to predicting protein interaction hot spots. *Proteins* **2007**, *68*, 813–823.

22.   Lise, S.; Archambeau, C.; Pontil, M.; Jones, D.T. Prediction of hot spot residues at protein-protein interfaces by combining machine learning and energy-based methods. *BMC Bioinform.* **2009**, *10*, 365.

23.   Wang, L.; Liu, Z.P.; Zhang, X.S.; Chen, L. Prediction of hot spots in protein interfaces using a random forest model with hybrid features. *Protein Eng. Des. Sel.* **2012**, *25*, 119–126.

24.   Tuncbag, N.; Gursoy, A.; Keskin, O. Identification of computational hot spots in protein interfaces: Combining solvent accessibility and inter-residue potentials improves the accuracy. *Bioinformatics* **2009**, *25*, 1513–1520.

25.   Guney, E.; Tuncbag, N.; Keskin, O.; Gursoy, A. HotSprint: Database of computational hot spots in protein interfaces. *Nucleic Acids Res.* **2008**, *36*, D662–D666.

26.   Cho, K.I.; Kim, D.; Lee, D. A feature-based approach to modeling protein-protein interaction hot spots. *Nucleic Acids Res.* **2009**, *37*, 2672–2687.

27.   Tuncbag, N.; Keskin, O.; Gursoy, A. HotPoint: Hot spot prediction server for protein interfaces. *Nucleic Acids Res.* **2010**, *38*, W402–W406.

28.   Lise, S.; Buchan, D.; Pontil, M.; Jones, D.T. Predictions of hot spot residues at protein-protein interfaces using support vector machines. *PLoS ONE* **2011**, *6*, e16774.

29.   Chen, P.; Li, J.; Wong, L.; Kuwahara, H.; Huang, J.Z.; Gao, X. Accurate prediction of hot spot residues through physicochemical characteristics of amino acid sequences. *Proteins* **2013**, *81*, 1351–1362.

30.   Ludmila, I.; Kuncheva, C.J.W. Measures of diversity in classifier ensembles and their relationship with the ensemble accuracy. *Mach. Learn.* **2003**, *51*, 181–207.

31.   Yachdav, G.; Kloppmann, E.; Kajan, L.; Hecht, M.; Goldberg, T.; Hamp, T.; Honigschmid, P.; Schafferhans, A.; Roos, M.; Bernhofer, M.; et al. PredictProtein—An open resource for online prediction of protein structural and functional features. *Nucleic Acids Res.* **2014**, *42*, W337–W343.

32.   Kawashima, S.; Pokarowski, P.; Pokarowska, M.; Kolinski, A.; Katayama, T.; Kanehisa, M. AAindex: Amino acid index database, progress report 2008. *Nucleic Acids Res.* **2008**, *36*, D202–D205.

33.   Altschul, S.F.; Madden, T.L.; Schäffer, A.A.; Zhang, J.; Zhang, Z.; Miller, W.; Lipman, D.J. Gapped BLAST and PSI-BLAST: A new generation of protein database search programs. *Nucleic Acids Res.* **1997**, *25*, 3389–3402.

34.   Chen, P.; Li, J. Sequence-based identification of interface residues by an integrative profile combining hydrophobic and evolutionary information. *BMC Bioinform.* **2010**, *11*, 402.

35.   Chen, P.; Wong, L.; Li, J. Detection of outlier residues for improving interface prediction in protein heterocomplexes. *IEEE/ACM Trans. Comput. Biol. Bioinform.* **2012**, *9*, 1155–1165.

36.   Chen, P.; Hu, S.; Zhang, J.; Gao, X.; Li, J.; Xia, J.; Wang, B. A sequence-based dynamic ensemble learning system for protein ligand-binding site prediction. *IEEE/ACM Trans. Comput. Biol. Bioinform.* **2016**, *13*, 901–912.

37.   Ritter, H.; Kohonen, T. Self-organizing semantic maps. *Biol. Cybern.* **1989**, *61*, 241.

38.   Papadimitriou, C.H.; Raghavan, P.; Tamaki, H.; Vempala, S. Latent semantic indexing: A probabilistic analysis. *J. Comput. Syst. Sci.* **2000**, *61*, 217–235.

39.   Kaski, S. Dimensionality reduction by random mapping: Fast similarity computation for clustering. In Proceedings of the IEEE International Joint Conference on Neural Networks Proceedings, World Congress on Computational Intelligence, Anchorage, AK, USA, 4–9 May 1998; Volume 1, pp. 413–418.

40.   Chen, P.; Huang, J.Z.; Gao, X. LigandRFs: Random forest ensemble to identify ligand-binding residues from sequence information alone. *BMC Bioinform.* **2014**, *15* (Suppl. S15), S4.

41.   Kuncheva, L.; Whitaker, C.; Shipp, C.; Duin, R. Limits on the majority vote accuracy in classifier fusion. *Pattern Anal. Appl.* **2003**, *6*, 22–31.

42.   Wang, B.; Chen, P.; Huang, D.S.; Li, J.j.; Lok, T.M.; Lyu, M.R. Predicting protein interaction sites from residue spatial sequence profile and evolution rate. *FEBS Lett.* **2006**, *580*, 380–384.

10

CytoCluster: A Cytoscape Plugin for Cluster Analysis and Visualization of Biological Networks

Min Li [1], Dongyan Li [2], Yu Tang [1], Fangxiang Wu [1,3] and Jianxin Wang [1,*]

[1] School of Information Science and Engineering, Central South University, Changsha 410083, China; limin@csu.edu.cn (M.L.); tangyu@csu.edu.cn (Y.T.); faw341@mail.usask.ca (F.X.W.)

[2] School of software, Central South University, Changsha 410083, China; dongyanli@csu.edu.cn

[3] Department of Mechanical Engineering and Division of Biomedical Engineering, University of Saskatchewan, Saskatoon, SK S7N 5A9, Canada

* Correspondence: jxwang@mail.csu.edu.cn

Abstract: Nowadays, cluster analysis of biological networks has become one of the most important approaches to identifying functional modules as well as predicting protein complexes and network biomarkers. Furthermore, the visualization of clustering results is crucial to display the structure of biological networks. Here we present CytoCluster, a cytoscape plugin integrating six clustering algorithms, HC-PIN (Hierarchical Clustering algorithm in Protein Interaction Networks), OH-PIN (identifying Overlapping and Hierarchical modules in Protein Interaction Networks), IPCA (Identifying Protein Complex Algorithm), ClusterONE (Clustering with Overlapping Neighborhood Expansion), DCU (Detecting Complexes based on Uncertain graph model), IPC-MCE (Identifying Protein Complexes based on Maximal Complex Extension), and BinGO (the Biological networks Gene Ontology) function. Users can select different clustering algorithms according to their requirements. The main function of these six clustering algorithms is to detect protein complexes or functional modules. In addition, BinGO is used to determine which Gene Ontology (GO) categories are statistically overrepresented in a set of genes or a subgraph of a biological network. CytoCluster can be easily expanded, so that more clustering algorithms and functions can be added to this plugin. Since it was created in July 2013, CytoCluster has been downloaded more than 9700 times in the Cytoscape App store and has already been applied to the analysis of different biological networks. CytoCluster is available from http://apps.cytoscape.org/apps/cytocluster.

Keywords: biological networks; cluster analysis; cytoscape; visualization

1. Introduction

In recent years, people have paid more and more attention to recognizing life activities within a cell by protein interactions and protein complexes [1–3] in the field of systems biology. Proteins are one of the most important biological molecules in a cell. Within a cell, a protein cannot work alone, but rather works together with other proteins to perform cellular functions. Proteins are involved in a life process through protein complexes. Protein complexes can help us to understand certain biological processes and to predict the functions of proteins. Also, they can realize the cell signaling regulation functions by allosteric, competitive binding, interaction, and post-translational modification [4]. Protein-protein interaction (PPI) networks are powerful models that represent the pairwise protein interactions of organisms. Clustering PPI networks can be useful for isolating groups of interacting proteins that participate in the same biological processes or that, together, perform specific biological functions.

Up to now, many clustering algorithms, which are used to predict protein complexes from proteomics data, have been proposed and applied to biological networks. Out of these methods, the graph-based approaches are the most popular, which includes the partition-based clustering method, the density-based clustering method, the hierarchical-based clustering method and the spectral-based clustering method.

The partition-based clustering algorithms detect protein complexes by finding an optimal network partition, and making sure that the divided objects in the same cluster are as close as possible and the objects in different clusters are as far away as possible, such as HCS (Highly Connected Subgraph) [5], RNSC (Restricted Neighborhood Search Clustering) [6], MSCF (Minimal Seed Cover for Finding protein complexes) [7]. These partition-based clustering algorithms need to know the partition number, which is albeit generally unknown to us. What is more, partition-based methods cannot predict overlapping clusters.

The density-based clustering algorithms identify protein complexes by mining dense subgraphs from biological networks, such as MCL (Markov CLuster) [8], MCODE (Molecular COmplex DEtection) [9], CPM (Clique Percolation Method) [10], LCMA (Local Clique Merging Algorithm) [11], Dpclus (Density-periphery based clustering) [12], IPCA (Identifying Protein Complex Algorithm) [13], CMC (Clustering based on Maximal Cliques) [14], MCL-Caw (a refinement of MCL for detecting yeast complexes) [15], ClusterONE (Clustering with Overlapping Neighborhood Expansion) [16], and so on. These clustering algorithms have the advantage of recognizing dense subgraphs. However, it is difficult to predict the clusters which are non-dense subgraphs with these methods, such as the subgraph of "star" and "cycle."

The basic idea of the hierarchical clustering method is measuring the possibility that any two proteins are located in the same cluster according to their similarity or the distance between them. Hierarchical clustering methods can be further divided into divisive methods and agglomerative methods. A divisive method is a top-down approach, whose main action regards the total PPI network as a cluster first, then divides the network according to a rule until all nodes belong to different clusters. An agglomerative method is a bottom-up approach, whose main action regards each protein in the PPI network as a cluster first, then merges any two clusters according to their similarity value until all nodes are assigned to clusters. For example, G-N (Girvan-Newman) [17], MoNet (Modular organization of protein interaction Networks) [18], FAG-EC (Fast AGglomerate algorithm for mining functional modules based on the Edge Clustering coefficients) [19], EAGLE (agglomerativE hierarchicAl clusterinG based on maximaL cliquE) [20], HC-PIN (Hierarchical Clustering algorithm in Protein Interaction Networks) [21] are all hierarchical clustering algorithms. Hierarchical clustering methods can be used for mining arbitrary shape clusters, and can render the hierarchical organization of the entire PPI network based on a tree structure. However, this type of method is very sensitive to noise data and cannot obtain overlapping clusters. Some researchers extend the hierarchical clustering method to detect overlapping clusters by initializing a triangle with three interacting proteins instead of a single protein, such as OH-PIN (identifying Overlapping and Hierarchical modules in Protein Interaction Networks) [22].

The spectral-based clustering algorithms predict protein complexes based on the spectrum theory, such as QCUT (Combines spectral graph partitioning and a local search to optimize the modularity Q) [23], ADMSC (Adjustable Diffusion Matrix-based Spectral Clustering) [24], and SSCC (Semi-Supervised Consensus Clustering) [25]. These spectral-based clustering methods can be a simple and fast approach to a certain extent. These clustering algorithms depend on the feature vector, which determines the final clustering results. In addition, many other kinds clustering algorithms can be found in survey papers [26,27].

With the developments of clustering methods, the visualization of clusters becomes more and more important. Several tools [28–33] have been developed to help researchers to better recognize positive protein complexes. Cytoscape [34] is a friendly and open bioinformatics platform, which shows an exceptional performance both in virtualizations and manipulation of biological networks. Cytoscape also has the advantage of formidable extensibility of integrating a vast amount of plugins with diverse functions over other platforms. There are 33 apps concerning clustering based on Cytoscape described in our supplement, many of which aim to find meaningful pathways, or visualize networks by semantic similarities, or construct dynamic networks. Among all of the apps, there are several apps, such as ClusterViz [35], clusterMake [36], and ClusterONE [16], which are used to detect and visualize protein complexes in PPI networks. They are all useful tools with different clustering methods, which have been used in different areas of life sciences in recent years. However, a great deal of newly developed clustering algorithms has lost favor with the Cytoscape platform and do not implement visualization. Also, several plugins with old versions cannot work on the new Cytoscape platform any more. In order to solve the above limitations, we developed a new plugin named CytoCluster, which integrates six new clustering algorithms in total. In our plugin, five new approaches named IPCA, OH-PIN, HC-PIN, DCU (Detecting Complexes based on Uncertain graph model) [37], IPC-MCE (Identifying Protein Complexes based on Maximal Complex Extension) [38] were added, which are not integrated in any existing apps, but are important methods used to predict protein complexes. Our CytoCluster plugin also contains the BinGO function, which is used to determine which Gene Ontology (GO) categories are statistically overrepresented in a set of genes or a subgraph of a biological network. So, our app becomes a versatile tool that offers such comprehensive clustering algorithms, in addition to the BinGO function for biological networks.

2. Architecture

In this paper, we adopt Cytoscape 3.x to develop our app. Cytoscape 3.x has notable advantages over Cytoscape 2.x, which can be described in the following two aspects. First, the platform of Cytoscape 3.x adopts the OSGI (Open Service Gateway Initiative) framework, which allows developers to dynamically install, load, update, unload, and uninstall the newly developed bundles in an easy way. Second, Cytoscape 3.x employs Maven, which can help developers manage many jar files. In Cytoscape 3.x, both core modules and apps are called OSGI bundles, and they can significantly reduce complexity in app development to some extent. Also, two methods can be used for developing apps in Cytoscape 3.x. The first way is to develop apps as bundles, which can both register a service in the OSGI framework and withdraw its service from the registry. The second way is to implement the apps with Simplified CyApp API (Application Programming Interface), just like in Cytoscape 2.x.

The architecture of CytoCluster is shown in Figure 1, which includes three main bundles: the interface of CytoCluster bundles, the cluster algorithm bundles, and the visualization, BinGO, and export bundles. The interface of CytoCluster bundles is made up of a graphic user interface and a data exchange system, which allows the users to obtain different forms of bioinformatics networks including .txt and .csv files, and send the clustering results to Cytoscape. The six clustering algorithms bundles play an important role in our plugin CytoCluster, and we have defined the abstract Java class named clustering algorithms, making it is easy for us to integrate more clustering algorithms in CytoCluster. The BinGO bundles are the core functionality in analyzing the GO terms, which can be used to determine which GO categories are statistically overrepresented in a set of genes or a subgraph of a biological network. The visualization of BinGO and export bundles provide a way to intuitively visualize the clustering results in Cytoscape, determine which GO categories are statistically overrepresented, and export the clustering results to .txt or .cvs files.

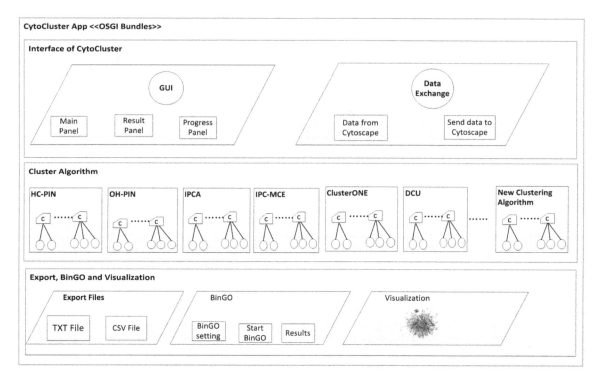

Figure 1. Architecture of CytoCluster.

3. Implementation

A user-friendly clustering software system to detect clusters is very important for biologists. By running the software, users can easily detect and analyze the protein complexes participating in the different life activities. Based on this basic idea, we developed our plugin CytoCluster by adopting the OSGI framework and the Cytoscape Maven archetypes. These frameworks and archetypes can create a maven-based project that builds an initial OSGI bundle-based Cytoscape app. The design is guided by the following three goals: first, to extend new clustering algorithms and add more functions; second, to dispatch the interface of CytoCluster and the algorithms; third, to respond quickly when the user operates the GUI (Graphical User Interface).

CyActivator class is an abstract class, which plays an important role in connecting Cytoscape with CytoCluster. All of the functions of CyActivator start to work as soon as you install the CytoCluster.jar for Cytoscape. The Analyze Action, as one of the service bundles, is the most important function in CytoCluster. Once the network is imported into Cytoscape, then our plugin CytoCluster is able to obtain these data from Cytoscape for further analysis. Two parts can be seen in the main panel. The top part mainly contains the two kinds of the clustering algorithms, overlap clustering algorithms and nonoverlap clustering algorithms. The bottom panel mainly provides six clustering algorithm panels, which are the IPCA panel, HC-PIN panel, OH-PIN panel, DCU panel, ClusterONE panel, and IPC-MCE panel. The user can choose different parameters according to their needs from these clustering algorithm panels. The result panel and the "export to .txt" function must be contained in CytoCluster, which provides an easy way to further analyze the results produced by different clustering algorithms. In addition, the progress panel is included in our app, which is used to visualize the progression of the running clustering algorithms.

Finally, we constructed this CytoCluster app containing four parts: Open, Close, About, and BinGO. Each part has its own function. Six clustering algorithms are included in the Open part. When users want to terminate this app, they should select the Close part. Here, BinGO plays an important role in determining which GO categories are statistically overrepresented in biological networks. Lastly, if you want to learn more information about the app, you cannot miss the About part.

3.1. Calculation and Basic Analysis

When users open the CytoCluster plugin, six clustering algorithms are provided, which are HC-PIN, OH-PIN, IPCA, IPC-MCE, ClusterONE, and DCU. In the following, these six clustering algorithms are briefly described.

3.1.1. HC-PIN (Hierarchical Clustering Algorithm in Protein Interaction Networks)

The HC-PIN algorithm [21] is a fast, hierarchical clustering algorithm, which can be used in a weighted graph or an unweighted graph. The main processes can be described as follows. First, all vertices in the PPI network are regarded as singleton clusters. Then, HC-PIN [21] calculates the clustering value of each edge and queues all of the edges into a queue Sq in non-increasing order according to their clustering values. The higher clustering value the edge has, the more likely its two vertices will be in the same module. In the process of adding edges in the queue Sq to cluster, λ-modules are formed. Finally, λ-modules can be outputted when the number of its proteins is no less than a threshold s.

3.1.2. OH-PIN (Identifying Overlapping and Hierarchical Modules in Protein Interaction Networks)

The OH-PIN algorithm [22] is an improved hierarchical clustering method, which can identify overlapping clusters. The basic idea of OH-PIN can be summarized as follows. At the beginning, the cluster set C_set is empty. For each edge in the protein interaction network, its B_Cluster is generated and the B_Cluster is added to the C_set, if B_Cluster is not already included in the C_set, until every B_Cluster is included. Then, OH-PIN [22] merges all highly overlapping cluster pairs in the C_set in terms of the threshold overlapping value. After the above step, OH-PIN assembles all of the clusters in the C_set into λ-modules by gradually merging the cluster pair with the maximum clustering coefficient.

3.1.3. IPCA (Identifying Protein Complex Algorithm)

The IPCA algorithm [13] is a density-based clustering algorithm, which can identify dense subgraphs in protein interaction networks. IPCA has four major sub-algorithms: weighting vertex, selecting weed, extending cluster, and extend-judgment. First, IPCA [13] calculates the weight of each edge by counting the common neighbors of its connected two nodes and computes the weight of each node by summing up the weights of its incident edges. The higher weight one node has, the more likely the node is regarded as the seed. At the beginning, a seed is initialed as a cluster. IPCA extends a cluster by adding vertices recursively from its neighbors in terms of the nodes' priority. Whether a node can be added to a cluster is determined by two conditions: its interaction probability and the shortest path between it and the nodes in the cluster.

3.1.4. IPC-MCE (Identifying Protein Complexes based on Maximal Complex Extension)

The IPC-MCE algorithm [38] is a maximal clique-based clustering algorithm. The basic idea of IPC-MCE can be described as follows. First, IPC-MCE removes all the nodes which have only one neighbor. Then IPC-MCE enumerates all the maximal cliques in the remained PPI network and puts them into the set MCS (Maximal Clique Sets). For each neighborhood vertex v of the maximal clique K in set MCS, if IP_{vk} is no less than the threshold t, the vertex v can be added to the maximal clique K. The definition of IP_{vk} is as follows:

$$IP_{vk} = \frac{|E_{vk}|}{|V_k|} \qquad (1)$$

E_{vK} is the number of the edges between the vertex v and K, and $|V_k|$ is the number of nodes in K. Finally, IPC-MCE [38] filters the repeated maximal clique according to a pre-defined overlapping value.

3.1.5. ClusterONE (Clustering with Overlapping Neighborhood Expansion)

The ClusterONE algorithm [16] mainly contains three steps. First, groups are grown by adding or removing vertices with high cohesiveness from selected seed proteins. At the beginning, the protein with the highest degree is regarded as the first seed and grows a cohesive group from it using a greedy procedure. ClusterONE repeats this grown process to form overlapping complexes until there are no proteins remaining in the PPI network. Then ClusterONE merges the highly overlapping pairs of locally optimal cohesive groups according to a pre-defined overlapping score. Finally, ClusterONE outputs protein complexes that contain no less than three proteins or whose density is larger than a given threshold ∂ (its default value is 0.8).

3.1.6. DCU (Detecting Complexes Based on Uncertain Graph Model)

The DCU algorithm [37] is a clustering algorithm, which detects protein complexes based on an uncertain graph model. First, DCU [37] starts from a seed vertex and adds other vertices by using a greedy procedure to form a candidate core with high cohesion and low coupling. Then, DCU uses a core-attachment strategy to add attachments to core sets to form complexes. Specifically, for each protein of a candidate set, if its internal absolute degree is less than its external absolute degree, which consists of neighbors of protein vertices in the candidate set, the protein must be removed from the candidate set. Finally, DCU needs to solve the problem of the repeated protein complexes by controlling their overlapping value. Users can select any kind of clustering algorithms they want in the main panel and input the parameters of the algorithm, which decide the creation of a specific clustering algorithm object in memory. Our CytoCluster plugin also provides the visualization of clustering results after running each of these six clustering algorithms, which can be seen in the result panel in the form of a thumbnail list. They can be sorted by the score, the size, or the modularity. In the result panel, the "Export" button and "Discard Result" button are included. The "Export" button is used for exporting results to a .txt file, including the name of algorithm, the parameters, and the clusters, while the "Discard Result" button is used for closing the result panel. Users can close the visualization of clustering results after running these six clustering algorithms with default parameters. In addition, users can see the visualization of cluster results after running a clustering algorithm. Therefore, CytoCluster is a convenient and fast app to obtain smaller networks from a large network.

3.2. BinGO

Here, we integrate the BinGO function to be the part of the CytoCluster. All this is done for the convenience of the users. When they install a cytocluster jar, users can not only choose different clustering algorithms, but also use BinGO. Once the BinGO part is opened, a panel will appear in the center of the computer monitor. Users can make a choice from this setting panel according to their need. The main function of BinGO is to determine the overrepresentation of Gene Ontology (GO) categories in a subgraph of a biological network or a set of genes. Once given a set of genes or a subgraph of a network on the GO hierarchy, BinGO can map the predominant functional themes and output this map in the form of a Cytoscape graph. The BinGO function has the same features as the BiNGO [39] plugin. These features contain graphs or genes list inputs; make and use custom annotations, ontologies, and reference sets; save the extensive results in a tab-delimited text file format; and so on. Selecting the "Start BiNGO" button is required after users have chosen their basic parameters. Then, the visualization of GO can be seen from a chosen network. The result can also be saved in a .bgo, which can be used for further studies.

In the BinGO part, two modes are included for selecting the set of genes to be functionally recommended. One is the default mode, and the other is the flexible mode. In the default mode, nodes can be chosen from a Cytoscape network, either manually or by other plugins. In the flexible mode, nodes can be selected from other sources, for example a set of nodes that are obtained from an experiment and pasted in a text input box. Here, the relevant GO annotations can be retrieved and propagated upwards through the GO hierarchy; namely, any genes related to a certain GO category can be predicted explicitly and included in all parental categories. Two statistical tests are also concerned so as to assess the enrichment of a GO term better. The most important characteristic of the BinGO part is its interactive use for molecular interaction networks, such as protein interaction networks. Furthermore, it is very flexible for BinGO to use ontologies and annotations. Both the traditional GO ontologies and the GOSlim ontologies are supported by BinGO. Then, the Cytoscape graph produced by BinGO can be seen, altered, and saved in a variety of ways.

4. Cases Studies

CytoCluster integrates different types of clustering algorithms including density-based clustering algorithms, hierarchical clustering algorithms, and maximal clique-based methods. Many researchers have downloaded and used the plugin since CytoCluster was released. So far, CytoCluster has been downloaded more than 9700 times since it was released in July 2013. Several important scientific articles indicated that CytoCluster can help scholars with their studies on the mechanisms of biological networks. There are several generic stages of how to run the clustering algorithms in our CytoCluster plugin, which include installing the CytoCluster app, loading the network, setting the data scope and parameters of clustering algorithms, running the cluster algorithm, and receiving or exporting the information of clustering results. The "CytoCluster" menu appears in the "App" menu, after installing the CytoCluster app. In this paper, we present a case to illustrate the use of our plugin. In addition, more cases on these six clustering algorithms can be seen in Table 1.

Table 1. More applications of CytoCluster and the six clustering algorithms integrated in it.

Algorithms	Application	Network	Description	Reference
IPCA	Exploring tomato gene functions	The tomato co-expression network was chosen and 465 complexes were found	IPCA was used to identity a densely connected network	[40]
	Unravelling gene function	The tomato co-expression network was chosen and 465 complexes were found	IPCA was choosen to identify thick connected nodes	[41]
	Predicting colon adenocarcinoma	The networks from IntAct and reactome were merged	IPCA was used to identify highly connected subnetworks	[42]
	The correlation between cold and heat patterns	The network from RA 18 was diagnosed with deficiency pattern and 15 others were diagnosed with nondeficiency pattern	IPCA was used to analyze the characteristics of networks	[43]
	Evidence-based complementary and alternative medicine	PPI network from genes was chosen so that the ratio of cold patterns to heat patterns in patients with RA was more or less than 1:1.4	IPCA was used to detect highly connected subnetworks	[44]
	Cold and heat patterns of rheumatoid arthritis	PPI network from these genes was chose that the ratio of cold patterns to heat patterns in patients with RA was more or less than 1:2	Highly connected regions associated with typical TCM cold patterns and heat patterns were identified	[45]
	Cold and heat pattern of rheumatoid arthritis	Network for differentially expressed genes between RA patients with TCM cold and heat patterns	IPCA was used to infer significant complexes or pathways in the PPI network	[46]
	Functional networks	Network contained some gene expressions or regulated proteins	Then eight highly connected regions were found by IPCA to infer complexes or pathways	[47]
	The molecular mechanism of interventions	PPI networks of biomedical combination was chosen and 11 complexes were found	IPCA was used to analyze the characteristics of the network	[48]
	The synergistic sechanisms	Network associated with Salvia miltiorrhiza and Panax notoginseng	Significant complexes or pathways were inferred	[49]
	Constraints on community	Associations between bacteria OTUs and four subnetworks were found	Subnetworks of OTUs were detected	[50]
HC-PIN	Strategies between two reef building cold-water coral species	Association network of the cold-water scleractinian corals bacterial communities	HC-PIN was used to identify OTUs	[51]
	Biomarkers	The network was extracted from the TCGA database	miRNA-gene clusters were identified	[52]
	Finding the candidate biomarkers for POAG disease	Network was extracted from previous studies with 474 proteins and nine subnetworks were found	HC-PIN was choosen to perform the clustering with a complex size threshold of 3	[53]
OH-PIN	Bacterial associations	Bulk soil DNA was extracted	The subnetworks were partitioned into modulars	[54]
ClusterONE	A census of human soluble protein complexes	Network was extracted from human HeLa S3 and HEK293 cells grown	ClusterONE was used to detect protein complexes	[55]
	An arabidopsis	A network with 8900 nodes and 6882 edges was chosen and 701 clusters were found	ClusterONE was used to obtain subnetworks	[56]
	Fndinge disease-drug modules	Disease-gene and drug-target associations were found from drug-target data	Overlapping subnetworks were identified	[57]

PPI: Protein-protein interaction; IPCA: Identifying Protein Complex Algorithm; TCM:Traditional Chinese Medicine; RA:Rheumatoid Arthritis; POAG: Primary Open Angle Glaucoma; OTU: Opearating Taxonomic Unit; TCGA:The Cancer Genome Atlas; OH-PIN: Identifying Overlapping and Hierarchical Modules in Protein Interaction Networks.

The case of CytoCluster was applied in botany [58]. This paper was published in Plant Physiology by Baute et al. The co-expression network was generated by Cytoscape 3.2.0 [59] according to the nodes and edges [60,61] at first. Then, the newly co-expression network was loaded, which incorporated 185 genes and 943 edges. Third, the main panel of the CytoCluster was opened and the HC-PIN clustering algorithm was chosen with standard settings and a complex size threshold of 10. In this case, 185 genes and 943 edges were included after dealing with the whole network. The identified subnetworks were further filtered, so as to only include the co-expression networks based on PCCs (Pearson Correlation Coefficients) of 0.7 and higher, as well as protein-protein interactions between query genes based on both experimental and predicted data from CORNET, when the users clicked on the "Analysis" button. Then, four subnetworks were formed after using our plugin for analysis, which can be seen from Figure 2. Each circle in Figure 2 shows a subnetwork. What is more, the generated co-expression network achieved by the HC-PIN algorithm can be seen in the result panel or exported to a .txt, so users can output the results from the different algorithms for further analysis. The table panel can list proprieties of clustering results when users select the corresponding clustering. The progress panel is used to visualize the progression of a specific cluster algorithm.

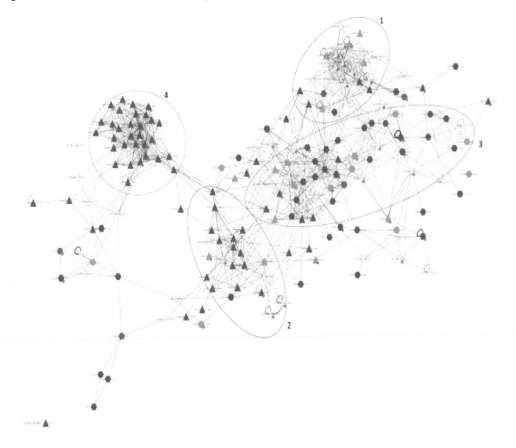

Figure 2. Four subnetworks achieved in the first case [58].

5. Conclusions

Our CytoCluster plugin is a platform-independent app for Cytoscape, which is also a functional diversity tool to offer different types of clustering algorithms, including IPCA, DCU, HC-PIN, OH-PIN, IPC-MCE, and ClusterONE. OH-PIN and HC-PIN are both hierarchical-based clustering algorithms, HC-PIN generates non-overlapping clusters, and on the contrary, OH-PIN produces overlapping clusters. IPCA, DCU, IPC-MCE, and ClusterONE are all density-based clustering algorithms, but the clusters generated by them also have some differences. Moreover, the same method will produce different results by changing the values of parameters. Users can both choose different clustering algorithms and analyze which GO categories are statistically overrepresented in a set of genes or a subgraph of a biological network. Our CytoCluster plugin is not only convenient for researchers to use, but also renders the investigated biological process easy to understand. Because our app has the advantage of expandability, more clustering algorithms such as those reported in References [62–65] as well as modules can be added to CytoCluster. Owing to such features, we firmly believe our app will be of great help in biology research.

Acknowledgments: This work was supported in part by the National Natural Science Foundation of China under Grants (No. 61622213, No. 61370024 and No. 61420106009).

Author Contributions: Min Li, Dongyan Li and Yu Tang conceived and designed the software, test and experiments; Dongyan Li and Yu Tang implemented the software and performed the experiments; Min Li and Dongyan Li wrote the paper. Min Li, Dongyan Li, Yu Tang, FangXiang Wu and Jianxin Wang revised the manuscript.

References

1. Wang, J.; Liang, J.; Zheng, W. A graph clustering method for detecting protein complexes. *J. Comput. Res. Dev.* **2015**, *52*, 1784–1793.

2. Alberts, B. The cell as a collection of protein machines: Preparing the next generation of molecular biologists. *Cell* **1998**, *92*, 291–294. [CrossRef]

3. Lasserre, J.P.; Beyne, E.; Pyndiah, S.; Pyndiah, S.; Lapaillerie, D.; Claverol, S.; Bonneu, M. A complexomic study of *Escherichia coli* using two-dimensional blue native/SDS polyacrylamide gel electrophoresis. *Electrophoresis* **2006**, *27*, 3306–3321. [CrossRef] [PubMed]

4. Gibson, T.J. Cell regulation: Determined to signal discrete cooperation. *Trends Biochem. Sci.* **2009**, *3410*, 471–482. [CrossRef] [PubMed]

5. Pržulj, N.; Wigle, D.A.; Jurisica, I. Functional topology in a network of protein interactions. *Bioinformatics* **2004**, *203*, 340–348. [CrossRef] [PubMed]

6. King, A.D.; Pržulj, N.; Jurisica, I. Protein complex prediction via cost-based clustering. *Bioinformatics* **2004**, *2017*, 3013–3020. [CrossRef] [PubMed]

7. Ding, X.; Wang, W.; Peng, X.; Wang, J. Mining protein complexes from PPI networks using the minimum vertex cut. *Tsinghua Sci. Technol.* **2012**, *176*, 674–681. [CrossRef]

8. Enright, A.J.; Dongen, S.V.; Ouzounis, C.A. An efficient algorithm for large-scale detection of protein families. *Nucleic Acids Res.* **2002**, *30*, 1575–1584. [CrossRef] [PubMed]

9. Bader, G.D.; Hogue, C.W.V. An automated method for finding molecular complexes in large protein interaction networks. *BMC Bioinform.* **2003**, *41*, 2.

10. Palla, G.; Derényi, I.; Farkas, I.; Vicsek, T. Uncovering the overlapping community structure of complex networks in nature and society. *Nature* **2005**, *435*, 814–818. [CrossRef] [PubMed]

11. Li, X.L.; Foo, C.S.; Tan, S.H.; Ng, S.K. Interaction graph mining for protein complexes using local clique merging. *Genome Inform.* **2005**, *16*, 260–269. [PubMed]

12. Altaf-Ul-Amin, M.; Shinbo, Y.; Mihara, K.; Kurokawa, K.; Kanaya, S. Development and implementation of an algorithm for detection of protein complexes in large interaction networks. *BMC Bioinform.* **2006**, *7*, 207. [CrossRef] [PubMed]

13. Li, M.; Chen, J.; Wang, J.; Hu, B.; Chen, G. Modifying the DPClus algorithm for identifying protein complexes based on new topological structures. *BMC Bioinform.* **2008**, *9*, 398. [CrossRef] [PubMed]

14. Liu, G.; Wong, L.; Chua, H.N. Complex discovery from weighted PPI networks. *Bioinformatics* **2009**, *25*, 1891–1897. [CrossRef] [PubMed]

15. Srihari, S.; Ning, K.; Leong, H.W. MCL-CAw: A refinement of MCL for detecting yeast complexes from weighted PPI networks by incorporating core-attachment structure. *BMC Bioinform.* **2010**, *11*, 504. [CrossRef] [PubMed]

16. Nepusz, T.; Yu, H.; Paccanaro, A. Detecting overlapping protein complexes in protein-protein interaction networks. *Nat. Methods* **2012**, *9*, 471–472. [CrossRef] [PubMed]

17. Girvan, M.; Newman, M.E.J. Community structure in social and biological networks. *Proc. Natl. Acad. Sci. USA* **2002**, *99*, 7821–7826. [CrossRef] [PubMed]

18. Luo, F.; Yang, Y.; Chen, C.F.; Chang, R.; Zhou, J.; Scheuermann, R.H. Modular organization of protein interaction networks. *Bioinformatics* **2007**, *23*, 207–214. [CrossRef] [PubMed]

19. Li, M.; Wang, J.; Chen, J. A fast hierarchical clustering algorithm for functional modules in protein interaction networks. In Proceedings of the IEEE 2008 International Conference on BioMedical Engineering and Informatics (BMEI), Sanya, China, 27–30 May 2008; Volume 1, pp. 3–7.

20. Shen, H.; Cheng, X.; Cai, K.; Hu, M.B. Detect overlapping and hierarchical community structure in networks. *Phys. A Stat. Mech. Appl.* **2009**, *388*, 1706–1712. [CrossRef]

21. Wang, J.; Li, M.; Chen, J.; Pan, Y. A fast hierarchical clustering algorithm for functional modules discovery in protein interaction networks. *IEEE/ACM Trans. Comput. Biol. Bioinform.* **2011**, *8*, 607–620. [CrossRef] [PubMed]

22. Wang, J.; Ren, J.; Li, M.; Wu, F.X. Identification of hierarchical and overlapping functional modules in PPI networks. *IEEE Trans. Nanobiosci.* **2012**, *11*, 386–393. [CrossRef] [PubMed]

23. Chen, D.; Fu, Y.; Shang, M. A fast and efficient heuristic algorithm for detecting community structures in complex networks. *Phys. A Stat. Mech. Appl.* **2009**, *388*, 2741–2749. [CrossRef]

24. Inoue, K.; Li, W.; Kurata, H. Diffusion model based spectral clustering for protein-protein interaction networks. *PLoS ONE* **2010**, *5*, e12623. [CrossRef] [PubMed]

25. Wang, Y.; Pan, Y. Semi-supervised consensus clustering for gene expression data analysis. *BioData Min.* **2014**, *7*, 1–13. [CrossRef] [PubMed]

26. Li, M.; Wu, X.; Wang, J.; Pan, Y. Progress on graph-based clustering methods for the analysis of protein-protein interaction networks. *Comput. Eng. Sci.* **2012**, *34*, 124–136.

27. Ji, J.; Liu, Z.; Liu, H.; Liu, C. An overview of research on functional module detection for protein-protein interaction networks. *Acta Autom. Sin.* **2014**, *40*, 577–593.

28. Protein-Protein Interaction Networks Co-Clustering. Available online: http://wwwinfo.deis.unical.it/rombo/co-clustering/ (accessed on 21 April 2017).

29. Batagelj, V.; Mrvar, A. Pajek-program for large network analysis. *Connections* **1998**, *21*, 47–57.

30. Adamcsek, B.; Palla, G.; Farkas, I.J.; Derényi, I.; Vicsek, T. CFinder: Locating cliques and overlapping modules in biological networks. *Bioinformatics* **2006**, *22*, 1021–1023. [CrossRef] [PubMed]

31. Moschopoulos, C.N.; Pavlopoulos, G.A.; Schneider, R.; Likothanassis, S.D.; Kossida, S. GIBA: A clustering tool for detecting protein complexes. *BMC Bioinform.* **2009**, *10*, S11. [CrossRef] [PubMed]

32. Zheng, G.; Xu, Y.; Zhang, X.; Liu, Z.P.; Wang, Z.; Chen, L.; Zhu, X.G. CMIP: A software package capable of reconstructing genome-wide regulatory networks using gene expression data. *BMC Bioinform.* **2016**, *17*, 137. [CrossRef] [PubMed]

33. Li, M.; Tang, Y.; Wu, X.; Wang, J.; Wu, F.X.; Pan, Y. C-DEVA: Detection, evaluation, visualization and annotation of clusters from biological networks. *Biosystems* **2016**, *150*, 78–86. [CrossRef] [PubMed]

34. Shannon, P.; Markiel, A.; Ozier, O.; Baliga, N.S.; Wang, J.T.; Ramage, D.; Amin, N.; Schwikowski, B.; Ideker, T. Cytoscape: A software environment for integrated models of biomolecular interaction networks. *Genome Res.* **2003**, *13*, 2498–2504. [CrossRef] [PubMed]

35. Wang, J.; Zhong, J.; Chen, G.; Li, M.; Wu, F.X.; Pan, Y. ClusterViz: A Cytoscape APP for cluster analysis of biological network. *IEEE/ACM Trans. Comput. Biol. Bioinform.* **2015**, *12*, 815–822. [CrossRef] [PubMed]

36. Morris, J.H.; Apeltsin, L.; Newman, A.M.; Baumbach, J.; Wittkop, T.; Su, G.; Bader, G.D.; Ferrin, T.E. clusterMaker: A multi-algorithm clustering plugin for Cytoscape. *BMC Bioinform.* **2011**, *12*, 436. [CrossRef] [PubMed]

37. Zhao, B.; Wang, J.; Li, M.; Wu, F.X.; Pan, Y. Detecting protein complexes based on uncertain graph model. *IEEE/ACM Trans. Comput. Biol. Bioinform.* **2014**, *11*, 486–497. [CrossRef] [PubMed]

38. Li, M.; Wang, J.X.; Liu, B.B.; Chen, J.E. An algorithm for identifying protein complexes based on maximal clique extension. *J. Cent. South Univ.* **2010**, *41*, 560–565.

39. Maere, S.; Heymans, K.; Kuiper, M. BiNGO: A Cytoscape plugin to assess overrepresentation of gene ontology categories in biological networks. *Bioinformatics* **2005**, *21*, 3448–3449. [CrossRef] [PubMed]

40. Fukushima, A.; Nishizawa, T.; Hayakumo, M.; Hikosaka, S.; Saito, K.; Goto, E.; Kusano, M. Exploring tomato gene functions based on coexpression modules using graph clustering and differential coexpression approaches. *Plant Physiol.* **2012**, *158*, 1487–1502. [CrossRef] [PubMed]

41. Schaefer, R.J.; Michno, J.M.; Myers, C.L. Unraveling gene function in agricultural species using gene co-expression networks. *Biochim. Biophys. Acta (BBA)-Gene Regul. Mech.* **2017**, *1860*, 53–63. [CrossRef] [PubMed]

42. Wang, Y.; Zhang, J.; Li, L.; Xu, X.; Zhang, Y.; Teng, Z.; Wu, F. Identification of molecular targets for Predicting Colon Adenocarcinoma. *Med. Sci. Monit. Int. Med. J. Exp. Clin. Res.* **2016**, *22*, 460–468. [CrossRef]

43. Wang, M.; Chen, G.; Lu, C.; Xiao, C.; Li, L.; Niu, X.; He, X.; Jiang, M.; Lu, A. Rheumatoid arthritis with deficiency pattern in traditional Chinese medicine shows correlation with cold and hot patterns in gene expression profiles. *Evid.-Based Complement. Altern. Med.* **2013**, *2013*, 248650. [CrossRef] [PubMed]

44. Lu, C.; Niu, X.; Xiao, C.; Chen, G.; Zha, Q.; Guo, H.; Jiang, M.; Lu, A. Network-based gene expression biomarkers for cold and heat patterns of rheumatoid arthritis in traditional Chinese medicine. *Evid.-Based Complement. Altern. Med.* **2012**, *2012*, 203043. [CrossRef] [PubMed]

45. Lu, C.; Xiao, C.; Chen, G.; Jiang, M.; Zha, Q.; Yan, X.; Kong, W.; Lu, A. Cold and heat pattern of rheumatoid arthritis in traditional Chinese medicine: Distinct molecular signatures indentified by microarray expression profiles in CD4-positive T cell. *Rheumatol. Int.* **2012**, *32*, 61–68. [CrossRef] [PubMed]

46. Chen, G.; Lu, C.; Zha, Q.; Xiao, C.; Xu, S.; Ju, D.; Zhou, Y.; Jia, W.; Lu, A. A network-based analysis of traditional Chinese medicine cold and hot patterns in rheumatoid arthritis. *Complement. Ther. Med.* **2012**, *20*, 23–30. [CrossRef] [PubMed]

47. Chen, G.; Liu, B.; Jiang, M.; Tan, Y.; Lu, A.P. Functional networks for Salvia miltiorrhiza and Panax notoginseng in combination explored with text mining and bioinformatical approach. *J. Med. Plants Res.* **2011**, *5*, 4030–4040.

48. Jiang, M.; Lu, C.; Chen, G.; Xiao, C.; Zha, Q.; Niu, X.; Chen, S.; Lu, A. Understanding the molecular mechanism of interventions in treating rheumatoid arthritis patients with corresponding traditional Chinese medicine patterns based on bioinformatics approach. *Evid.-Based Complement. Altern. Med.* **2012**, *2012*, 129452. [CrossRef] [PubMed]

49. Chen, G.; Liu, B.; Jiang, M.; Aiping, L. System Analysis of the Synergistic Mechanisms between Salvia Miltiorrhiza and Panax Notoginseng in Combination. *World Sci. Technol.* **2010**, *12*, 566–570.

50. Kalenitchenko, D.; Fagervold, S.K.; Pruski, A.M.; Vétion, G.; Yücel, M.; Le Bris, N.; Galand, P.E. Temporal and spatial constraints on community assembly during microbial colonization of wood in seawater. *ISME J.* **2015**, *9*, 2657–2670. [CrossRef] [PubMed]

51. Meistertzheim, A.L.; Lartaud, F.; Arnaud-Haond, S.; Kalenitchenko, D.; Bessalam, M.; Le Bris, N.; Galand, P.E. Patterns of bacteria-host associations suggest different ecological strategies between two reef building cold-water coral species. *Deep Sea Res. Part I Oceanogr. Res. Pap.* **2016**, *114*, 12–22. [CrossRef]

52. Guo, H.; Chen, J.; Meng, F. Identification of novel diagnosis biomarkers for lung adenocarcinoma from the cancer genome atlas. *Orig. Artic.* **2016**, *9*, 7908–7918.

53. Atan, N.A.D.; Yekta, R.F.; Nejad, M.R.; Nikzamir, A. Pathway and network analysis in primary open angle glaucoma. *J. Paramed. Sci.* **2014**, *5*. [CrossRef]

54. Wang, H.; Wei, Z.; Mei, L.; Gu, J.; Yin, S.; Faust, K.; Raes, J.; Deng, Y.; Wang, Y.; Shen, Q.; Yin, S. Combined use of network inference tools identifies ecologically meaningful bacterial associations in a paddy soil. *Soil Biol. Biochem.* **2017**, *105*, 227–235. [CrossRef]

55. Havugimana, P.C.; Hart, G.T.; Nepusz, T.; Yang, H.; Turinsky, A.L.; Li, Z.; Wang, P.I.; Boutz, D.R.; Fong, V.; Phanse, S.; et al. A census of human soluble protein complexes. *Cell* **2012**, *150*, 1068. [CrossRef] [PubMed]

56. Van Landeghem, S.; de Bodt, S.; Drebert, Z.J.; Inzé, D.; van de Peer, Y. The potential of text mining in data integration and network biology for plant research: A case study on Arabidopsis. *Plant Cell* **2013**, *25*, 794–807. [CrossRef] [PubMed]

57. Wu, C.; Gudivada, R.C.; Aronow, B.J.; Jegga, A.G. Computational drug repositioning through heterogeneous network clustering. *BMC Syst. Biol.* **2013**, *7*, S6. [CrossRef] [PubMed]

58. Baute, J.; Herman, D.; Coppens, F.; de Block, J.; Slabbinck, B.; dell'Aqcua, M.; Pè, M.E.; Maere, S.; Nelissen, H.; Inzé, D. Combined large-scale phenotyping and transcriptomics in maize reveals a robust growth regulatory network. *Plant Physiol.* **2016**, *170*, 1848–1867. [CrossRef] [PubMed]

59. Czerwinska, U.; Calzone, L.; Barillot, E.; Zinovyev, A. DeDaL: Cytoscape 3 app for producing and morphing data-driven and structure-driven network layouts. *BMC Syst. Biol.* **2015**, *9*, 46. [CrossRef] [PubMed]

60. Kerrien, S.; Aranda, B.; Breuza, L.; Bridge, A.; Broackes-Carter, F.; Chen, C.; Duesbury, M.; Dumousseau, M.; Feuermann, M.; Hinz, U.; et al. The IntAct molecular interaction database in 2012. *Nucleic Acids Res.* **2012**, *40*, D841–D846. [CrossRef] [PubMed]

61. Croft, D.; O'Kelly, G.; Wu, G.; Haw, R.; Gillespie, M.; Matthews, L.; Caudy, M.; Garapati, P.; Gopinath, G.; Jassal, B.; et al. Reactome: A database of reactions, pathways and biological processes. *Nucleic Acids Res.* **2010**, *39*, D691–D697. [CrossRef] [PubMed]

62. Li, M.; Wang, J.; Chen, J.; Cai, Z.; Chen, G. Identifying the overlapping complexes in protein interaction networks. *Int. J. Data Min. Bioinform.* **2010**, *4*, 91–108. [CrossRef] [PubMed]

63. Li, X.; Wang, J.; Zhao, B.; Wu, F.X.; Pan, Y. Identification of protein complexes from multi-relationship protein interaction networks. *Hum. Genom.* **2016**, *10*, 17. [CrossRef] [PubMed]

64. Lei, X.; Ding, Y.; Wu, F.X. Detecting protein complexes from DPINs by density based clustering with Pigeon-Inspired Optimization Algorithm. *Sci. China Inf. Sci.* **2016**, *59*, 070103. [CrossRef]

65. Zhao, B.; Wang, J.; Li, M.; Li, X.; Li, Y.; Wu, F.X.; Pan, Y. A new method for predicting protein functions from dynamic weighted interactome networks. *IEEE Trans. Nanobiosci.* **2016**, *15*, 131–139. [CrossRef] [PubMed]

11

UltraPse: A Universal and Extensible Software Platform for Representing Biological Sequences

Pu-Feng Du [1], Wei Zhao [1], Yang-Yang Miao [1,2], Le-Yi Wei [1] and Likun Wang [3,*]

[1]   School of Computer Science and Technology, Tianjin University, Tianjin 300350, China;
      PufengDu@gmail.com (P.-F.D.); wzhao_cstju@yeah.net (W.Z.); miaoyangyang1998@163.com (Y.-Y.M.);
      weileyi@tju.edu.cn (L.-Y.W.)
[2]   School of Chemical Engineering, Tianjin University, Tianjin 300350, China
[3]   Institute of Systems Biomedicine, Beijing Key Laboratory of Tumor Systems Biology, Department of
      Pathology, School of Basic Medical Sciences, Peking University Health Science Center, Beijing 100191, China
*   Correspondence: wanglk@hsc.pku.edu.cn

Abstract: With the avalanche of biological sequences in public databases, one of the most challenging problems in computational biology is to predict their biological functions and cellular attributes. Most of the existing prediction algorithms can only handle fixed-length numerical vectors. Therefore, it is important to be able to represent biological sequences with various lengths using fixed-length numerical vectors. Although several algorithms, as well as software implementations, have been developed to address this problem, these existing programs can only provide a fixed number of representation modes. Every time a new sequence representation mode is developed, a new program will be needed. In this paper, we propose the UltraPse as a universal software platform for this problem. The function of the UltraPse is not only to generate various existing sequence representation modes, but also to simplify all future programming works in developing novel representation modes. The extensibility of UltraPse is particularly enhanced. It allows the users to define their own representation mode, their own physicochemical properties, or even their own types of biological sequences. Moreover, UltraPse is also the fastest software of its kind. The source code package, as well as the executables for both Linux and Windows platforms, can be downloaded from the GitHub repository.

Keywords: pseudo-amino acid compositions; pseudo-k nucleotide compositions; extensible software

1. Introduction

Over the last two decades, huge numbers of biological sequences have been deposited in public databases. Until today, the number of these sequences is still increasing exponentially. However, the cellular and functional attributes of these sequences, no matter whether they are nucleotide sequences or protein sequences, remain largely unknown. It is a very important task for computational biology to predict the functional and cellular attributes of these sequences.

In the view of machine learning, most of these prediction tasks can be formulated as pattern classification problems. As elaborated in a series of publications [1–8], one of the most challenging parts is to represent a biological sequence with a fixed-length numerical vector, yet still keep a considerable amount of the sequence-order information. This is because almost every existing algorithm for these tasks can only handle fixed-length vectors, but not the sequences.

For protein and peptide sequences, Chou proposed pseudo-amino acid compositions (PseAAC) [9] and amphiphilic pseudo-amino acid compositions (AmPseAAC) [10]. Ever since the concepts of pseudo-factors were introduced, they have rapidly penetrated into almost every area of computational proteomics [11–20]. As elaborated in a review article, the form of classic pseudo-amino acid compositions has been generalized to contain various types of information [21], which is known as the general-form pseudo-amino acid compositions. The applications of PseAAC concepts have been summarized in the review papers [22,23].

Recently, the concept of PseAAC has been extended to represent nucleotide sequences [24]. Chen et al. developed pseudo-dinucleotide compositions (PseDNC) to predict DNA recombination hostspots [25]. This formulation was then extended as pseudo-k nucleotide compositions (PseKNC), which have been applied in predicting splicing sites [26], predicting translation initiation sites [27], predicting nucleosome positions [28], predicting promoters [29], predicting DNA methylation sites [30], predicting microRNA precursors [31] and many others [32–41].

In the early days of pseudo-amino acid compositions, every study had to implement PseAAC independently. Although the algorithms in every implementation are identical, different implementations may introduce computational discrepancies due to technical details. For example, different implementations may give results with different precisions. This kind of differences may be amplified by machine-learning based predictors, which may eventually produce different prediction results. For another example, different implementations may have very different computational efficiencies. This means one implementation may only use a second to process a dataset, while another program may require over an hour to achieve the same results on the same dataset with the same parameters.

To solve these problems, a universal implementation of the algorithm should be provided. Many efforts have been made for this purpose [42–52]. The first program focus on the PseAAC formulation is the PseAAC server [43], which was brought online in the year 2008. The PseAAC server can compute Type-I and Type-II PseAAC using six different kinds of physicochemical properties of amino acids. The PseAAC server has a friendly user interface, which is convenient and efficient for small datasets. However, for large datasets and the repeatedly parameter scanning process, the computational efficiency of the PseAAC server is not ideal. The PseAAC-Builder [45], which was released in the year 2012, is dedicated to improving the efficiency. Unlike the PseAAC server, the PseAAC-Builder is a stand-alone program that can be executed locally. It has a simple graphical user interface (GUI) for the users' convenience. It can also be executed in a command line environment. The computational efficiency of PseAAC-Builder is much higher than the PseAAC server, especially in the command line environment. Although the PseAAC-Builder includes over 500 different types of physicochemical properties, it did not provide the ability to compute general form PseAAC. PseAAC-General [46], which is a major upgrade to the PseAAC-Builder, was developed to solve this problem. PseAAC-General provides the ability to compute several commonly used general forms of PseAAC, such as the GO mode, the functional domain mode and the evolutionary mode. The users of PseAAC-General can slightly extend its ability by using Lua scripts.

After Chen et al. proposed the PseKNC representations for nucleotide sequences, similar software and services were needed for DNA and RNA sequences. Chen et al. released the PseKNC [48] and PseKNC-General [49] packages for converting DNA/RNA sequences into its PseKNC or general form of PseKNC representations. Liu et al. developed the repDNA [50], repRNA [51], and Pse-In-One [52] services for more types of descriptors. The Pse-In-One service attempts to be a universal online service that can be applied on both protein and nucleotide sequences.

However, all existing software packages and online services suffer from three problems. (1) Lack of extensibility. Most of the existing software can only be used to produce existing modes of representation. The users cannot extend the software to handle their own novel representation modes. Although PseAAC-General can be extended by using Lua script, it can only be used for protein sequences; (2) Lack of flexibility. Most of the existing software can only handle one type of biological sequences, either nucleotide sequences or protein sequences. Pse-In-One is the only existing service that can handle protein sequence as well as nucleotide sequences. However, no program can handle user-defined sequence types. For example, when studying the protein phosphorylation sites, the modified residues should have different notations of sequences, which are not in the standard 20 letters. The users need to define the extra letters to represent the modified residues. As far as we know, no program can handle this kind of sequence; (3) Lack of computational efficiency on large datasets. Most of the existing programs are not designed to handle large datasets. They may need many minutes to process a million sequences. If a user needs to repeatedly scan parameters of a representation, the processing time may be days or even weeks.

In this paper, we proposed the UltraPse program, which is a universal and extensible software platform for all possible sequence representation modes. The UltraPse program unified the processing of nucleotide and protein sequence in one program, as well as the user-defined sequence types. UltraPse supports two forms of extension modules, the BSOs (Binary Shared Objects) and the Lua scripts, which are called the TDFs (Task Definition Files) in UltraPse. The users can develop their own modes by just writing several lines of Lua scripts. UltraPse has very high computational efficiency. It is even faster than the PseAAC-General, which used to be the fastest program of its kind. For the users' convenience, we have integrated many existing modes within the UltraPse. We expect that the UltraPse program can be a useful platform which simplifies all future programming works in developing novel sequence representation modes. All source codes of UltraPse, including some extension modules can be downloaded freely under the term of GNU GPL (GNU General Public License) v3 from the GitHub repository: https://github.com/pufengdu/UltraPse.

2. Results and Discussion

2.1. Computational Efficiency Analysis

We compared the computational efficiency of UltraPse to that of PseAAC-General and Pse-In-One under the same conditions. As in Figure 1, the UltraPse can process over 120 thousand sequences per second, while PseAAC-General can process about 85 thousand sequences per second. Unfortunately, the Pse-In-One can process only less than one thousand sequences per second. According to these results, the computational efficiency of UltraPse is roughly 1.5 times of the PseAAC-General, and about 185 times of Pse-In-One. Since the algorithms of the three programs are essentially the same, the reason for the efficiency differences resides in the technical details of the implementations.

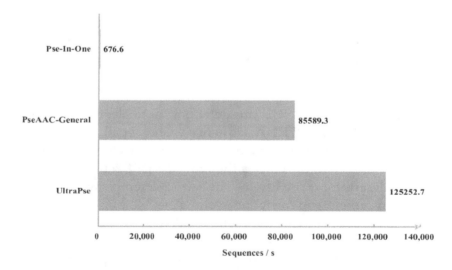

Figure 1. Computational efficiency comparisons. Three programs are compared. The comparison was carried out by letting the three programs compute amino acid compositions on the same dataset on the same machine. Every program was executed with the same parameters for three times. The average execution time was applied in calculating the computational efficiency. The computational efficiency is measured by the average number of sequences that are processed every second. Pse-In-One: A program in literature [52]; PseAAC-General: A program in literature [46]; UltraPse: A program of this work.

2.2. Flexibility and Extensibility

We integrated 35 sequence representation modes within the UltraPse. The representation modes can be organized hierarchically as in Figure 2. The integrated modes can be used to represent protein, as well as DNA and RNA sequences. The modes cover most of the representation modes that can be generated by PseAAC-General, PseKNC-General, and Pse-In-One. Moreover, UltraPse can generate even more modes, for example, the commonly used one-hot encoding mode [53–55]. The sequence representation modes of UltraPse can be extended by using BSOs and TDFs. According to our own works, using UltraPse in developing novel representation modes can save over half of the programing labor.

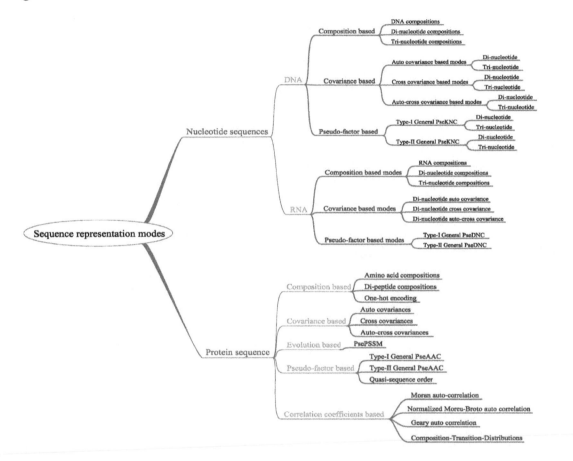

Figure 2. Hierarchical organization of integrated sequence representation modes. UltraPse integrated the sequence representation modes in its distribution package. Most of these modes can also be applied in user-defined sequence types, as long as the users provide proper definitions of the physicochemical properties.

Besides the user-defined representation modes of protein and nucleotide sequences, the users of UltraPse can define their own sequence types using TDFs. They are allowed to choose a set of letters other than the standard ones to represent additional information. For example, a user can use C for cytidines on a DNA sequence, and M for methylated cytidines. The choice of the letter M totally depends on the users. Even more, the users of UltraPse can define their own physicochemical properties with TDFs.

The TDFs of UltraPse is written using Lua language, which is a simple, powerful and extensible programming language which has been applied in bioinformatics software previously [56]. We provide over 20 UltraPse specific functions and interfaces. Users can access and modify UltraPse internal data structures using these functions in TDFs. We compared the flexibility and the extensibility of different software in Table 1.

Table 1. Software function comparison in terms of flexibility and extensibility.

Software Functions	Sequence Types	Extensibility
UltraPse	DNA, RNA, Protein, User-defined types	Users can define their own sequence types, representation modes and physicochemical properties
PseAAC-General [46]	Protein	Users can define their own representation modes
PseAAC-Builder [45]	Protein	No extensibility
Pse-In-One [52]	DNA, RNA, Protein	Users can define their own physicochemical properties
PseKNC [48]	DNA, RNA	Users can define their own physicochemical properties
PseKNC-General [49]	DNA, RNA	Users can define their own physicochemical properties

2.3. Compatibility and Robustness

UltraPse can recognize FASTA format files that are directly downloaded from one of the following five databases: GenBank, UniProt, EMBL, DDBJ, and RefSeq. The sequence identifiers and comments in these public databases can be automatically recognized. For FASTA file that are not from these public databases, UltraPse can also recognize them as long as the comment line of every sequence is unique in the FASTA file. Besides the FASTA format requirements, there is no additional restriction on input data format. As indicated in Table 2, this is a unique advantage of UltraPse.

According to Chou's five step rule [12,21,57–60], before converting biological sequences into numerical vectors, a high-quality benchmark dataset must be constructed. The construction of a dataset usually includes a step to filter out the sequences containing non-standard letters. For example, B, J, or X appear in protein sequences in the UniProt database. However, the sequences containing these letters are hardly suitable for further analysis in many cases. As indicated in Table 2, UltraPse provides a user-controllable data fault tolerant ability. According to users' choice, when one of these sequences is encountered, UltraPse can automatically skip the sequence or abort all further computations. This function is useful in adopting third-party datasets in practical works, because filtering out the sequences usually requires tedious programming work.

Table 2. Software function comparison in terms of data processing ability.

Software	Output Formats	Input Formats	Data Fault Tolerant [a]
UltraPse	SVM [b], TSV [c], CSV [d]	Multi-line FASTA (Automatic ID recognition for UniProt, GenBank, EMBL, DDBJ and RefSeq)	User-controllable behavior on data faults
PseAAC-General [46]	SVM, TSV, CSV	Single-line FASTA (With restrictions on comment line) [e]	Automatically ignore and report data faults
PseAAC-Builder [45]	SVM, TSV, CSV	Single-line FASTA (With restrictions on comment line)	Automatically ignore and report data faults
Pse-In-One [52]	SVM, TSV, CSV	Mutlti-line FASTA	Abort processing on data faults
PseKNC [48]	SVM, TSV, CSV	Mutlti-line FASTA	Abort processing on data faults
PseKNC-General [49]	SVM, TSV, CSV	Mutlti-line FASTA	Abort processing on data faults

[a] Data fault tolerant: The behavior of a software when it encounters some invalid data records. Here, the invalid data records include the sequences with non-standard letter and the sequence without sufficient length; [b] SVM: data format for libSVM [61]; [c] TSV: tab separated vector; [d] CSV: comma separated vector; [e] Single-line FASTA: the sequence of a record in the file must not spread to multiple lines. Both PseAAC-General and PseAAC-Builder have the same restrictions.

2.4. Technical Detail Comparison

Most state-of-the art software is written in Python, while PseAAC-General and UltraPse are written in C++. This difference eventually made the difference in computational efficiency. Since the computational efficiencies of PseAAC-General and UltraPse are comparable, we can compare several technical details of them.

PseAAC-General is a program that can be extended by using Binary Extension Modules (BEMs). However, it should be noted that, the BEMs of PseAAC-General are completely different to the BSOs in UltraPse. A BEM of PseAAC-General is just a compressed data block. However, how this data block should be used, was still implemented by the PseAAC-General main program. In the UltraPse, a BSO is actually a dynamically loaded library, which contains all the information and instructions for constructing one or more sequence representation modes. Therefore, the BSOs of UltraPse are much more flexible than the BEMs of PseAAC-General.

We have seen that UltraPse has roughly 1.5 times the efficiency of PseAAC-General. This advantage is achieved by an internal representation scheme and a pre-computing mechanism of UltraPse. In PseAAC-General, the sequences are converted to a series of physicochemical properties. The sequence descriptors are then computed according to the corresponding algorithms. However, this intuitive implementation requires repeatedly computing dot-product or Euclidean distance between physicochemical vectors of different amino acids. Since the combination of two different amino acids is limited, we pre-compute the dot-product and Euclidean distance for all possible combinations in UltraPse. The sequences in UltraPse are not converted into a series of physicochemical properties. They are converted into UltraPse internal indices, which can be used to quickly find correct values that have been pre-computed. When computing only the amino acids compositions, the implementations of PseAAC-General and UltraPse are similar. However, UltraPse still benefits from converting all sequences into internal indices first. Because, the amino acids counting procedure becomes simpler, this allows the compiler to do more optimization for speed. This is why UltraPse is faster than PseAAC-General.

2.5. Future Works in Plan

Besides the practical application of UltraPse program in research projects, there is still much work to do in terms of software development. The work at first priority is to add an automated unit-testing facility in the source code of UltraPse. Unit-testing is good practice in software engineering to ensure robustness of large scale software. It will be very important for the future versions of UltraPse. The next work in plan is to enable UltraPse support more data formats as input files. As far as we can tell, no existing program in representing biological sequences can handle file formats other than FASTA. We will make the next version of UltraPse handle FASTA, FASTQ, and several other formats of input file.

2.6. Availability

The UltraPse software is provided as source codes and binary packages. All the source codes can be downloaded from the GitHub repository (Available online: https://github.com/pufengdu/UltraPse). The binary distribution packages can also be downloaded from the Release sub-directory in the GitHub repository. Currently, there are binary packages for Windows and Linux platforms. The Windows binary program can be executed directly. The Linux binary package has been tested on a freshly installed Ubuntu Linux Server 16.04.3.

3. Methods

3.1. Efficiency Comparison Protocols

We performed computational efficiency comparisons on a server with an Intel Xeon X3470 processor and 32 GB memory. To perform a fair comparison, we installed Pse-In-One locally on the server. We also locally compiled and installed PseAAC-General and UltraPse on the same server.

The testing dataset is the "huge" testing dataset that can be obtained from the official website of PseAAC-General. This dataset contains 516,081 protein sequences. Since the Pse-In-One keeps complaining about non-standard letters and too short sequences in the dataset, we excluded all the sequences that have non-standard letters. The remaining 513,536 protein sequences were fed into three programs independently. All three programs are configured to compute only amino acid compositions. The computational times are measured by the "real" time value of the standard Linux time command. To eliminate random errors, every program was executed consecutively with exactly the same configuration three times. The average computational time was used in calculating computational efficiency.

3.2. Abstracted Software Design

We illustrate the internal structure and the data-flows of UltraPse in Figure 3. There are four major parts within UltraPse. They are the FASTA parser, sequence preprocessor, computing engine, and the result writer. The FASTA parser is responsible for loading FASTA format sequences into the memory from a hard drive. It also organizes the sequences according to their identifiers and their sequence types. These sequences are then sent to the sequence preprocessor, where the sequences are converted to UltraPse internal indices according to the sequence type definitions. The computing engine is composed of several mode modules, which are configured according to user requirements. The internal indices go through all mode modules. Eventually, sequence descriptors are generated. The result writer exports these descriptors on the hard drive according to the format requirements.

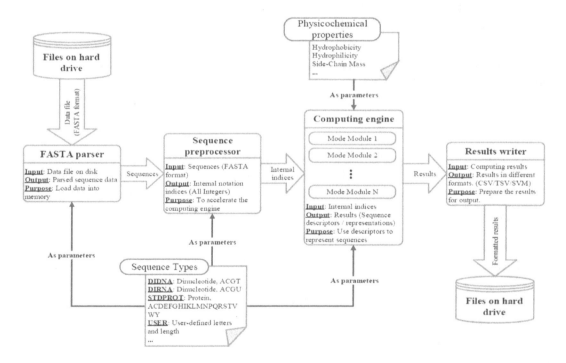

Figure 3. The abstracted software design and data flow chart of UltraPse.

3.3. Implementation Technology

The UltraPse main program is written using standard C++ language, following C++14 standard. The destination hardware architecture is x86-64. The dependencies of the UltraPse main program include GNU standard C library and the embedded interpreter of Lua scripting language. The BSOs of UltraPse are also written using C++, following the same rules as the main program.

On the Linux platform, the compiler for producing binary executables is the GNU g++ version 7.2. The users should first install Lua scripting language. The configuration and compilation of UltraPse need the library provided by the Lua package. On the Windows platform, the MinGW64 version g++ compiler is applied. Several independent libraries are required to compile the codes. For the convenience of Windows users, we provide a binary executable package for the Windows platform.

The TDFs are provided as platform-independent Lua scripts, which can be viewed, edited, and loaded as their original form. The internal data structures of UltraPse can be accessed by Lua scripts using UltraPse specific functions and interfaces. The details on how to write TDFs can be found in the software manual.

3.4. A Practical Example

Figure 4 demonstrate a practical example. The classic pseudo-amino acid composition modes, including type-I and type-II, are implemented using a TDF in UltraPse. The TDF for classic pseudo-amino acid compositions can be found in the "tdfs" subdirectory of UltraPse. The right part of Figure 4 is a part of this TDF. With this TDF, the users only need to specify some parameters on the command line. For example, the "–l 10 –w 0.05" on the command line indicate the value of λ and ω in the PseAAC formulations. Unlike PseAAC-General, where the meanings of all command line options are fixed, the meanings of command line options can be altered by the TDFs in the UltraPse. This is to simplify the development of novel sequence representation modes, where parameters are required to perform correct and efficient computations.

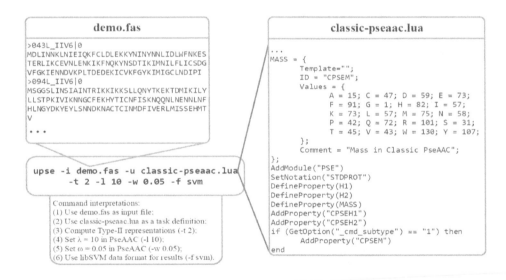

Figure 4. An example on using UltraPse. UltraPse was used to implement classic pseudo-amino acid compositions. A TDF: classic-pseaac.lua, was applied. The FASTA format sequences are stored in the demo.fas file. The command options indicate that the Type 2 PseAAC will be computed with parameters: λ = 10 and ω = 0.05. The output format is compatible to libSVM.

4. Conclusions

In this paper, we described our new software, the UltraPse (Available online: https://github.com/pufengdu/UltraPse). UltraPse is a universal and extensible software platform for generating biological sequence representations. Since many programs have already been released for various sequence representations, UltraPse has no intention to be a new competitor on the same playground. We expect that UltraPse can work side-by-side with other existing programs, such as PseAAC-General, PseAAC-Builder and Pse-In-One, to accelerate the process of generating sequence representations under various working environments.

Although we have integrated many existing sequence modes within the UltraPse, it should be noted that the major advantage of UltraPse is its flexibility and extensibility. It was designed to be a software platform rather than a program with specific functions. It aims at simplifying all future programming works in developing novel sequence representations.

Web servers have already been proved to be a good method in releasing software. However, presenting UltraPse with a web server will severely damage its computational efficiency. Therefore, we do not provide an online web server for UltraPse. We would rather provide it as a local program. The users need to compile and install it on their own servers. The graphical user interfaces (GUI) is useful on platforms like Microsoft Windows. We will develop a GUI for UltraPse on the Windows platform in future.

Acknowledgments: This work is funded by the National Natural Science Foundation of China (NSFC 31401132); National Natural Science Foundation of China (NSFC 61005041); Tianjin Natural Science Foundation (No. 12JCQNJC02300).

Author Contributions: Pu-Feng Du designed the software, wrote most of the codes and the paper in part. Wei Zhao partially wrote the codes and in part tested the software. Yang-Yang Miao partially tested the software and in part wrote the paper. Le-Yi Wei partially wrote the paper. Likun Wang provided technical discussions, and in part wrote the code and the paper.

Abbreviations

AmPseAAC	Amphiphilic pseudo amino acid composition
BEM	Binary extension module
BSO	Binary shared objects
GPL	General public license
GUI	Graphical user interfaces
PseAAC	Pseudo-amino acid composition
PseDNC	Pseudo-dinucleotide composition
PseKNC	Pseudo-k nucleotide composition
TDF	Task definition file

References

1. Jiao, Y.-S.; Du, P.-F. Predicting Golgi-resident protein types using pseudo amino acid compositions: Approaches with positional specific physicochemical properties. *J. Theor. Biol.* **2016**, *391*, 35–42. [CrossRef] [PubMed]
2. Jiao, Y.-S.; Du, P.-F. Predicting protein submitochondrial locations by incorporating the positional-specific physicochemical properties into Chou's general pseudo-amino acid compositions. *J. Theor. Biol.* **2017**, *416*, 81–87. [CrossRef] [PubMed]
3. Nanni, L.; Brahnam, S.; Lumini, A. High performance set of PseAAC and sequence based descriptors for protein classification. *J. Theor. Biol.* **2010**, *266*, 1–10. [CrossRef] [PubMed]
4. Nanni, L.; Brahnam, S.; Lumini, A. Prediction of protein structure classes by incorporating different protein descriptors into general Chou's pseudo amino acid composition. *J. Theor. Biol.* **2014**, *360*, 109–116. [CrossRef] [PubMed]
5. Li, L.; Yu, S.; Xiao, W.; Li, Y.; Hu, W.; Huang, L.; Zheng, X.; Zhou, S.; Yang, H. Protein submitochondrial localization from integrated sequence representation and SVM-based backward feature extraction. *Mol. Biosyst.* **2014**, *11*, 170–177. [CrossRef] [PubMed]
6. Lin, H.; Chen, W.; Yuan, L.-F.; Li, Z.-Q.; Ding, H. Using Over-Represented Tetrapeptides to Predict Protein Submitochondria Locations. *Acta Biotheor* **2013**, *61*, 259–268. [CrossRef] [PubMed]
7. Zuo, Y.-C.; Peng, Y.; Liu, L.; Chen, W.; Yang, L.; Fan, G.-L. Predicting peroxidase subcellular location by hybridizing different descriptors of Chou' pseudo amino acid patterns. *Anal. Biochem.* **2014**, *458*, 14–19. [CrossRef] [PubMed]
8. Nanni, L.; Lumini, A.; Gupta, D.; Garg, A. Identifying Bacterial Virulent Proteins by Fusing a Set of Classifiers Based on Variants of Chou's Pseudo Amino Acid Composition and on Evolutionary Information. *IEEE-ACM Trans. Comput. Biol. Bioinform.* **2012**, *9*, 467–475. [CrossRef] [PubMed]

9. Chou, K.-C. Prediction of protein cellular attributes using pseudo-amino acid composition. *Proteins* **2001**, *43*, 246–255. [CrossRef] [PubMed]

10. Chou, K.-C. Using amphiphilic pseudo amino acid composition to predict enzyme subfamily classes. *Bioinformatics* **2005**, *21*, 10–19. [CrossRef] [PubMed]

11. Chou, K.-C. Pseudo Amino Acid Composition and its Applications in Bioinformatics, Proteomics and System Biology. *Curr. Proteom.* **2009**, *6*, 262–274. [CrossRef]

12. Qiu, W.-R.; Sun, B.-Q.; Xiao, X.; Xu, Z.-C.; Chou, K.-C. iHyd-PseCp: Identify hydroxyproline and hydroxylysine in proteins by incorporating sequence-coupled effects into general PseAAC. *Oncotarget* **2016**, *7*, 44310–44321. [CrossRef] [PubMed]

13. Xu, Y.; Ding, Y.-X.; Ding, J.; Wu, L.-Y.; Deng, N.-Y. Phogly–PseAAC: Prediction of lysine phosphoglycerylation in proteins incorporating with position-specific propensity. *J. Theor. Biol.* **2015**, *379*, 10–15. [CrossRef] [PubMed]

14. Jia, J.; Zhang, L.; Liu, Z.; Xiao, X.; Chou, K.-C. pSumo-CD: Predicting sumoylation sites in proteins with covariance discriminant algorithm by incorporating sequence-coupled effects into general PseAAC. *Bioinformatics* **2016**, *32*, 3133–3141. [CrossRef] [PubMed]

15. Ahmad, K.; Waris, M.; Hayat, M. Prediction of Protein Submitochondrial Locations by Incorporating Dipeptide Composition into Chou's General Pseudo Amino Acid Composition. *J. Membr. Biol.* **2016**, *249*, 293–304. [CrossRef] [PubMed]

16. Feng, P.-M.; Chen, W.; Lin, H.; Chou, K.-C. iHSP-PseRAAAC: Identifying the heat shock protein families using pseudo reduced amino acid alphabet composition. *Anal. Biochem.* **2013**, *442*, 118–125. [CrossRef] [PubMed]

17. Lin, W.-Z.; Fang, J.-A.; Xiao, X.; Chou, K.-C. iLoc-Animal: A multi-label learning classifier for predicting subcellular localization of animal proteins. *Mol. Biosyst.* **2013**, *9*, 634–644. [CrossRef] [PubMed]

18. Mohabatkar, H.; Mohammad Beigi, M.; Esmaeili, A. Prediction of GABAA receptor proteins using the concept of Chou's pseudo-amino acid composition and support vector machine. *J. Theor. Biol.* **2011**, *281*, 18–23. [CrossRef] [PubMed]

19. Jiao, Y.-S.; Du, P.-F. Prediction of Golgi-resident protein types using general form of Chou's pseudo-amino acid compositions: Approaches with minimal redundancy maximal relevance feature selection. *J. Theor. Biol.* **2016**, *402*, 38–44. [CrossRef] [PubMed]

20. Du, P.; Wang, L. Predicting human protein subcellular locations by the ensemble of multiple predictors via protein-protein interaction network with edge clustering coefficients. *PLoS ONE* **2014**, *9*, e86879. [CrossRef] [PubMed]

21. Chou, K.-C. Some remarks on protein attribute prediction and pseudo amino acid composition. *J. Theor. Biol.* **2011**, *273*, 236–247. [CrossRef] [PubMed]

22. Chou, K.-C. Some remarks on predicting multi-label attributes in molecular biosystems. *Mol. Biosyst.* **2013**, *9*, 1092–1100. [CrossRef] [PubMed]

23. Chou, K.-C. Impacts of bioinformatics to medicinal chemistry. *Med. Chem.* **2015**, *11*, 218–234. [CrossRef] [PubMed]

24. Chen, W.; Lin, H.; Chou, K.-C. Pseudo nucleotide composition or PseKNC: An effective formulation for analyzing genomic sequences. *Mol. Biosyst.* **2015**, *11*, 2620–2634. [CrossRef] [PubMed]

25. Chen, W.; Feng, P.-M.; Lin, H.; Chou, K.-C. iRSpot-PseDNC: Identify recombination spots with pseudo dinucleotide composition. *Nucleic. Acids Res.* **2013**, *41*, e68. [CrossRef] [PubMed]

26. Chen, W.; Feng, P.-M.; Lin, H.; Chou, K.-C. iSS-PseDNC: Identifying splicing sites using pseudo dinucleotide composition. *Biomed. Res. Int.* **2014**, *2014*, 623149. [CrossRef] [PubMed]

27. Chen, W.; Feng, P.-M.; Deng, E.-Z.; Lin, H.; Chou, K.-C. iTIS-PseTNC: A sequence-based predictor for identifying translation initiation site in human genes using pseudo trinucleotide composition. *Anal. Biochem.* **2014**, *462*, 76–83. [CrossRef] [PubMed]

28. Guo, S.-H.; Deng, E.-Z.; Xu, L.-Q.; Ding, H.; Lin, H.; Chen, W.; Chou, K.-C. iNuc-PseKNC: A sequence-based predictor for predicting nucleosome positioning in genomes with pseudo k-tuple nucleotide composition. *Bioinformatics* **2014**, *30*, 1522–1529. [CrossRef] [PubMed]

29. Lin, H.; Deng, E.-Z.; Ding, H.; Chen, W.; Chou, K.-C. iPro54-PseKNC: A sequence-based predictor for identifying sigma-54 promoters in prokaryote with pseudo k-tuple nucleotide composition. *Nucleic Acids Res.* **2014**, *42*, 12961–12972. [CrossRef] [PubMed]

30. Chang, C.-C.; Lin, C.-J.; Chen, W.; Feng, P.; Ding, H.; Lin, H.; Chou, K.-C. iRNA-Methyl: Identifying N^6-methyladenosine sites using pseudo nucleotide composition. *Anal. Biochem.* **2015**, *490*, 26–33. [CrossRef]

31. Liu, B.; Fang, L.; Liu, F.; Wang, X.; Chou, K.-C. iMiRNA-PseDPC: MicroRNA precursor identification with a pseudo distance-pair composition approach. *J. Biomol. Struct. Dyn.* **2016**, *34*, 223–235. [CrossRef] [PubMed]

32. Chen, W.; Tang, H.; Ye, J.; Lin, H.; Chou, K.-C. iRNA-PseU: Identifying RNA pseudouridine sites. *Mol. Ther. Nucleic Acids* **2016**, *5*, e332. [CrossRef] [PubMed]

33. Liu, B.; Long, R.; Chou, K.-C. iDHS-EL: Identifying DNase I hypersensitive sites by fusing three different modes of pseudo nucleotide composition into an ensemble learning framework. *Bioinformatics* **2016**, *32*, 2411–2418. [CrossRef] [PubMed]

34. Liu, B.; Yang, F.; Huang, D.-S.; Chou, K.-C. iPromoter-2L: A two-layer predictor for identifying promoters and their types by multi-window-based PseKNC. *Bioinformatics* **2017**. [CrossRef] [PubMed]

35. Iqbal, M.; Hayat, M. "iSS-Hyb-mRMR": Identification of splicing sites using hybrid space of pseudo trinucleotide and pseudo tetranucleotide composition. *Comput. Methods Programs Biomed.* **2016**, *128*, 1–11. [CrossRef] [PubMed]

36. Kabir, M.; Iqbal, M.; Ahmad, S.; Hayat, M. iTIS-PseKNC: Identification of Translation Initiation Site in human genes using pseudo k-tuple nucleotides composition. *Comput. Biol. Med.* **2015**, *66*, 252–257. [CrossRef] [PubMed]

37. Zhang, M.; Sun, J.-W.; Liu, Z.; Ren, M.-W.; Shen, H.-B.; Yu, D.-J. Improving N(6)-methyladenosine site prediction with heuristic selection of nucleotide physical-chemical properties. *Anal. Biochem.* **2016**, *508*, 104–113. [CrossRef] [PubMed]

38. Dong, C.; Yuan, Y.-Z.; Zhang, F.-Z.; Hua, H.-L.; Ye, Y.-N.; Labena, A.A.; Lin, H.; Chen, W.; Guo, F.-B. Combining pseudo dinucleotide composition with the Z curve method to improve the accuracy of predicting DNA elements: A case study in recombination spots. *Mol. Biosyst.* **2016**, *12*, 2893–2900. [CrossRef] [PubMed]

39. Liu, B.; Liu, Y.; Huang, D. Recombination Hotspot/Coldspot Identification Combining Three Different Pseudocomponents via an Ensemble Learning Approach. *Biomed. Res. Int.* **2016**, *2016*, 8527435. [CrossRef] [PubMed]

40. Qiu, W.-R.; Jiang, S.-Y.; Xu, Z.-C.; Xiao, X.; Chou, K.-C. iRNAm5C-PseDNC: Identifying RNA 5-methylcytosine sites by incorporating physical-chemical properties into pseudo dinucleotide composition. *Oncotarget* **2017**, *8*, 41178–41188. [CrossRef] [PubMed]

41. Xu, Z.-C.; Wang, P.; Qiu, W.-R.; Xiao, X. iSS-PC: Identifying Splicing Sites via Physical-Chemical Properties Using Deep Sparse Auto-Encoder. *Sci. Rep.* **2017**, *7*, 8222. [CrossRef] [PubMed]

42. Li, Z.R.; Lin, H.H.; Han, L.Y.; Jiang, L.; Chen, X.; Chen, Y.Z. PROFEAT: A web server for computing structural and physicochemical features of proteins and peptides from amino acid sequence. *Nucleic Acids Res.* **2006**, *34*, W32–W37. [CrossRef] [PubMed]

43. Shen, H.-B.; Chou, K.-C. PseAAC: A flexible web server for generating various kinds of protein pseudo amino acid composition. *Anal. Biochem.* **2008**, *373*, 386–388. [CrossRef] [PubMed]

44. Cao, D.-S.; Xu, Q.-S.; Liang, Y.-Z. Propy: A tool to generate various modes of Chou's PseAAC. *Bioinformatics* **2013**, *29*, 960–962. [CrossRef] [PubMed]

45. Du, P.; Wang, X.; Xu, C.; Gao, Y. PseAAC-Builder: A cross-platform stand-alone program for generating various special Chou's pseudo-amino acid compositions. *Anal. Biochem.* **2012**, *425*, 117–119. [CrossRef] [PubMed]

46. Du, P.; Gu, S.; Jiao, Y. PseAAC-General: Fast building various modes of general form of Chou's pseudo-amino acid composition for large-scale protein datasets. *Int. J. Mol. Sci.* **2014**, *15*, 3495–3506. [CrossRef] [PubMed]

47. Xiao, N.; Cao, D.-S.; Zhu, M.-F.; Xu, Q.-S. protr/ProtrWeb: R package and web server for generating various numerical representation schemes of protein sequences. *Bioinformatics* **2015**, *31*, 1857–1859. [CrossRef] [PubMed]

48. Chen, W.; Lei, T.-Y.; Jin, D.-C.; Lin, H.; Chou, K.-C. PseKNC: A flexible web server for generating pseudo K-tuple nucleotide composition. *Anal. Biochem.* **2014**, *456*, 53–60. [CrossRef] [PubMed]

49. Chen, W.; Zhang, X.; Brooker, J.; Lin, H.; Zhang, L.; Chou, K.-C. PseKNC-General: A cross-platform package for generating various modes of pseudo nucleotide compositions. *Bioinformatics* **2015**, *31*, 119–120. [CrossRef] [PubMed]

50. Liu, B.; Liu, F.; Fang, L.; Wang, X.; Chou, K.-C. repDNA: A Python package to generate various modes of feature vectors for DNA sequences by incorporating user-defined physicochemical properties and sequence-order effects. *Bioinformatics* **2015**, *31*, 1307–1309. [CrossRef] [PubMed]

51. Liu, B.; Liu, F.; Fang, L.; Wang, X.; Chou, K.-C. repRNA: A web server for generating various feature vectors of RNA sequences. *Mol. Genet. Genom.* **2016**, *291*, 473–481. [CrossRef] [PubMed]

52. Liu, B.; Liu, F.; Wang, X.; Chen, J.; Fang, L.; Chou, K.-C. Pse-in-One: A web server for generating various modes of pseudo components of DNA, RNA, and protein sequences. *Nucleic Acids Res.* **2015**, *43*, W65–W71. [CrossRef] [PubMed]

53. Li, T.; Du, P.; Xu, N. Identifying human kinase-specific protein phosphorylation sites by integrating heterogeneous information from various sources. *PLoS ONE* **2010**, *5*, e15411. [CrossRef] [PubMed]

54. Chen, Q.-Y.; Tang, J.; Du, P.-F. Predicting protein lysine phosphoglycerylation sites by hybridizing many sequence based features. *Mol. Biosyst.* **2017**, *13*, 874–882. [CrossRef] [PubMed]

55. Lei, G.-C.; Tang, J.; Du, P.-F. Predicting *S*-sulfenylation Sites Using Physicochemical Properties Differences. *Lett. Org. Chem.* **2017**, *14*, 665–672. [CrossRef]

56. Steinbiss, S.; Gremme, G.; Schärfer, C.; Mader, M.; Kurtz, S. AnnotationSketch: A genome annotation drawing library. *Bioinformatics* **2009**, *25*, 533–534. [CrossRef] [PubMed]

57. Jia, J.; Liu, Z.; Xiao, X.; Liu, B.; Chou, K.-C. iCar-PseCp: Identify carbonylation sites in proteins by Monte Carlo sampling and incorporating sequence coupled effects into general PseAAC. *Oncotarget* **2016**, *7*, 34558–34570. [CrossRef] [PubMed]

58. Qiu, W.-R.; Xiao, X.; Lin, W.-Z.; Chou, K.-C. iMethyl-PseAAC: Identification of protein methylation sites via a pseudo amino acid composition approach. *Biomed. Res. Int.* **2014**, *2014*, 947416. [CrossRef] [PubMed]

59. Liu, B.; Xu, J.; Lan, X.; Xu, R.; Zhou, J.; Wang, X.; Chou, K.-C. iDNA-Prot|dis: Identifying DNA-binding proteins by incorporating amino acid distance-pairs and reduced alphabet profile into the general pseudo amino acid composition. *PLoS ONE* **2014**, *9*, e106691. [CrossRef] [PubMed]

60. Xu, Y.; Wen, X.; Wen, L.-S.; Wu, L.-Y.; Deng, N.-Y.; Chou, K.-C. iNitro-Tyr: Prediction of nitrotyrosine sites in proteins with general pseudo amino acid composition. *PLoS ONE* **2014**, *9*, e105018. [CrossRef] [PubMed]

61. Chang, C.-C.; Lin, C.-J. LIBSVM: A Library for Support Vector Machines. *ACM Trans. Intell. Syst. Technol.* **2011**, *2*, 27. [CrossRef]

12

3D-QSAR and Molecular Docking Studies on the *TcPMCA1*-Mediated Detoxification of Scopoletin and Coumarin Derivatives

Qiu-Li Hou †, Jin-Xiang Luo †, Bing-Chuan Zhang, Gao-Fei Jiang, Wei Ding and Yong-Qiang Zhang *

Laboratory of Natural Products Pesticides, College of Plant Protection, Southwest University, Chongqing 400715, China; houqiuli2000@126.com (Q.-L.H.); xiangxiangnx@sohu.com (J.-X.L.); zhbichting@163.com (B.-C.Z.); Gaofei.Jiang@toulouse.inra.fr (G.-F.J.); dwing818@163.com (W.D.)
* Correspondence: zyqiang@swu.edu.cn
† These authors contributed equally to this work.

Abstract: The carmine spider mite, *Tetranychus cinnabarinus* (Boisduval), is an economically important agricultural pest that is difficult to prevent and control. Scopoletin is a botanical coumarin derivative that targets Ca^{2+}-ATPase to exert a strong acaricidal effect on carmine spider mites. In this study, the full-length cDNA sequence of a plasma membrane Ca^{2+}-ATPase 1 gene (*TcPMCA1*) was cloned. The sequence contains an open reading frame of 3750 bp and encodes a putative protein of 1249 amino acids. The effects of scopoletin on *TcPMCA1* expression were investigated. *TcPMCA1* was significantly upregulated after it was exposed to 10%, 30%, and 50% of the lethal concentration of scopoletin. Homology modeling, molecular docking, and three-dimensional quantitative structure-activity relationships were then studied to explore the relationship between scopoletin structure and *TcPMCA1*-inhibiting activity of scopoletin and other 30 coumarin derivatives. Results showed that scopoletin inserts into the binding cavity and interacts with amino acid residues at the binding site of the *TcPMCA1* protein through the driving forces of hydrogen bonds. Furthermore, CoMFA (comparative molecular field analysis)- and CoMSIA (comparative molecular similarity index analysis)-derived models showed that the steric and H-bond fields of these compounds exert important influences on the activities of the coumarin compounds.Notably, the C3, C6, and C7 positions in the skeletal structure of the coumarins are the most suitable active sites. This work provides insights into the mechanism underlying the interaction of scopoletin with *TcPMCA1*. The present results can improve the understanding on plasma membrane Ca^{2+}-ATPase-mediated (PMCA-mediated) detoxification of scopoletin and coumarin derivatives in *T. cinnabarinus*, as well as provide valuable information for the design of novel PMCA-inhibiting acaricides.

Keywords: *Tetranychus cinnabarinus*; plasma membrane Ca^{2+}-ATPase; scopoletin; coumarin derivatives; molecular docking; three-dimensional quantitative structure activity relationship (3D-QSAR); interaction mechanism

1. Introduction

The plasma membrane Ca^{2+}-ATPase (PMCA) pumps Ca^{2+} out of the cell to maintain cytosolic Ca^{2+} concentration at a level that is compatible with messenger function. The concentration of nerve membrane Ca^{2+} is normally higher in the cytoplasm than that in the extracellular matrix;furthermore, Ca^{2+} is sequestered by sarco/endoplasmic reticulum Ca^{2+} pumps (SERCA) or by Ca^{2+}-binding proteins, or else extruded by Na^+/Ca^{2+} exchangers or PMCAs [1–3]. PMCAs exhibit cell-specific expression patterns and play an essential role in Ca^{2+} homeostasis in various cell types, including sensory neurons [4–7]. The inhibition of PMCAs in rat and fire salamander cilia by specific drugs, such as vanadate or carboxyeosin, suggests that PMCAs play a predominant role in Ca^{2+} clearance [8,9]. In mammals, four genes encode PMCAs [10]. PMCA isoforms 1 and 4 are ubiquitously expressed and considered as housekeeping isoforms, whereas PMCA isoforms 2 and 3 exhibit limited expression in tissues [4–7]. Through quantitative analysis, human PMCA1 is shown to be more abundant than PMCA4 at mRNA and protein levels [11]. Numerous methods, such as transient transfection, the use of stable cell lines, and use of the vaccinia viral vector, are used to advance knowledge on the differential properties of these isoforms [12–14].

The carmine spider mite, *Tetranychus cinnabarinus* (Boisduval), is a global agricultural pest that parasitizes more than 100 plant species, including beans, cotton, eggplants, tomatoes, and peppers. *T. cinnabarinus* infestations significantly reduce the quality and yield of these crops. These mites are difficult to prevent and control given its high fecundity, short developmental duration, small individual size, limited territory, and high inbreeding rate [15,16]. The control and prevention of *T. cinnabarinus* are currently dependent on chemical insecticides and acaricides, such as spiromesifen, pyridaben, and etoxazole, which introduce a high amount of chemical residues to the environment and induce drug resistance in the target species [17]. Therefore, a novel, environmentally friendly acaricidal compound should be identified and developed to manage these problems.

Scopoletin (7-hydroxy-6-methoxychromen-2-one) is an important coumarin phytoalexin found in many herbs [18]. Scopoletin displays a wide array of pharmacological and biochemical activities [19]. In addition, scopoletin exerts insecticidal, acaridal, antibacterial, and allelopathic activities that are useful in agricultural applications [20–22]. A previous study found that scopoletin extracted from *Artemisia annua* L. exhibits strong acaricidal activity against carmine spider mites and inhibits oviposition [22]. Furthermore, many studies on the effects of scopoletin on various protective enzymes in the nervous system of *T. cinnabarinus* indicated that scopoletin inhibits Ca^{2+}-ATPase [23]. Thus, scopoletin is has increasingly attracting interest as a potential botanical acaricide because it is more environmentally friendly compared with chemical and physical agents. However, the interaction between Ca^{2+}-ATPase and scopoletin in *T. cinnabarinus* remains unclear.

The objective of this study is to investigate the PMCA-meditated detoxification mechanism of scopoletin. Molecular docking and three-dimensional quantitative structure activity relationship (3D-QSAR) analyses were performed to achieve this aim. The full-length cDNA that encodes the PMCA 1 gene (*TcPMCA1*) was obtained from *T. cinnabarinus*. The expression profiles of *TcPMCA1* at the various life stages of carmine spider mites were then reported. The effects of scopoletin on *TcPMCA1* expression during the adult stage of *T. cinnabarinus* were also investigated. The results of the molecular docking and 3D-QSAR studies were used to investigate the mechanism underlying the interaction between scopoletin and *TcPMCA1*, as well as the active site of coumarin compounds. This work provides an insight into the detoxification mechanism of scopoletin at the active site for future studies on the optimized structural design of scopoletin and other coumarin derivatives.

2. Results

2.1. Cloning and Sequence Analysis

The partial cDNA sequence that codes for PMCA1 was identified through the use of transcriptome data and alignment with nucleotide sequences from the genome datasets of *Tetranychus urticae* [24]. The remaining 5′ and 3′ ends were amplified through a RACE (rapid amplification of cDNA ends)/PCR (Polymerase Chain Reaction)-based strategy. The full-length cDNA sequence, which was designated as *TcPMCA1*, was deposited in the GenBank database and with the accession number of KP455490. The full-length cDNA of *TcPMCA1* is 4369 bp in length and contains a 3750-bp open reading frame (ORF), a 456-bp 5′-untranslated region (UTR), and a 163-bp 3′-UTR with a putative polyadenylation signal upstream of the *poly(A)* (Figure 1). The ORF encodes 1249 amino acid residues with a predicted molecular mass of 137.7 kDa and an isoelectric point of 8.10 (Figure 1).

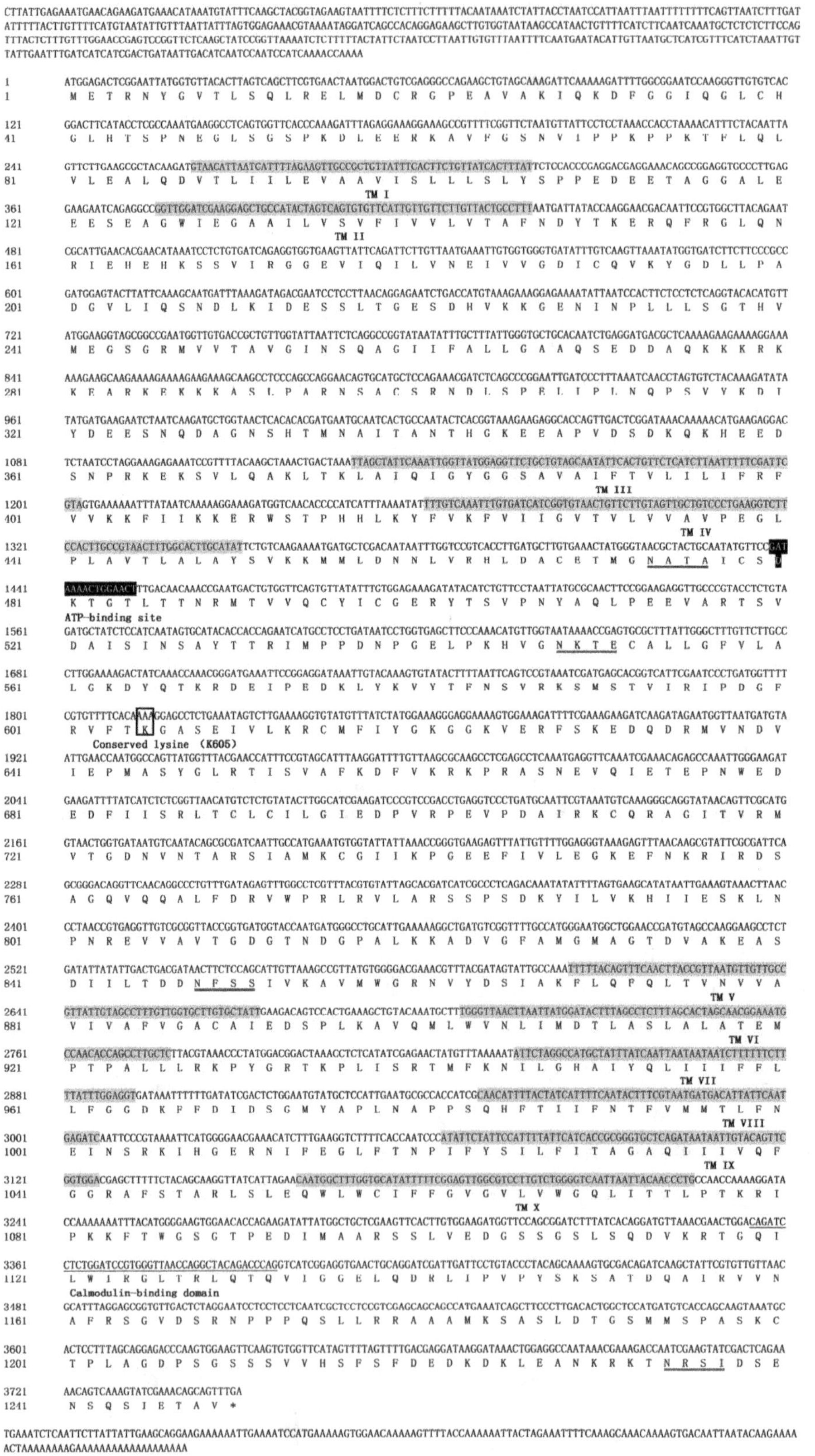

Figure 1. Nucleotide and deduced amino acid sequences of Ca^{2+}-ATPase 1 gene (*TcPMCA1*) from the carmine spider mite (*Tetranychus cinnabarinus* (Boisduval)). Nucleotide numbers are provided on the left. The 10 transmembrane (TM) domains, which are denoted as TM I to TM X, are shaded. The ATP (Adenosine Triphosphate)-binding site, together with phosphorylable aspartate (D480), is shaded black, whereas the conserved lysine (K605) is boxed. The calmodulin-binding domain is indicated by a single line and the four *N*-glycosylation sites are indicated by double lines. * represents the termination signal.

The analysis of the deduced amino acid sequence of *TcPMCA1* revealed the presence of ten membrane-spanning segments (TM), which were denoted as TM I to TM X, as well as four main cytosolic domains located between TM II and TM III, between TM IV and TM V, and at the *N*- and *C*-terminal regions. Some characteristic segments also were predicted. *TcPMCA1* contains an ATP-binding site (from amino acid D480 to T484) and a calmodulin-binding domain (Q1119 to Q1130) (Figure 1).

The multiple protein alignments of the C-terminal conserved catalytic domains of the PMCAs from Arachnida and insects showed that *TcPMCA1* exhibits 99.7% amino acid sequence identity with *T. urticae* PMCA1. *TcPMCA1* also showed nearly 70% similarity with the PMCA genes of *Ixodes scapularis*, and 60–75% similarity with the PMCA genes of insects and nematodes (Figure 2).

```
TcPMCA1    ····PTKRIPKKFTWGSGTPEDIMAARSSLVEDGSSGSLSQDVKRTGQILWIRGLTRLQTQVIG
TuPMCA1    ····PTKRIPKKFTWGSGTPEDIMAARSSLVEDGSSGSLSQDVKRTGQILWIRGLTRLQTQVIG
ApPMCA1    ····PTRKIPKLLSWGRGHPEEYTNAIN-LGEENRYDPDSGQKPRAGQILWIRGLTRLQTQVIG
CbPMCA1    ····PTGSLPANMTIGSGEAPTNDPLMPDYEDSDTHE------KRSGQILWVRGLTRLQTQVIG
           **   :*   :: * * .         :..        *:*****:************
```
```
TcPMCA1    GELQDRLIPVPYSKSATDQAIRVVNAFRSGVDSRNPPPQSLLRRAAAMKSASLDTGSMMSPAS
TuPMCA1    GELQDRLIPVPYSKSATDQAIRVVNAFRSGVDSRNPPPQSLLRRAAAMKSASLDTGSMMSPAS
ApPMCA1    GELQERLIPVPYSKSSTDQAIRVVNAFRQGLDARYGESLAEVLRKQPCFSKRLSAATGGGGGS
CbPMCA1    GERSDHLIPVPLSSAPTDQAIRVVKAFQAGLDRREPSLTGQSAARLREISRQLRLQVDSENRA
           ** .::***** *.:.*********:**: *:* *      .      *   *        :
```
```
TcPMCA1    KCTPLAGDPSGSSSVVHSFSFDEDKDKLEANKRKTNRSIDSENSQSIETAV-
TuPMCA1    KCTPLAGDPSGSSSVVHSFSFDEDKDKLEANKRKTNRSIDSENSQSIETAV-
ApPMCA1    SSKTGGGGGTGTGGSIEYADSDNAIGVPHIDVERLSSHSHTETAV------
CbPMCA1    RSTSRG---------------------NIKETNNL-----------
           ··· .                    : .. .
```

Figure 2. ClustalW alignment of the C-terminal sequence comparison of plasma membrane Ca²⁺-ATPase 1 (*PMCA1*) obtained from different species. Alignment of the sequences of the *PMCAs*, starting after the last (10th) putative membrane-spanning domain and ending at the last residue. Residues that are completely conserved are marked with an asterisk (*); those that are highly conserved are indicated by colon (:); while similar residues are indicated by a dot (.). "-" represents interval. *PMCA1* sequences used in the alignment are as follows: *TcPMCA1*, *Tetranychus cinnabarinus*; *TuPMCA1*, *Tetranychus urticae*; *ApPMCA1*, *Acythosiphon pisum*; and *CbPMCA1*, *Caenorhabditis briggsae*. The *PMCA2* sequences used in the alignment are as follows: *TcPMCA2*, *Tetranychus cinnabarinus*; *TuPMCA2*, *Tetranychus urticae*; *IsPMCA*, *Ixodes scapularis*; and *CbPMCA1*, *Caenorhabditis briggsae*.

2.2. Phylogenetic Analysis

A neighbor-joining phylogenetic tree was constructed by comparing the amino acid sequence of *TcPMCA1* with those of PMCA genes from other animal species. Phylogenetic analysis showed that *TcPMCA1* belongs to the cluster of *Ixodes* PMCA. The PMCA genes of *T. cinnabarinus* and *T. urticae* clustered into the PMCA family and apparently share a single clade. These results suggested that the PMCA genes of *T. cinnabarinus* and *T. urticae* are evolutionarily related and share similar physiological functions (Figure 3).

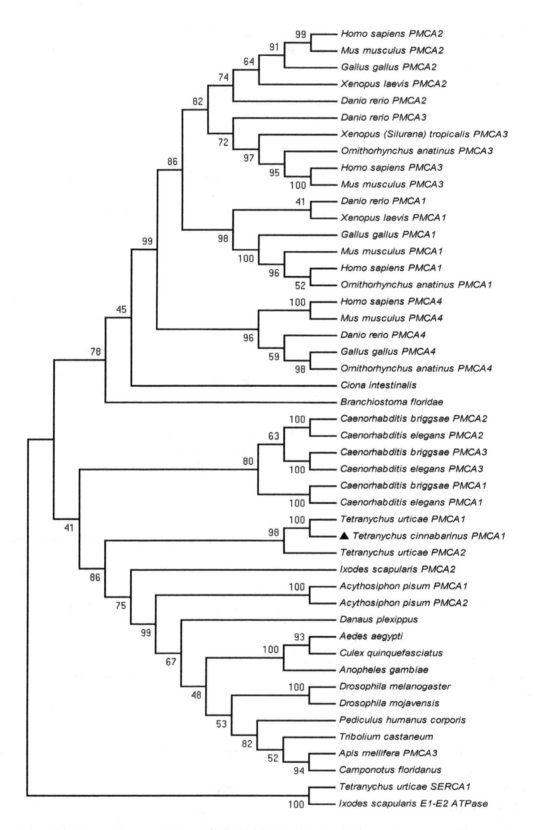

Figure 3. Phylogenetic analysis of *TcPMCA1* obtained from the carmine spider mite (*Tetranychus cinnabarinus* (Boisduval)). The phylogenetic tree was constructed using Molecular Evolutionary Genetics Analysis (MEGA) 5.04 using the neighbor-joining method based on amino acid sequences. *TcPMCA1* was indicated by "▲". Bootstrap support values derived from 1000 replicates are shown on the branches. Sequence accession numbers are given in Electronic Supplementary Material, Table S1.

2.3. Developmental Expression Patterns

To gain insights into the potential role of *TcPMCA1*, the expression levels of *TcPMCA1* in female individuals at various life stages were quantified through Real-time Quantitative polymerase chain reaction (RT-qPCR). The results showed that *TcPMCA1* mRNA was detected at all developmental stages, including the larval, nymphal, and adult stages. More specifically, the *TcPMCA1* transcript was slightly detectable at the egg stage, was highly expressed at the larval, nymphal, and adult stages, and was the highest at the nymphal stage (Figure 4).

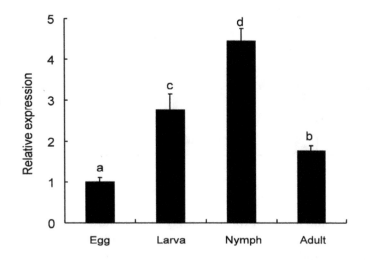

Figure 4. Expression levels of the plasma membrane Ca^{2+}-ATPase 1 gene (*TcPMCA1*) at different developmental stages of *Tetranychus cinnabarinus* were evaluated using Real-time Quantitative polymerase chain reaction (RT-qPCR). The egg, larval, nymphal, and adult stages were analyzed. Relative expression was calculated according to the value of the lowest expression level, which was assigned with an arbitrary value of 1. Letters above the bars indicate significant differences among different developmental stages. *RPS18* was used as reference gene. Data were presented as the means ($\pm$SE) of three biological replications per developmental stage. Different letters on the error bars indicate significant differences revealed by ANOVA test ($p < 0.05$).

2.4. Effects of Scopoletin Exposure on TcPMCA Expression

Scopoletin exposure caused spasms and high mortality among adult *T. cinnabarinus*. The results of induction showed that exposure to scopoletin significantly changed the *TcPMCA1* expression. *TcPMCA1* was significantly upregulated following exposure to low lethal (LC_{10}), sublethal (LC_{30}), and median lethal (LC_{50}) scopoletin concentrations for 12, 24, 36, or 48 h. The relative expression levels of *TcPMCA1* were upregulated by more than 100-fold of that of the control following 24 or 36 h of exposure to scopoletin at LC_{30} dose. However, *TcPMCA1* activation by scopoletin weakened gradually with the extension of time (Figure 5).

2.5. Homology Modeling

Bell Labs Layered Space-Time (BLAST) analysis revealed that the primary sequence of the target enzyme had a high sequence identity of 73% with the template 3BA6. BLAST analysis guarantees that the model structure is of a high quality. Further energy minimization was performed to remove geometric restraints prior to model construction [25]. The homology modeling of *TcPMCA1* is shown in Figure 6. The 3D structure of this enzyme was further checked by Procheck to evaluate the stereo-chemical quality. Ramachandran plot analysis showed that most residues are present at the most favored regions. In particular, 90.3% of the residues were in the most favored regions, 9.0% residues in the additional allowed regions, giving a total of 99.3%. Other 0.4% residues in the generously allowed regions and 0.4% residues in the disallowed regions. The results of the procheck analysis demonstrated that the 3D-modeling structure exhibits reasonable and reliable stereo-chemical properties and is thus appropriate for subsequent molecular docking study.

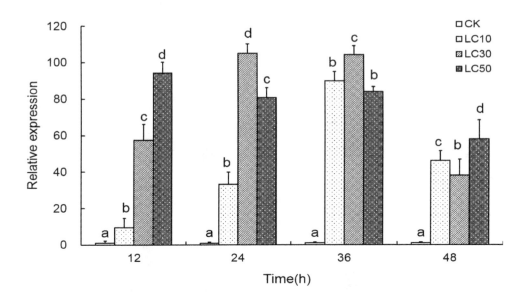

Figure 5. Relative expression levels of the *TcPMCA1* gene in adult female *Tetranychus cinnabarinus* exposed LC_{10} (0.219 mg mL^{-1}), LC_{30} (0.581 mg mL^{-1}), and LC_{50} (1.142 mg mL^{-1}) scopoletin. Expression levels were quantified using qPCR after 12, 24, 36, and 48 h of treatment through leaf-dip bioassay ($n = 3$). Scopoletin was mixed with acetone and Tween-80 (scopoletin: Tween-80 = 3:1; acetone was added until scopoletin dissolved, generally limited within 5%). *T. cinnabarinus* treated with double distilled water containing 0.5% acetone and Tween-80 were used as controls (CK). The mRNA levels in the control and in each treatment were normalized to the expression of the reference gene *RPS18*. The mean expression in each treatment was shown as fold change compared with the mean expression in the control, which was assigned with a basal value of 1. Letters on the error bar indicate significant difference between scopoletin treatment and control ($p < 0.05$).

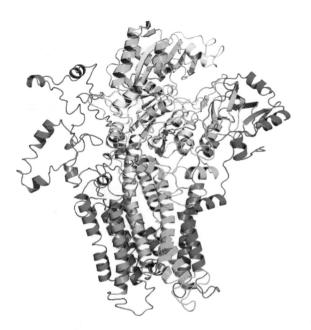

Figure 6. Homology modeling 3D-structure of *TcPMCA1*.

2.6. Molecular Docking

To comprehend the interaction between the ligand scopoletin and TcPMCA1, molecular docking was performed to investigate the binding mode of scopoletin within the binding pocket of TcPMCA1, and to further understand their structure–activity relationship. The ligand structure of scopoletin is shown in Figure 7. The result showed that scopoletin docked with high affinity to the nucleotide-binding pocket of TcPMCA1 and amino acid residues Ser297 and 300, Thr144, Cys299, Glu83, Gln86, Asp87, and Lys301 surrounded scopoletin. Furthermore, five hydrogen bonds (the red dash lines) formed between the 7-hydroxy with Sre297, 6-methoxy with Ala298, oxygen at position 1 with Lys301, and oxygen at position 2 with Lys301 and Ser300 (Figure 8).

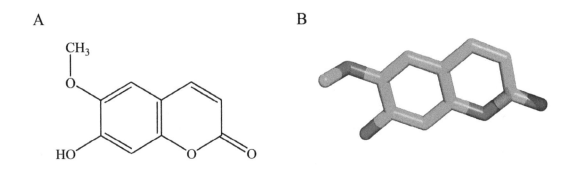

Figure 7. (**A**) Chemical structural formula and (**B**) the cartoon representation of scopoletin. Red regions represent oxygen atoms; green regions represent carbon atoms.

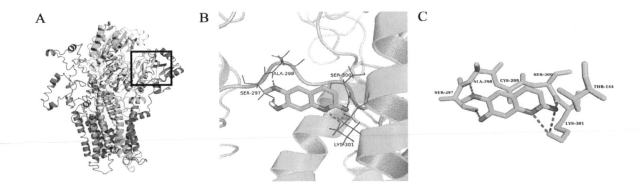

Figure 8. (**A**) Binding pocket of *TcPMCA1* was indicated by the black frame; (**B**) best conformation of scopoletin docked to binding pocket of *TcPMCA1*; (**C**) cartoon representation of residues involved in the binding of scopoletin to *TcPMCA1*. The black box represents the binding cavity. Short, red dashed lines represent hydrogen bonds. Red regions represent oxygen atoms of scopoletin; green regions represent the carbon atoms of scopoletin; the others represent the amino acid residue of the protein.

The 30 coumarin derivatives (Table 1) were also subjected to molecular docking calculations. The derivatives all docked with high affinity to the nucleotide-binding domain (NBD). These results appeared promising and encouraged the calculation of molecular docking at the NBD for all compounds. Defined molecular docking (DMD) at the nucleotide-binding pocket revealed that all compounds showed low binding energy values. The lowest binding energy of -6.03 kcal/mol was exhibited by compound 2 (Table 1). Therefore, compound 2 appears to be the most stable compound.

Table 1. Docking results of coumarins with Ca^{2+}-ATPase 1 gene of *Tetranychus cinnabarinus* (*TcPMCA1*).

Compound	AutoDock				Compound	AutoDock			
	Einter	Eintra	Etors	ΔG		Einter	Eintra	Etors	ΔG
1	−6.87	−0.47	1.19	−5.64	16	−4.64	−0.15	0.3	−4.35
2	−7.22	−0.59	1.19	−6.71	17	−4.77	−0.37	0.6	−5.01
3	−4.55	−0.56	0.3	−4.65	18	−5.95	−0.86	0.89	−5.03
4	−4.95	−0.02	0.3	−5.07	19	−4.84	−1.45	0.89	−4.33
5	−4.65	−0.1	0.3	−4.38	20	−4.67	−1.13	0.6	−4.69
6	−4.95	0.03	0.3	−4.41	21	−4.61	−1.27	0.6	−4.23
7	−6.01	−0.55	0.89	−5.14	22	−4.29	0.02	0.3	−4.32
8	−4.86	−0.09	0.3	−5.04	23	−3.97	0	0	−4.47
9	−6.56	−1.73	0.89	−5.24	24	−5.89	−0.38	0.89	−6.08
10	−4.66	0.03	0.3	−3.83	25	−4.89	−0.25	0.6	−6.1
11	−4.35	−0.06	0.3	−4.59	26	−4.12	0	0	−4.59
12	−4.79	0.01	0.3	−4.58	27	−5.59	−0.59	0.89	−5.25
13	−4.56	−0.26	0.6	−4.84	28	−4.91	−0.11	0.3	−5.13
14	−4.49	0	0	−5.28	29	−4.54	−0.68	0.3	−5.35
15	−4.56	−0.14	0.6	−4.36	30	−4.42	−0.89	0.89	−4.6

1, 3-(2-benzimidazolyl)-7-(diethylamino)coumarin; 2, 3-(2-benzothiazolyl)-7-(diethylamino)coumarin; 3, 3-Aminocoumarin; 4, 3-Acetylcoumarin; 5, 4-Methoxycoumarin; 6, 4-Hydroxycoumarin; 7, 5,7-dihydroxy-4-phenyl coumarin; 8, 6-Nitrocoumarin; 9, 7,8-dihydroxy-4-phenyl coumarin; 10, 7-amino-4-phenyl coumarin; 11, 7-methoxycoumarin(herniarin); 12, 7-mercapto-4-methyl coumarin; 13, 6,7-dimethoxy coumarin(Scoparone); 14, Psoralen; 15, 7-Hydroxy-6-methoxycoumarin(Scopoletin); 16, Xanthotoxin; 17, Pimpinellin; 18, Imperatorin; 19, Fraxetin; 20, Esculetin; 21, Daphnetin; 22, Umbelliferone; 23, Coumarin; 24, Oxypeucedanin; 25, Isopimpinellin; 26, 6-Methylcoumarin; 27, Osthole; 28, Bergapten; 29, Xanthotol; 30, Isofraxidin.

2.7. CoMFA and CoMSIA Statistical Result

The same training (24 compounds) and test sets (six compounds) (Table 2) were used to derive models through CoMFA and CoMSIA. The statistical details were summarized in Table 3. The results showed that the optimal CoMFA model provided a leave-one-out q^2 of 0.75 (>0.5) with an optimal number of principal components (ONC) of 7. A correlation coefficient R^2 of 0.993 with a low standard error of the estimate (SEE) of 0.042, and an F-statistic value of 383.856 were also obtained. In contribution, the CoMFA steric field and electrostatic field contributed 72.6% and 27.4%, respectively. The best CoMSIA model provided a q^2 of 0.71 with an ONC of 6. An R^2 of 0.975 with a low SEE of 0.080 and an F value of 124.834 were obtained. In CoMSIA model, the contributions of the steric, electrostatic, hydrophobic, H-bond donor and acceptor were 14.0%, 33.4%, 23.9%, 19.7% and 9.0%, respectively (Table 3). Based on these field contributions, the steric field is the most important field in the CoMFA model, whereas the electrostatic field is the most important field in the CoMSIA model.

The test set (six compounds) was used to evaluate the predictive accuracy of the CoMFA and CoMSIA models. Table 4 showed the experimentally determined and predicted activitiesand the training and test sets residual values. The residual values obtained by calculating the difference between the predicted and actual pLC_{50} are below one logarithmic unit for all the compounds (Figure 9). Therefore, the predictive abilities of the optimal CoMFA/CoMSIA models are excellent.

Table 2. Structures and acaricidal activities (LC$_{50}$ values) of the compounds tested in this study.

Compound	Structure	LC$_{50}$ (mmol/L)	Compound	Structure	LC$_{50}$ (mmol/L)
1a		1.2175	16a		6.0313
2a		0.8638	17a		5.188
3a		2.971	18a		5.3789
4a		3.52	19a		6.2036
5b		2.2563	20a		12.6973
6b		61.2926	21b		3.8273
7a		22.784	22a		20.0142
8a		3.319	23a		14.1447
9b		5.4987	24a		4.876
10b		14.1318	25a		5.0816
11a		33.8571	26a		15.4398
12b		22.269	27a		1.9186
13a		1.3813	28a		15.1358
14a		25.6564	29a		3.8
15a		6.4698	30a		2.5798

a, Training compounds; b, test set compounds. The others are the same as those in Table 1.

Table 3. Summary of the results obtained from CoMFA (comparative molecular field analysis) and CoMSIA (comparative molecular similarity index analysis) analyses.

Statistical Parameter	CoMFA Model	CoMSIA Model
q^2	0.750	0.710
ONC	7	6
R^2	0.993	0.975
SEE	0.042	0.080
F	383.856	124.834
R^2pred	0.6465	0.931
	Contribution	
Steric	0.726	0.140
Electrostatic	0.274	0.334
Hydrophobic		0.239
H-bond donor		0.197
H-bond acceptor		0.090

Table 4. Observed and predicted activities of the test compounds.

Compound	pLC_{50}	CoMFA		CoMSIA	
		Predicted pLC_{50}	Residual	Predicted pLC_{50}	Residual
1a	2.915	2.868	0.047	2.924	−0.009
2a	3.064	3.097	−0.033	3.021	0.043
3a	2.527	2.514	0.013	1.83	0.697
4a	2.453	2.487	−0.034	2.465	−0.012
5b	2.647	1.651	0.996	1.92	0.727
6b	1.213	2.328	−1.115	1.916	−0.703
7a	1.642	1.394	0.248	1.65	−0.008
8a	2.479	2.894	−0.415	2.493	−0.014
9b	2.260	1.67	0.59	1.917	0.343
10b	1.850	2.097	−0.247	1.857	−0.007
11a	1.470	2.184	−0.714	1.716	−0.246
12b	1.652	2.245	−0.593	1.756	−0.104
13a	2.860	2.258	0.602	2.779	0.081
14a	1.591	2.271	−0.68	1.739	−0.148
15a	2.189	1.84	0.349	2.18	0.009
16a	2.220	1.703	0.517	2.127	0.093
17a	2.285	1.931	0.354	2.344	−0.059
18a	2.269	2.304	−0.035	2.263	0.006
19a	2.207	2.309	−0.102	2.311	−0.104
20a	1.896	1.947	−0.051	1.769	0.127
21b	2.417	2.74	−0.323	2.063	0.354
22a	1.699	1.841	−0.142	1.641	0.058
23a	1.849	2.583	−0.734	1.806	0.043
24a	2.312	2.008	0.304	2.298	0.014
25a	2.294	1.967	0.327	2.265	0.029
26a	1.811	2.122	−0.311	1.765	0.046
27a	2.717	1.759	0.958	2.697	0.02
28a	1.820	1.697	0.123	1.683	0.137
29a	2.420	2.152	0.268	1.832	0.588
30a	2.588	1.681	0.907	3.694	−1.106

a, Training compounds; b, test set compounds. The others are the same as those in Table 1. CoMFA, comparative molecular field analysis; CoMSIA, comparative molecular similarity index analysis; pLC_{50}, −log(LC_{50}).

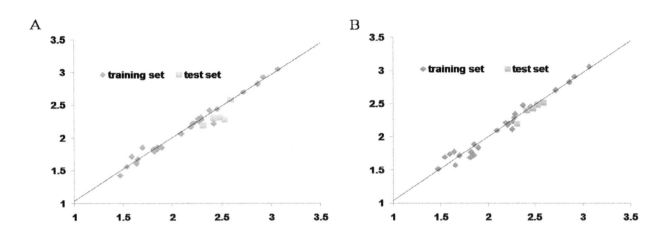

Figure 9. Plots of experimental activity [log $(1/LC_{50})$] against activity as predicted using CoMFA- (**A**) and CoMSIA-derived (**B**) models.

2.8. Contour Maps of CoMFA-Derived Models

Stdev * Coeff contour maps were plotted on the basis of the optimal CoMFA/CoMSIA-derived models. Core structure of these test compounds were shown in Figure 10A. Compound 2 was employed as the template molecule for the analysis of contour maps (Figure 10B) because of it had the highest acaricidal effect and its lowest binding energy among all compounds. Figure 11 presents the steric and electrostatic contour maps for the optimal CoMFA-derived models. The green and yellow contours in the contour maps indicated default 80% and 20% contribution levels, respectively. From Figure 11A, a medium-sized green contour near the R5-position of ring B indicated that inhibitory activity could be improved with a bulky substituent introduced in this region. Correspondingly, other compounds have bulky substituents at this position. Another green contour occurred around the R1-position of ring A, suggesting that inserting a bulky group into ring A increases inhibitory activity. By contrast, a large yellow contour near the R5-position of ring B implied that the introduction of a bulky group at this position negatively affects inhibitory activity. Another large yellow contour around the R2 and R3 positions suggested that inserting a bulky group in these positions decreases inhibitory activity. Indeed, the inhibitory activities of compounds 1–4 (with a group at R1-or R5-position) are higher than that of compound 23 (with an H atom at this position; Table 2).

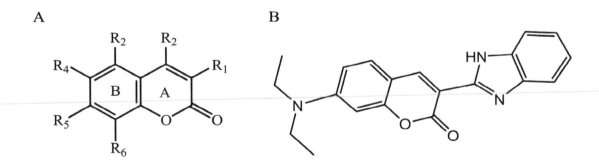

Figure 10. (**A**) Core structure of the test compounds and (**B**) the chemical structure of compound 2.

A   B

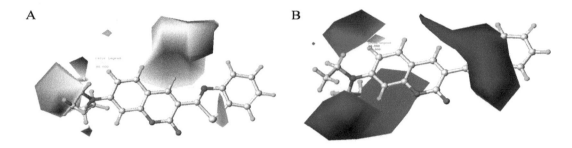

Figure 11. Steric (**A**) and electrostatic (**B**) contour maps obtained using CoMFA-derived models based on molecule 2. Green regions (**A**) indicates regions where the introduction of a bulky group would increase activity. Yellow regions (**A**) indicates regions where the introduction of a bulky group would decrease activity. Red regions (**B**) indicates regions where the introduction of electronegative groups is favored. Blue regions (**B**) indicates regions where the introduction of electropositive groups is favored. The others in Figure A and B represent the compound 2 (Red, oxygen atoms; yellow and blue, nitrogen atom; cyan, hydrogen atom; gray, carbon atoms).

Figure 11B showed the electrostatic contour maps obtained from CoMFA-derived models. Red contour indicates electronegative groups are favored; blue contour indicates electropositive groups are favored. These contours depict default contribution levels. A large blue contour near the R5 and R6 positions of ring B suggested that the introduction of electronegative groups in this position will decrease inhibitory activity. Another large blue contour near the R1-position of ring A indicated that the introduction of electropositive groupsenhances inhibitory activity. A large red contour near the R4-positions of ring B suggested that replacing the original groups with electronegative groups at these positions could improve inhibitory activity. For example, the inhibitory activities of compounds 3 (R1 = $-NH_2$) and 4 (R1 = $-COCH_3$) are greater than that of compound 23 (R1 = $-H$), and the inhibitory activity of compound 8 (R4 = $-NO_2$) is greater than that of compound 23 (R1 = $-H$) (Table 2).

2.9. Contour Maps of CoMSIA-derived Models

The steric, electrostatic, hydrophobic, and H-bond contour maps for the optimal CoMSIA-derived models are shown in Figure 12. Figure 12A,B show the steric and electrostatic contour maps, respectively, which were obtained from the optimal CoMSIA model. The CoMSIA steric and electrostatic contour maps are similar to the corresponding CoMFA contour map. Therefore, the preceding discussion also applies to the steric and electrostatic contour maps from the CoMFA model.

Figure 12C shows the hydrophobic contour map of the CoMSIA model is displayed. In the CoMSIA-derived hydrophobic field, a medium-sized cyan contour near the ring B indicated that introducing hydrophilic groups to that position could improve the inhibitory activity of the molecule. Another two yellow contours around the R1-position of ring A suggested that hydrophobic groups preferentially localize at these positions. Figure 12D shows the H-bond contour map for the optimal CoMSIA model. In this figure, the cyan color indicated regions that favor H-bond donors, whereas the red color indicated regions that disfavor H-bond donors. A medium-sized cyan contour occurred at the 2-position on ring A, thus indicating that the inhibitory activity would be improved with an H-bond acceptor group introduced at this position. A large red contour near the 4-position of ring B implied that introducing an H-bond donor group in this position could decrease inhibitory activity.

The detailed analysis of the contour maps obtained using the optimal CoMFA- and CoMSIA-derived models may facilitate the design of a novel selective TcPMCA1 inhibitors. Introducing an electropositive, hydrophobic, or H-accepting group in region A (R1- and R2-positions of ring A) can increase inhibitory activity, and introducing a hydrophobic group in region B (R3-position of ring B) can increase activity. Meanwhile, introducing an electronegative group in region C (R4-position of ring B) is favorable, and introducing a bulky, hydrophobic, or electropositive group in region D (R5- and R6-position of ring B) can increase activity (Figure 13).

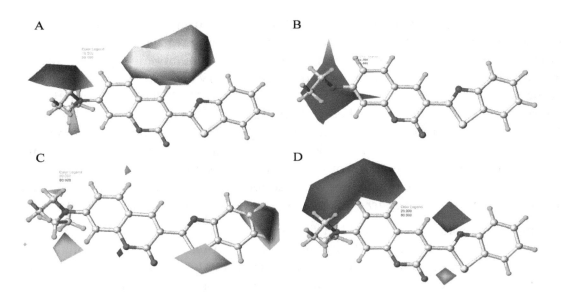

Figure 12. Steric (**A**), electrostatic (**B**), hydrophobic (**C**), and H-bond (**D**) contour maps obtained using CoMSIA-derived models based on molecule 2. Green (**A**) indicates regions where the introduction of a bulky group would increase activity. Yellow (**A**) indicates regions where the introduction of a bulky group would decrease activity. Blue (**B**) indicates regions where the introduction of electropositive groups is favored. Cyan (**C**) indicates regions where the introduction of hydrophobic is favored. Purple (**D**) indicates regions where the introduction of H-bond acceptors is favored. Red (**D**) indicates regions where the introduction of H-bond acceptors is disfavored. The others in Figure **A**–**D** represent the compound 2 (Red, oxygen atoms; yellow and blue, nitrogen atom; cyan, hydrogen atom; gray, carbon atoms).

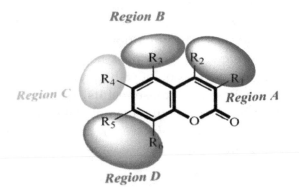

Figure 13. Diagram of structure–activity relationship based on the core structure of the tested compounds. Blue (region A) indicates regions where the introduction of electropositive group, hydrophobic group, or H-accepting groups would increase activity. Red (region B) indicates regions where the introduction of hydrophobic group is favored. Cyan (region C) indicates regions where the introduction of electronegative group is favored. Magenta (region D) indicates regions where the introduction of a bulky group, hydrophobic group, or electropositive group would increase the activity. Dark indicates the core structure of the test compounds.

3. Discussion

Scopoletin is a naturally occurring, low-molecular-weight alloleochemical that is ubiquitous in the plant kingdom. Moreover, scopoletin is present in some foods and plant species used in traditional medicine. Scopoletin extracted from *Artemisia annua* L. exhibits strong activity against the carmine spider mite; in addition, it affects ATPase activity and is possibly a neurotoxin [22].

In the present study, full-length cDNA encoding PMCA1 from *T. cinnabarinus* was characterized and designated as *TcPMCA1*. The predicted amino acid sequences of *TcPMCA1* consists of three major regions: the first intracellular loop region located between transmembrane segments TM II and TM III; the second large intracellular loop region located between TM IV and TM V; which possesses a putative ATP-binding site; the third part extended "tail" found next to TM X. This conformation is consistent with the structure of previously described PMCAs [26–29]. The putative CaM-binding domain of *TcPMCA1* binds to the C-terminal region downstream of the last transmembrane domain and shares a common pattern with those in vertebrates [30]. Alternative splicing expands the diversity of mRNA transcripts and augments the functions of modulatory genes [31]. Previous efforts to discriminate *TcPMCA1* splice variants failed, this failure was also reported in *Spodoptera littoralis* [32]. By contrast, mammals and *Drosophila melanogaster* possess a large number of splice variants [28].

The expression profiles of *TcPMCA1* in *T. cinnabarinus* were similar to that in *S. littoralis*, which is present at all investigated stages and exhibits maximal expression at the nymphal stage [32]. This expression pattern is correlated to the massive synthesis of TcPMCA1 during the developmental stages, thereby confirming that TcPMCA1 is essential for the functions of *T. cinnabarinus*.

The reported pharmacological effects of scopoletin presuppose some interactions with membrane-bound enzymes, such as Ca^{2+}-ATPase, which is vital in nervous signal conduction [33–35]. Oliveira [36] reported that in rats, scopoletin inhibits Ca^{2+}-ATPase activity by inhibiting the mobilization of intracellular calcium from noradrenaline-sensitive Ca stores. Ca^{2+}-ATPase is a major neurotransmitter, and PMCA extrudes Ca^{2+} from the postsynaptic region of the nerve [37]. In insects, PMCA inhibition results in internal Ca^{2+} flow, causing neurotransmitter accumulation [38]. In the present study, the results of scopoletin induction indicated that *TcPMCA1* in *T. cinnabarinus* was significantly upregulated after exposure to scopoletin within 36 h. Scopoletin also increases the expression of both peroxisome proliferator-activated receptor γ2 and adipocyte-specific fatty acid binding protein [39]. Moreover, scopoletin inhibits the expression of cyclooxygenase in a concentration-dependent manner [40]. These results implicated *TcPMCA1* in the detoxification metabolism of scopoletin in *T. cinnabarinus*. The inhibition of Ca^{2+}-ATPase activity or increase in PMCA expression possibly indicates the existence of a feedback regulatory mechanism that compensates for enzyme content. The decrease of gene expression at 48 h may related to the organism damage caused by continuous scopoletin exposing. Basing on these results, we surmise that *TcPMCA1* inhibition in *T. cinnabarinus* causes intra- and extracellular calcium ion imbalance and thus blocks the transmission of neural activity, causing the death of mites [41,42].However, the influence of scopoletin on Ca^{2+}-ATPase mechanism in the carmine spider mite requires extensive exploration because of the intricacy of PMCA-mediated detoxification.

Scopoletin is also designated as 7-hydroxy-6-methoxy coumarin and is a coumarin derivative. Coumarin is a leading molecule in biopesticides. Given the pesticidal potential of this class of compounds, the toxic effects of coumarin derivatives against mosquito species *Culex quinquefasciatus* and *Aedes aegypti* were evaluated, and the results showed that modifying the 7-OH position remarkably enhances the ovicidal activity of coumarin [43]. The antitermiticidal activity of scopoletin and coumarin derivatives were investigated against *Coptotermes formosanus*, and the results suggested that scopoletin has the highest activity among the tested compounds [44]. To investigate the structure–activity relationship of the methoxy and hydroxy groups at the C-6 and C-7 positions of the coumarin skeleton, 6-alkoxycoumarin derivatives and 7-alkoxycoumarins and related analogs were synthesized. The findings indicated that the presence of alkenyloxy and alkynyloxy groups at the C-6 position, as well as the cyclohexyloxy and aryloxy groups at the C-7 position, are important for the termiticidal and antifeedant activities of coumarin [45,46]. These results revealed that scopoletin actually inserts into the binding cavity and interacts with the active sites of TcPMCA1, suggesting that the microenvironments and conformation of the enzymes change because of these interactions [47]. Furthermore, these results indicated that the C-6 and C-7 positions of scopoletin are important for acaricidal activity.

Molecular docking and the homology modeling of the 3D structure of the target protein were used to identify conformational protein–ligand interaction patterns [48,49]. Pharmacophore have been used to develop 3D-QSAR models over the past the decade [50]. Combined information on protein–ligand interactions from a pharmacophore and accurate binding conformations from molecular docking offers the potential for enhanced prediction accuracy [51]. In the present study, the crystallographic structure of sarco/endoplasmic reticulum Ca^{2+}-ATPases (SERCA) was defined in rabbit [52]. The BLAST analysis performed showed that TcPMCA1 shares 73% sequence identity with the SERCA Ca^{2+}-ATPase of rabbit, indicating the validity of homologous protein structure [53,54]. The homologous 3D structure of TcPMCA1 allowed the evaluation of the binding energies and docking positions of scopoletin on TcPMCA1 protein. In our docking results, the hydrophobic environment of the active site is favorable for interactions with scopoletin, and the special arrangements at the C6 and C7 sites are assumed to be favorable for the acaricidal activity of scopoletin. Furthermore, the 3D-CoMFA and CoMSIA models indicating that C3, C6, and C7 regions of coumarins appear to be important acaricidal active sites of coumarins. This result is in agreement with the results of the acaricidal activity assay, which showed that coumarins substituted with methoxy at C6 or C7 have significantly better activity than coumarins substituted with other compounds at the same positions. Furthermore, coumarins with C3 substitutions also demonstrated enhanced acaricidal activity. Nakamura [55] previously investigated the structure–activity relationship between 63 natural oxycoumarin derivatives and their effects on the expression of inducible nitric oxide synthase, which showed that the C-5, C-6 and C-7 positions of oxycoumarin derivatives are essential for potent activities. In addition, the discovery and structure–activity relationship of a novel series of coumarin-based tumor necrosis factor α (TNF-α) inhibitors showed that substitution at the C-3 and C-6 position of the coumarin ring system most dramatically influences inhibitory activity against TNF-α [56]. The docking results and the detailed analysis of the contour maps obtained by 3D-CoMFA and CoMSIA-derived models will encourage the design of novel, selective *TcPMCA* inhibitors.

4. Materials and Methods

4.1. Test Mites

The carmine spider mite culture was collected from cowpea *Vigna unguiculata* (L.) grown in Beibei, Chongqing, China. The mites were maintained on potted cowpea seedlings (30–40 cm tall) in a walk-in insect rearing room at 26 ± 1 °C under 75 to 80% RH and 16L:8D photoperiod. The colony was maintained for more than 12 years without any contact with insecticides/acaricides. The voucher specimens of *T. cinnabarinus* were deposited at the Insect Collection of Southwest University, Chongqing, China.

4.2. Leaf-Dip Bioassay

More than 600 leaf discs were prepared to obtain uniform individuals at different developmental stages. Fresh cowpea leaves that had not been exposed to pesticides were washed thoroughly. Leaf discs with 3 cm diameters were placed on a 4 mm water-saturated sponge in a Petri dish (9 cm in diameter) [57]. Approximately 30 adult females were transferred to each leaf disc, allowed to lay eggs, and removed after 12 h. After a batch of uniform eggs had hatched, the offspring was maintained until the progeny had developed into 3- to 5-d-old females [58].

For the leaf-dip bioassay, female adult mites were treated with scopoletin (provided by Southwest University, Beibei, Chongqing, China). The responses of *TcPMCAs* in mites to scopoletin were investigated by exposing the adult female mites to 10% of the lethal concentration (LC_{10}), LC_{30}, and LC_{50} of scopoletin for 12, 24, 36, and 48 h. The LC_{10} (0.219 mg·mL^{-1}), LC_{30} (0.581 mg·mL^{-1}), and LC_{50} (1.142 mg·mL^{-1}) of *T. cinnabarinus* to scopoletin were determined using leaf-dip bioassays prior to acaricide treatments. Each leaf disc, which contained 30 mites on its surface, was soaked for 5 s in acaricide solutions. For each treatment, more than 500 surviving mites were collected and three biological replicates were performed. A total of 200 mites were dipped in distilled water for 5 s and used as the control. All of the surviving mites were collected and stored at −80 °C for RNA extraction.

4.3. RNA Isolation and Reverse Transcription

Total RNA was isolated using RNeasy® Plus Micro Kit (Qiagen, Hilden, Germany), and genomic DNA was removed using a gDNA elimination column in accordance with the manufacturer's instructions. The quantities of total RNA were assessed at 260 nm using Nanovue UV-Vis spectrophotometer (GE Healthcare, Fairfield, CT, USA). RNA purities were quantified at an absorbance ratio of OD260/280. RNA integrity was evaluated via 1% agarose gel electrophoresis. cDNA was synthesized using total RNA and the rapid amplification of cDNA ends (RACE) method. First-strand cDNA was synthesized from 0.5 µg of RNA in a 10 µL reaction mixture by using PrimeScript® 1st strand cDNA Synthesis Kit (TaKaRa, Dalian, China) and oligo (dT)18 primers. The synthesized samples were then stored at −20 °C.

4.4. Sequencing and Phylogenetic Analysis

To obtain the full-length DNA sequences of *TcPMCA* genes, specific primers were designed using Primer 5.0 (Available online: http://www.premierbiosoft.com/) based on the transcript unigene sequences obtained from the transcriptome (Table S2). A set of gene-specific primers and nested primers were designed to amplify the fragments. The rapid amplification of cDNA ends (RACE) methodwas amplified using the SMARTer™ RACE cDNA Amplification Kit (Clontech, Palo Alto, CA, USA). The total PCR volume was 25 µL and contained 2.5 µL of 10× PCR buffer (Mg^{2+} free), 2.0 µL of dNTPs (2.5 mM), 2.0 µL of Mg^{2+} (2.5 mM), 1 µL of cDNA templates, 1 µL of each primer (10 mM), 0.25 µL of rTaq™ polymerase (TaKaRa), and 15.5 µL of ddH$_2$O. The PCR program was performed as follows: initial denaturation for 3 min at 94 °C, followed by 34 cycles of 94 °C for 30 s, 55 to 60 °C (depending on gene specific primers) for 30 s, and 72 °C extension for 2 min, and final extension for 10 min at 72 °C. The PCR products were separated by agarose gel electrophoresis andpurified using Gel Extraction Mini Kit (Watson Biotechnologies, Shanghai, China). The purified PCR products were ligated into the pGEM-T vector (Promega, Fitchburg, MA, USA) and then sequenced (Invitrogen Life Technologies, Shanghai, China).

BLAST searching was performed using the NCBI BLAST website (Available online: http://www.ncbi.nlm.nih.gov/Blast.cgi). The molecular weight and isoelectric points of the deduced protein sequences were calculated by ExPASy Proteomics Server (Available online: http://cn.expasy.org/tools/pi_tool.html) [59]. The transmembrane domain positions and protein domain were estimated using Phobius (Available online: http://phobius.sbc.su.se/), Calmodulin Target Database (Available online: http://calcium.uhnres.utoronto.ca/ctdb/pub_pages/search/index.htm), and ATPint (Available online: http://www.imtech.res.in/raghava/atpint/submit.html) servers. Signal peptides were predicted using SignalP 3.0 (Available online: http://www.cbs.dtu.dk/service/SignalP/) [60]. N-glycosylation sites were predicted by NetNGlyc 1.0 Server (Available online: http://www.cbs.dtu.dk/services/NetNGlyc/). DNAMAN 6.0 (Lynnon BioSoft, Vaudreuil, QC, Canada) was used to edit *TcPMCA1* nucleotide sequences, and the corresponding phylogenetic trees were constructed using the neighbor-joining method, with 1000 bootstrap replicates, in MEGA5.01 [61].

4.5. Real-Time Quantitative PCR (qPCR)

Primers used for qPCR were designed by Primer 3.0 software [62]. qPCR was performed in 20 µL-reaction mixture that contained 10 µL of qSYBR Green Supermix (BIO-RAD laboratories, Hercules, CA, USA), 1 µL of cDNA template, 1 µL of each primer (0.2 mM) and 7 µL of ddH$_2$O. qPCR was performed on a Stratagene Mx3000P Thermal Cycler (Stratagene, La Jolla, CA, USA) as following protocol: an initial denaturation at 95 °C for 2 min, followed by 40 cycles at 95 °C for 15 s, 60 °C for 30 s, and elongation at 72 °C for 30 s. At the end of each reaction, a melt curve analys (from 60 to 95 °C) was generated to rule out the possibility of primer-dimer formation. *RPS18* was used as a stable housekeeping gene for the qPCR analysis [63]. Relative gene expression levels were calculated by $2^{-\Delta\Delta Ct}$ method [64]. Three biological and two technical replicates were performed.

Expression pattern of TcPMCA1 at different developmental stages. To investigate the expression patterns of *TcPMCA1* at different developmental stages, we collected mites in uniform developmental stages (2000 eggs, 1500 larvae, 1000 nymphs, and 500 adults). The samples were isolated and placed in a 1.5 mL diethyl pyrocarbonate (DEPC)-treated centrifuge tube containing RNA storage reagent (Tiangen, Beijing, China), immediately frozen in liquid nitrogen, and stored at $-80\ ^\circ$C for RNA extraction. Three independent biological replications were performed.

Expression levels of TcPMCA1 after scopoletin exposure. The differential expression levels of *TcPMCA1* in response to scopoletin were investigated by exposing adult female mites to LC_{10}, LC_{30}, and LC_{50} scopoletin, as in leaf bioassays. After 12, 24, and 36 h intervals, only the surviving adults obtained from the treated and control groups (at least 500 larvae) were collected and frozen at $-80\ ^\circ$C for RNA extraction. After scopoletin exposure, total RNA was isolated to analyze the expression levels of *TcPMCA* by TR-qPCR.

4.6. Homology Modeling

The homology modeling was conducted on the I-TASSER server (Available online: http:// zhanglab.ccmb.med.umich.edu/I-TASSER/) [65], and the 3D structure of TcPMCA1 protein was obtained. The details of I-TASSER protocol have been described previously [66–70]. Briefly, it consists of three steps: template identification, full-length structure assembly and structure-based function annotation. Firstly, starting from the query sequence, I-TASSER identifies homologous structure templates from the PDB library [71] using LOMETS [69,72], a meta-threading program that consists of multiple threading algorithms. Then, the topology of the full-length models is constructed by reassembling the continuously aligned fragment structures excised from the templates, where the structures of the unaligned regions are built from scratch by *ab initio* folding based on replica-exchange Monte Carlo simulations [73]. The low free-energy states are further identified by SPICKER [74]. To refine the structural models, a second round of structure reassembly is conducted starting from the SPICKER clusters. The low free-energy conformations refined by full-atomic simulations using FG-MD [75] and ModRefiner [76]. Finally, the biological functions of the target proteins were derived by matching the I-TASSER models with proteins in the BioLiP library [77–79].

Based on identity with the primary sequence of the target *TcPMCA1*, the crystal structure of the phosphoenzyme intermediate of the rabbit SERCA Ca^{2+}-ATPase (PDB ID code: 3BA6) was retrieved from the Protein Data Bank (PDB, Available online: http://www.rcsb.org/pdb/home/home.do) and used as the template for homology modeling (the amino acid sequences of the trmplate was shown in Figure S1). The Psi/Phi Ramachandran plot obtained from Procheck analysis was used to validate the modeled 3D structure of TcPMCA1 protein [80,81].

4.7. Dataset and Molecular Modeling

The acaricidal activities of the 30 collected compounds (Table S3) were obtained from a previous study [82]. These 30 compounds are natural or synthetic compounds that are readily available to coumarin, which were purchased from Chengdu Aikeda Chemical Reagent Co., Ltd. and Shanghai yuanye Bio-Technology Co., Ltd. The purity of these compounds was more than 98%. The structures and half-maximal inhibitory concentration (LC_{50}) of the compounds are shown in Table 2. These values were transformed into the corresponding pLC_{50} $[-\log(LC_{50})]$ as the expression of inhibitor potency. The 30 compounds were placed in a training set of 24 compounds (80%) and a test set of 6 compounds (20%).

The 3D structures of these ligand compounds were constructed in Sybyl 6.9 (Tripos Software, St. Louis, MO, USA). Structures were energy minimized by using the Gasteiger–Hückel charge [83], Tripos force field [84], and Powell methods [85] with a convergence criterion of 0.005 kcal/(mol Å). The iterations maximum number was set to 10,000, and multiple conformation search was used. Coumarin structure was used as the common scaffold for molecular alignment, and compound 2 with the highest acaricidal activity was used as the template molecule. All other compounds were aligned with the coumarin core using the "align database" command in Sybyl.

4.8. Molecular Docking

The protein model was prepared using Sybyl prior to docking simulations. All bound water molecules and ligands were removed from the protein, and hydrogen atoms and AM1-BCC charges [86] were added to the amino acid residues. The generated homology model of TcPMCA1 was used for molecular docking, and the binding pocket was defined using Discovery Studio 2.5 (Accelrys Software Inc., San Diego, CA, USA). The 3D structure of the compound was prepared as the ligand, and all of the hydrogen atoms and AM1-BCC charges were added [86]. Molecular docking was performed with AutoDock 4.0 [87]. The grid spacing was changed from 0.375 nm, and the cubic grid map was $40 \times 40 \times 40$ Å toward the TcPMCA binding site. The docking parameters were set as follows: the number of GA Runs was set as 10, population size was set as 150, the maximum number of evaluations was set as 25,000,000, and 250 runs were performed. All other parameters were set as the default. The docking process was performed as follows: first, molecular docking was performed to evaluate the docking poses. Then, defined docking was conducted on the binding pocket. Three to six independent docking calculations were conducted. The corresponding lowest binding energies and predicted inhibition constants (pK_i) were obtained from the docking log files (dlg). The mean $\pm$ SD of binding energies was calculated from the dockings. AutoDock Tools and Visual Molecular Dynamics (VMD, Theoretical and Computational Biophysics group at the Beckman Institute, University of Illinois at Urbana-Champaign) [88,89] was used to visualize the docking result. Surface representation images that show the binding pocket of TcPMCA1 were generated using VMD software.

4.9. 3D-QSAR Study

CoMFA and CoMSIA descriptor fields were employed in the present 3D-QSAR studies. The CoMFA fields were carried out to generate the steric and electrostatic fields with the default value of the energy cutoff at 30 kcal·mol^{-1} CoMSIA fields were carried out to calculate the steric, electrostatic, hydrophobic, hydrogen-bond donor and hydrogen-acceptor donor with a default attenuation factor of 0.3 for Gaussian function. Field type "Stdev * Coeff" was used as the coefficient to analysis the contour map of each field. The partial least squares (PLS) [90] was used to construct a linear correlation by setting the biological activity (pLC_{50} values) as the dependent variables and the CoMFA/CoMSIA descriptors as independent variables.

4.10. Statistical Analysis

All results were expressed as the mean $\pm$ standard error. The differences among the four developmental stages and time-dependent responses to scopoletin exposure were analyzed using one-way analysis of variance (ANOVA). The level of significance of the means was then separated by Fisher's LSD multiple comparison test ($p < 0.05$). The fold change in TcPMCA gene expression was analyzed using SPSS (v.16.0, SPSS Inc., Chicago, IL, USA), and significance was determined by independent sample t-test ($p < 0.05$).

5. Conclusions

The molecular characteristics of the TcPMCA1 gene were identified and described, and the gene expression levels of TcPMCA1 after scopoletin exposure were investigated. The TcPMCA1-mediated detoxification mechanism of scopoletin in T. cinnabarinus was preliminarily explored through the integrated study of homology modeling and molecular docking. Moreover, CoMFA and CoMSIA 3D-QSAR studies have been performed to put the pharmacophoric environment that will help future structure based drug design. The results of the present study showed that scopoletin forms hydrogen bonds with the active site of TcPMCA1, and that the C3, C6, and C7 positions in the skeletal structure of coumarins are the most suitable active sites. These results provide a better understanding of the *TcPMCA1*-mediated detoxification mechanisms of scopoletin and of other coumarin derivatives. These compounds can be structurally modified to increase their acaricidal and inhibitory effects. More detailed investigations of the mechanism of action and pharmacological activities of these compounds may provide novel anti-PMCA agents for spider mite control.

Acknowledgments: We are grateful to Yuwei Wang in School of Pharmacy, Lanzhou University for Molecular Docking and 3D-QSAR analysis. This research was partially supported by a combination of funding from the National Science Foundation of China (31272058, 31572041 and 31601674) and Chongqing social undertakings and people's livelihood guarantee scientific and technological innovation (cstc2015shms-ztzx0129).

Author Contributions: Qiu-Li Hou, Jin-Xiang Luo, Bing-Chuan Zhang and Yong-Qiang Zhang conceived and designed the experiments; Qiu-Li Hou, Yong-Qiang Zhang, Bing-Chuan Zhang, Jin-Xiang Luo, and Gao-Fei Jiang performed the experiments and analyzed the data; Qiu-Li Hou and Yong-Qiang Zhang wrote the paper; Wei Ding, Jin-Xiang Luo and Yong-Qiang Zhang revised the paper.

References

1. Glynn, I.M. *The Enzymes of Biological Membranes*; Martonosi, A.N., Ed.; New York Press: New York, NY, USA, 1985; pp. 35–114.

2. Carafoli, E. The calcium pumping ATPase of the plasma membrane. *Annu. Rev. Physiol.* **1991**, *53*, 531–547. [CrossRef] [PubMed]

3. Penniston, J.T.; Enyedi, A. Modulation of the plasma membrane Ca^{2+} pump. *J. Membr. Biol.* **1998**, *165*, 101–109. [CrossRef]

4. Krizaj, D.; Steven, J.D.; Johnson, J.; Strehler, E.E.; Copenhagen, D.R. Cell-specific expression of plasma membrane calcium ATPase isoforms in retinal neurons. *J. Comp. Neurol.* **2002**, *451*, 1–21. [CrossRef] [PubMed]

5. Ii, W.J.P.; Thayer, S.A. Transient rise in intracellular calcium produces a long-lasting increase in plasma membrane calcium pump activity in rat sensory neurons. *J. Neurochem.* **2002**, *83*, 1002–1008.

6. Zenisek, D.; Matthews, G. The role of mitochondria in presynaptic calcium handling at a ribbon synapse. *Neuron* **2000**, *25*, 229–237. [CrossRef]

7. Street, V.A.; Mckee-Johnson, J.W.; Fonseca, R.C.; Tempel, B.L.; Noben-Trauth, K. Mutations in a plasma membrane Ca^{2+}-ATPase gene cause deafness in deafwaddler mice. *Nat. Genet.* **1998**, *19*, 390–394. [PubMed]

8. Salome, A.; Hugh, J.R.; Matthews, R. Olfactory response termination involves Ca^{2+}-ATPase in vertebrate olfactory receptor neuron cilia. *J. Gen. Physiol.* **2010**, *135*, 367–378.

9. Castillo, K.; Delgado, R.; Bacigalupo, J. Plasma membrane Ca^{2+}-ATPase in the cilia of olfactory receptor neurons: Possible role in Ca^{2+} clearance. *Eur. J. Neurosci.* **2007**, *26*, 2524–2531. [CrossRef] [PubMed]

10. Strehler, E.E.; Filoteo, A.G.; Penniston, J.T.; Caride, A.J. Plasma membrane Ca^{2+}-pumps: Structural diversity as basis for functional versatility. *Biochem. Soc. Trans.* **2007**, *35*, 919–922. [CrossRef] [PubMed]

11. Stauffer, T.P.; Guerini, D.; Carafoli, E. Tissue distribution of the 4 gene-products of the plasma-membrane Ca^{2+} pump-a study using specific antibodies. *J. Biol. Chem.* **1995**, *270*, 12184–12190. [CrossRef] [PubMed]

12. Foletti, D.; Guerini, D.; Carafoli, E. Subcellular targeting of the endoplasmic reticulum and plasma membrane Ca^{2+} pumps: A study using recombinant chimeras. *FASEB J.* **1995**, *9*, 670–680. [PubMed]

13. Zvaritch, E.; Vellani, F.; Guerini, D.; Carafoli, E. A signal for endoplasmic reticulum retention located at the carboxyl terminus of the plasma membrane Ca^{2+}-ATPase isoform 4CI. *J. Biol. Chem.* **1995**, *270*, 2679–2688. [CrossRef] [PubMed]

14. Schwab, B.L.; Guerini, D.; Didszun, C.; Bano, D.; Ferrando-May, E.; Fava, E.; Tam, J.; Xu, D.; Xanthoudakis, S.; Nicholson, D.W. Cleavage of plasma membrane calcium pumps by caspases: A link between apoptosis and necrosis. *Cell Death Differ.* **2002**, *9*, 818–831. [CrossRef] [PubMed]

15. Cakmak, I.; Baspinar, H. Control of the Carmine Spider Mite *Tetranychus cinnabarinus* boisduval by the predatory mite *Phytoseinlus persimilis* (Athias-Henriot) in protected strawberries in Aydin, Turkey. *Turk. J. Agric. For.* **2005**, *29*, 259–265.

16. Hazan, A.; Gerson, U.; Tahori, A.S. Spider mite webbing. I. The production of webbing under various environmental conditions. *Acarologia* **1974**, *16*, 68–84.
17. Bi, J.L.; Niu, Z.M.; Yu, L.; Toscano, N.C. Resistance status of the carmine spider mite, *Tetranychus cinnabarinus* and the twospotted spider mite, *Tetranychus urticae* to selected acaricides on strawberries. *Insect Sci.* **2016**, *23*, 88–93. [CrossRef] [PubMed]
18. Tal, B.; Robeson, D.J. The induction, by fungal inoculation, of ayapin and scopoletin biosynthesis in *Helianthus annuus*. *Phytochemistry* **1985**, *25*, 77–79. [CrossRef]
19. Gnonlonfin, G.J.B.; Sanni, A.; Brimer, L. Review Scopoletin—A coumarin phytoalexin with medicinal properties. *Crit. Rev. Plant Sci.* **2012**, *31*, 47–56. [CrossRef]
20. Rollinger, J.M.; Hornick, A.; Langer, T.; Stuppner, H.; Prast, H. Acetylcholinesterase inhibitory activity of scopolin and scopoletin discovered by virtual screening of natural products. *J. Med. Chem.* **2013**, *47*, 6248–6254. [CrossRef] [PubMed]
21. Tripathi, A.K.; Bhakuni, B.H.; Upadhyay, S.; Gaur, R. Insect feeding deterrent and growth inhibitory activities of scopoletin isolated from *Artemisia annua* against *Spilarctia obliqua* (Lepidoptera: Noctuidae). *Insect Sci.* **2011**, *18*, 189–194. [CrossRef]
22. Zhang, Y.Q.; Wei, D.; Zhao, Z.M.; Jing, W.U.; Fan, Y.H. Studies on acaricidal bioactivities of *Artemisia annua* L. extracts against *Tetranychus cinnabarinus* Bois. (Acari: Tetranychidae). *Agric. Sci. China* **2008**, *7*, 577–584. [CrossRef]
23. Hou, Q.L.; Wang, D.; Zhang, B.C.; Ding, W.; Zhang, Y.Q. Biochemical evidences for scopoletin inhibits Ca^{2+}-ATPase activity in the Carmine spider mite, *Tetranychus cinnabarinus* (Boisduval). *Agric. Sci. Technol.* **2015**, *4*, 826–831.
24. Xu, Z.; Zhu, W.; Liu, Y.; Liu, X.; Chen, Q.; Peng, M.; Wang, X.; Shen, G.; He, L. Analysis of insecticide resistance-related genes of the Carmine spider mite *Tetranychus cinnabarinus* based on a de novo assembled transcriptome. *PLoS ONE* **2014**, *9*, e94779. [CrossRef] [PubMed]
25. Zhang, Q.Y.; Jian, W.; Xu, X.; Yang, G.F.; Ren, Y.L.; Liu, J.J.; Wang, H.; Yu, G. Structure-based rational quest for potential novel inhibitors of human HMG-CoA reductase by combining CoMFA 3D QSAR modeling and virtual screening. *J. Comb. Chem.* **2007**, *9*, 131–138. [CrossRef] [PubMed]
26. Carafoli, E.; Guerini, D. Molecular and cellular biology of plasma membrane calcium ATPase. *Trends Cardiovasc. Med.* **1993**, *3*, 177–184. [CrossRef]
27. Lnenicka, G.A.; Grizzaffi, J.; Lee, B.; Rumpal, N. Ca^{2+} dynamics along identified synaptic terminals in *Drosophila* larvae. *J. Neurosci.* **2006**, *26*, 12283–12293. [CrossRef] [PubMed]
28. Strehler, E.E.; Zacharias, D.A. Role of alternative splicing in generating isoform diversity among plasma membrane calcium pumps. *Physiol. Rev.* **2001**, *81*, 21–50. [PubMed]
29. Di, L.F.; Domi, T.; Fedrizzi, L.; Lim, D.; Carafoli, E. The plasma membrane Ca^{2+} ATPase of animal cells: Structure, function and regulation. *Arch. Biochem. Biophys.* **2008**, *476*, 65–74.
30. Brodin, P.; Falchetto, R.; Vorheer, T.; Carafoli, E. Identification of two domains which mediate the binding of activating phospholipids to the plasma-membrane Ca^{2+} pump. *Eur. J. Biochem.* **1992**, *204*, 939–946. [CrossRef] [PubMed]
31. Hicks, M.J.; Lam, B.J.; Hertel, K.J. Analyzing mechanisms of alternative pre-mRNA splicing using in vitro splicing assays. *Methods* **2005**, *37*, 306–313. [CrossRef] [PubMed]
32. François, A.; Bozzolan, F.; Demondion, E.; Montagné, N.; Lucas, P.; Debernard, S. Characterization of a plasma membrane Ca^{2+}-ATPase expressed in olfactory receptor neurons of the moth *Spodoptera littoralis*. *Cell Tissue Res.* **2012**, *350*, 239–250. [CrossRef] [PubMed]
33. Ezeokonkwo, C.A.; Obidoa, O.; Eze, L.C. Effects of scopoletin and aflatoxin B 1 on bovine erythrocyte membrane Na-K-ATPase. *Plant Physiol. Commun.* **2010**, *41*, 715–719. [CrossRef]
34. Ezeokonkwo, C.A.; Obidoa, O. Effect of scopoltin on erythrocyte membrane ion motive ATPases. *Niger. J. Nat. Prod. Med.* **2001**, *5*, 37–40.
35. Ojewole, J.A.; Adesina, S.K. Cardiovascular and neuromuscular actions of scopoletin from fruit of *Tetrapleura tetraptera*. *Planta Med.* **1983**, *49*, 99–102. [CrossRef] [PubMed]
36. Oliveira, E.J.; Romero, M.A.; Silva, M.S.; Silva, B.A.; Medeiros, I.A. Intracellular calcium mobilization as a target for the spasmolytic action of scopoletin. *Planta Med.* **2001**, *67*, 605–608. [CrossRef] [PubMed]
37. Palmgren, M.G.; Nissen, P. P-type ATPases. *Annu. Rev. Biophys.* **2011**, *40*, 243–266. [CrossRef] [PubMed]

38. Wang, Y.N.; Jin, Y.S.; Shi, G.L.; Bu, C.Y.; Zhao, L.; Du, J.; Liu, Y.B.; Zhao, L.L. Effects of *Kochia scoparia* extracts to activities of several enzymes of *Tetranychus viennensis*. *Sci. Silvae Sin.* **2008**, *44*, 1–5.

39. Zhang, W.Y.; Lee, J.J.; Kim, Y.; Kim, I.S.; Park, J.S.; Myung, C.S. Amelioration of insulin resistance by scopoletin in high-glucose-induced, insulin-resistant HepG2 cells. *Horm. Metab. Res.* **2010**, *42*, 930–935. [CrossRef] [PubMed]

40. Kim, H.J.; Jang, S.I.; Kim, Y.J.; Chung, H.T.; Yun, Y.G.; Kang, T.H.; Jeong, O.S.; Kim, Y.C. Scopoletin suppresses pro-inflammatory cytokines and PGE2 from LPS-stimulated cell line, RAW 264.7 cells. *Fitoterapia* **2004**, *75*, 261–266. [CrossRef] [PubMed]

41. Desaiah, D.; Cutkomp, L.K.; Koch, R.B. Inhibition of spider mite ATPases by plictran and three organochlorine acaricides. *Life Sci.* **1973**, *13*, 1693–1703. [CrossRef]

42. Jeyaprakash, A.; Hoy, M.A. The mitochondrial genome of the predatory mite *Metaseiulus occidentalis* (Arthropoda: Chelicerata: Acari: Phytoseiidae) is unexpectedly large and contains several novel features. *Gene* **2007**, *391*, 264–274. [CrossRef] [PubMed]

43. Deshmukh, M.; Pawar, P.; Joseph, M.; Phalgune, U.; Kashalkar, R.; Deshpande, N.R. Efficacy of 4-methyl-7-hydroxy coumarin derivatives against vectors *Aedes aegypti* and *Culex quinquefasciatus*. *Indian J. Exp. Biol.* **2008**, *46*, 788–792. [PubMed]

44. Adfa, M.; Yoshimura, T.; Komura, K.; Koketsu, M. Antitermite activities of coumarin derivatives and scopoletin from *Protium javanicum* Burm. f. *J. Chem. Educ.* **2010**, *36*, 720–726. [CrossRef] [PubMed]

45. Adfa, M.; Hattori, Y.; Yoshimura, T.; Komura, K.; Koketsu, M. Antifeedant and termiticidal activities of 6-alkoxycoumarins and related analogs against *Coptotermes formosanus* Shiraki. *J. Chem. Educ.* **2011**, *37*, 598–606. [CrossRef] [PubMed]

46. Adfa, M.; Hattori, Y.; Yoshimura, T.; Koketsu, M. Antitermite activity of 7-alkoxycoumarins and related analogs against *Coptotermes formosanus* Shiraki. *Int. Biodeter. Biodegr.* **2012**, *74*, 129–135. [CrossRef]

47. Lin, H.; Fang, C.; Zhu, R.; Qiang, P.; Ding, L.; Min, W. Inhibitory effect of phloretin on α-glucosidase: Kinetics, interaction mechanism and molecular docking. *Int. J. Biol. Macromol.* **2017**, *95*, 520–527.

48. Deb, P.K.; Sharma, A.; Piplani, P.; Akkinepally, R.R. Molecular docking and receptor-specific 3D-QSAR studies of acetylcholinesterase inhibitors. *Mol. Divers.* **2012**, *16*, 803–823. [CrossRef] [PubMed]

49. Sippl, W.; Contreras, J.M.; Parrot, I.; Rival, Y.M.; Wermuth, C.G. Structure-based 3D QSAR and design of novel acetylcholinesterase inhibitors. *J. Comput. Aided Mol. Des.* **2001**, *15*, 395–410. [CrossRef] [PubMed]

50. Verma, J.; Khedkar, V.M.; Coutinho, E.C. 3D-QSAR in drug design—A review. *Curr. Top. Med. Chem.* **2010**, *10*, 95–115. [CrossRef] [PubMed]

51. Katsamakas, S.; Bermperoglou, E.; Hadjipavloulitina, D. Considering autotaxin inhibitors in terms of 2D-QSAR and 3D-mapping-review and evaluation. *Curr. Med. Chem.* **2015**, *22*, 1428–1461. [CrossRef] [PubMed]

52. Myint, W.; Gong, Q.; Ahn, J.; Ishima, R. Characterization of sarcoplasmic reticulum Ca^{2+} ATPase nucleotide binding domain mutants using NMR spectroscopy. *Biochem. Biophys. Res. Commun.* **2011**, *405*, 19–23. [CrossRef] [PubMed]

53. Martí-Renom, M.A.; Stuart, A.C.; Fiser, A.; Sánchez, R.; And, F.M.; Šali, A. Comparative protien structure modeling of genes and genomes. *Annu. Rev. Biophys. Biomol. Struct.* **2000**, *29*, 291–325. [CrossRef] [PubMed]

54. Min, J.; Lin, D.; Zhang, Q.; Zhang, J.; Yu, Z. Structure-based virtual screening of novel inhibitors of the uridyltransferase activity of *Xanthomonas oryzae* pv. oryzae GlmU. *Eur. J. Med. Chem.* **2012**, *53*, 150–158. [CrossRef] [PubMed]

55. Nakamura, T.; Kodama, N.; Oda, M.; Tsuchiya, S.; Yu, A.; Kumamoto, T.; Ishikawa, T.; Ueno, K.; Yano, S. The structure—Activity relationship between oxycoumarin derivatives showing inhibitory effects on iNOS in mouse macrophage RAW264.7 cells. *J. Nat. Med.* **2009**, *63*, 15–20. [CrossRef] [PubMed]

56. Cheng, J.F.; Chen, M.; Wallace, D.; Tith, S.; Arrhenius, T.; Kashiwagi, H.; Ono, Y.; Ishikawa, A.; Sato, H.; Kozono, T. Discovery and structure-activity relationship of coumarin derivatives as TNF-α inhibitors. *Bioorg. Med. Chem. Lett.* **2004**, *14*, 2411–2415. [PubMed]

57. Hu, J.; Wang, C.; Wang, J.; You, Y.; Chen, F. Monitoring of resistance to spirodiclofen and five other acaricides in *Panonychus citri* collected from Chinese citrus orchards. *Pest Manag. Sci.* **2010**, *66*, 1025–1030. [CrossRef] [PubMed]

58. Michel, A.P.; Mian, M.A.R.; Davila-Olivas, N.H.; Cañas, L.A. Detached leaf and whole plant assays for *Soybean aphid* resistance: Differential responses among resistance sources and biotypes. *J. Econ. Entomol.* **2010**, *103*, 949–957. [CrossRef] [PubMed]

59. Bairoch, A. The PROSITE dictionary of sites and patterns in proteins, its current status. *Nucleic Acids Res.* **1993**, *21*, 3097–3103. [CrossRef] [PubMed]

60. Bendtsen, J.D.; Nielsen, H.; Von, H.G.; Brunak, S. Improved prediction of signal peptides: SignalP 3.0. *J. Mol. Biol.* **2004**, *340*, 783–795. [CrossRef] [PubMed]

61. Tamura, K.; Peterson, D.; Peterson, N.; Stecher, G.; Nei, M.; Kumar, S. MEGA5: Molecular evolutionary genetics analysis using maximum likelihood, evolutionary distance, and maximum parsimony methods. *Mol. Biol. Evol.* **2011**, *28*, 2731–2739. [CrossRef] [PubMed]

62. Rozen, S.; Skaletsky, H. Primer3 on the WWW for general users and for biologist programmers. *Methods Mol. Biol.* **2000**, *132*, 365–386. [PubMed]

63. Sun, W.; Jin, Y.; He, L.; Lu, W.; Li, M. Suitable reference gene selection for different strains and developmental stages of the carmine spider mite, *Tetranychus cinnabarinus*, using quantitative real-time PCR. *J. Insect Sci.* **2013**, *10*, 208. [CrossRef] [PubMed]

64. Livak, K.J.; Schmittgen, T.D. Analysis of relative gene expression data using real-time quantitative PCR and the $2^{-\Delta\Delta Ct}$ Method. *Methods* **2001**, *25*, 402–408. [CrossRef] [PubMed]

65. Zhang, Y. I-TASSER server for protein 3D structure prediction. *BMC Bioinform.* **2008**, *9*, 40. [CrossRef] [PubMed]

66. Yang, J.Y.; Zhang, Y. Protein structure and function prediction using I-TASSER. *Curr. Protoc. Bioinform.* **2016**, *52*, 5.8.1–5.8.15.

67. Yang, J.; Yan, R.; Roy, A.; Xu, D.; Poisson, J.; Zhang, Y. The I-TASSER Suite: Protein structure and function prediction. *Nat. Methods* **2015**, *12*, 7–8. [CrossRef] [PubMed]

68. Roy, A.; Kucukural, A.; Zhang, Y. I-TASSER: A unified platform for automated protein structure and function prediction. *Nat. Protoc.* **2010**, *5*, 725–738. [CrossRef] [PubMed]

69. Wu, S.; Zhang, Y. LOMETS: A local meta-threading-server for protein structure prediction. *Nucleic Acids Res.* **2007**, *35*, 3375–3382. [CrossRef] [PubMed]

70. Zhang, Y. Template-based modeling and free modeling by I-TASSER in CASP7. *Proteins* **2007**, *69*, 108–117. [CrossRef] [PubMed]

71. Dutta, S.; Berman, H.M.; Bluhm, W.F. Using the tools and resources of the RCSB protein data bank. *Curr. Protoc. Bioinform.* **2007**, *20*, 1–24.

72. Zhang, Y. Progress and challenges in protein structure prediction. *Curr. Opin. Struct. Biol.* **2008**, *18*, 342–348. [CrossRef] [PubMed]

73. Zhang, Y.; Kolinski, A.; Skolnick, J. TOUCHSTONE II: A new approach to ab initio protein structure prediction. *Biophys. J.* **2003**, *85*, 1145–1164. [CrossRef]

74. Zhang, Y.; Skolnick, J. Automated structure prediction of weakly homologous proteins on a genomic scale. *Proc. Natl. Acad. Sci. USA* **2004**, *101*, 7594–7599. [CrossRef] [PubMed]

75. Zhang, J.; Liang, Y.; Zhang, Y. Atomic-level protein structure refinement using fragment-guided molecular dynamics conformation sampling. *Structure* **2011**, *19*, 1784–1795. [CrossRef] [PubMed]

76. Xu, D.; Zhang, Y. Improving the physical realism and structural accuracy of protein models by a two-step atomic-level energy minimization. *Biophys. J.* **2011**, *101*, 2525–2534. [CrossRef] [PubMed]

77. Yang, J.; Roy, A.; Zhang, Y. BioLiP: A semi-manually curated database for biologically relevant ligand-protein interactions. *Nucleic Acids Res.* **2013**, *41*, D1096–D1103. [CrossRef] [PubMed]

78. Roy, A.; Zhang, Y. Recognizing protein-ligand binding sites by global structural alignment and local geometry refinement. *Structure* **2012**, *20*, 987–997. [CrossRef] [PubMed]

79. Yang, J.; Roy, A.; Zhang, Y. Protein-ligand binding site recognition using complementary binding-specific substructure comparison and sequence profile alignment. *Bioinformatics* **2013**, *29*, 2588–2595. [CrossRef] [PubMed]

80. Laskowski, R.A.; Macarthur, M.W.; Moss, D.S.; Thornton, J.M. PROCHECK: A program to check the stereochemical quality of protein structures. *J. Appl. Crystallogr.* **1993**, *26*, 283–291. [CrossRef]

81. Porter, L.L.; Englander, S.W. Redrawing the Ramachandran plot after inclusion of hydrogen-bonding constraints. *Proc. Natl. Acad. Sci. USA* **2011**, *108*, 109–113. [CrossRef] [PubMed]

82. Zhang, B.C.; Luo, J.X.; Lai, T.; Wang, D.; Ding, W.; Zhang, Y.Q. Study on acaricidal bioactivity and quantitative structure activity relationship of coumarin compounds against *Tetranychus cinnabarinus* Bois. (Acari: Tetranychidae). *Chin. J. Pestic. Sci.* **2016**, *18*, 37–48.

83. Gasteiger, J.; Marsili, M. Iterative partial equalization of orbital electronegativity-a rapid access to atomic charges. *Tetrahedron* **1980**, *36*, 3219–3228. [CrossRef]

84. Clark, M.; Cramer, R.D.; Van Opdenbosch, N. Validation of the general purpose tripos 5.2 force field. *J. Comput. Chem.* **1989**, *10*, 982–1012. [CrossRef]

85. Powell, M.J.D. Restart procedures for the conjugate gradient method. *Math. Program.* **1977**, *12*, 241–254. [CrossRef]

86. Araz, J.; David, B.J.; Christopher, I.B. Fast, efficient generation of high-quality atomic charges. AM1-BCC model: II. Parameterization and validation. *J. Comput. Chem.* **2002**, *23*, 1623–1641.

87. Morris, G.M.; Huey, R.; Lindstrom, W.; Sanner, M.F.; Belew, R.K.; Goodsell, D.S.; Olson, A.J. AutoDock4 and AutoDockTools4: Automated docking with selective receptor flexibility. *J. Comput. Chem.* **2009**, *30*, 2785–2791. [CrossRef] [PubMed]

88. Welch, W.; Ruppert, J.; Jain, A.N. Hammerhead: Fast, fully automated docking of flexible ligands to protein binding sites. *Cell Chem. Biol.* **1996**, *3*, 449–462. [CrossRef]

89. Kadioglu, O.; Saeed, M.E.M.; Valoti, M.; Frosini, M.; Sgaragli, G.; Efferth, T. Interactions of human P-glycoprotein transport substrates and inhibitors at the drug binding domain: Functional and molecular docking analyses. *Biochem. Pharmacol.* **2016**, *104*, 42–51. [CrossRef] [PubMed]

90. Wold, S.; Geladi, P.; Esbensen, K.; Ohman, J. Multi way principal components and PLS analysis. *J. Chemom.* **2005**, *1*, 41–56. [CrossRef]

13

Protein Complexes Prediction Method Based on Core—Attachment Structure and Functional Annotations

Bo Li *,† **and Bo Liao** †

College of Computer Science and Electronic Engineering, Hunan University, Changsha 410082, China;
dragonbw@163.com
* Correspondence: nonegenius@hnu.edu.cn
† These authors contributed equally to this work.

Abstract: Recent advances in high-throughput laboratory techniques captured large-scale protein–protein interaction (PPI) data, making it possible to create a detailed map of protein interaction networks, and thus enable us to detect protein complexes from these PPI networks. However, most of the current state-of-the-art studies still have some problems, for instance, incapability of identifying overlapping clusters, without considering the inherent organization within protein complexes, and overlooking the biological meaning of complexes. Therefore, we present a novel overlapping protein complexes prediction method based on core–attachment structure and function annotations (CFOCM), which performs in two stages: first, it detects protein complex cores with the maximum value of our defined cluster closeness function, in which the proteins are also closely related to at least one common function. Then it appends attach proteins into these detected cores to form the returned complexes. For performance evaluation, CFOCM and six classical methods have been used to identify protein complexes on three different yeast PPI networks, and three sets of real complexes including the Munich Information Center for Protein Sequences (MIPS), the Saccharomyces Genome Database (SGD) and the Catalogues of Yeast protein Complexes (CYC2008) are selected as benchmark sets, and the results show that CFOCM is indeed effective and robust for achieving the highest F-measure values in all tests.

Keywords: protein–protein interaction network; overlapping; clustering

1. Introduction

Most proteins in living organisms, performing their biological functions or involving with cellular processes, barely serve as single isolated entities, but rather via molecular interactions with other partners to form complexes [1]. In fact, protein complexes are the key molecular entities to perform cellular functions, such as signal transduction, post-translational modification, DNA transcription, and mRNA translation. Moreover, the damage of protein complexes is one of the main factors inducing severe diseases [2]. Identification of protein complexes, therefore, becomes a fundamental task in better understanding the biological functions in different cellular systems, uncovering regularities of cellular activities and contributing to interpreting the causes, diagnosis, and even the treatments of complex diseases. As a result, lots of techniques including laboratory-based and computational-based have been proposed to address this issue.

Up to now, significant progress in high-throughput laboratory techniques involving Tandem Affinity Purification (TAP) [3] and Mass Spectrometry (MS) [4] has been made to discover protein complexes on a large scale. However, laboratory experiments are expensive and time-consuming, resulting in poor coverage of the complete protein complexes. Fortunately, the genomic-scale protein–protein interaction (PPI) networks created from pairwise protein–protein interactions make it possible to automatically and computationally detect protein complexes. Given a PPI network, as the protein complexes are formed by physical aggregations of several binding proteins, they are assumed to be the functionally and structurally cohesive substructures, and thus graph clustering methods have been put forward to search densely connected regions in PPI networks as protein complexes.

Since some proteins have multiple functions, in other words, they may belong to more than one protein complex, so the ideal approaches need to be able to detect overlapping complexes. However, several types of graph clustering methods don't allow overlaps between detected protein complexes due to the confinements of the rationales behind them. For example, the partition-based clustering methods such as the Restricted Neighborhood Search Clustering algorithm (RNSC) [5], the Bayesian Nonnegative Matrix Factorization(NMF)-based weighted Ensemble Clustering algorithm (EC-BNMF) [6], obtain, however, some highly reliable protein complexes, since they need prior knowledge of the exact number of clusters that thus cannot detect overlapping functional modules, and, in addition, most of the hierarchy-based clustering methods [7–9] utilize hierarchical trees to represent the hierarchical module organization for a PPI network, but it is difficult to detect overlapping complexes as well. In addition, although some algorithms are capable of finding overlapping complexes, they still have some distinct shortcomings—for instance, the Molecular Complex Detection (MCODE) [10] predicts only quite a small number of protein complexes. CFinder [11] first discovers k-cliques by using the clique percolation method (CPM) [12], and then combines the adjacent k-cliques to get the functional modules, but may fail to detect some regular complexes. ClusterONE [13] requires one pre-determined parameter, which is depended on the quality of PPI network, and it is difficult to determine.

Furthermore, the aforementioned methods still have a common fatal weakness—ignorance of the inherent organization of the complexes—but actually experimental analysis has already reported that a protein complex generally consists of a core, in which proteins share similar functions and tend to be highly co-expressed, and other attach proteins surrounding to the core [14]. Based on these, several core–attachment based algorithms have been presented, and experimental results indicate that they can acquire better performance compared to traditional methods neglecting inherent organization. Among them, CORE [15] first calculates the probability of each pairwise proteins to be in the same core and then uses it to detect cores. COACH [16] detects cores from neighborhood graphs of the selected seed proteins, and then applies an outward growing strategy to generate protein complexes. Compared with CORE, COACH can find overlapping cores. Other methods including [17] predict complexes based on multi-structures in PPI network, and achieve significant performance. The complexes predicted by structure-based methods, in general, have been verified more in accordance with the known complexes.

In addition, to precisely predict more biological explainable complexes, some methods of fusing various types of prior knowledge including functional annotations [18–20], gene expression data [21–23], as well as sub-cellular location of proteins [24], are presented and have already been proved that can help to improve the performance to some extent. However, these kinds of valuable information are either used in data preprocessing or post-processing, such as filtering low-confidence edges, weighting edges, discarding some biological meaningless complexes, but seldom helps mining cores with better biological meaning, in which most proteins are co-subcellular or co-expression or with similar functions. Furthermore, since these data are undeniably incomplete and imprecise, how to generate a impartial and efficient model incorporating different types of data is still a hot topic in complex prediction [25–27].

In summary, we may come to the conclusion that a comparatively well-designed protein complexes identification method may need to meet the following conditions: capable of detecting overlapping complexes, fewer parameters, being easy to be determine, consideration of the inherent organization of protein complexes, particularly finding topological and biological meaningful cores, properly incorporating prior information as much as possible into the predicting model, and robust to PPI networks with false positives and false negatives. Unfortunately, even though many effective techniques have been proposed, as far as we know, few of them satisfy most of the above-mentioned requirements, which results in impeding further practical applications, and thus there is still urgent need for new approaches.

In this manuscript, we introduced a novel core–attachment based method to predict protein complexes, and the proteins in our detected cores are closely linked, share high similar topology that is highly connected to internal vertexes and relatively sparsely connected to outsides, and are more biologically significant, namely more likely to participate in one or more biological processes with the appliance of GO functional annotation. Furthermore, the detected complexes can be overlapping. We applied our algorithm to two PPI networks of yeast, and validated our predicted complexes using benchmark complexes collected from several public databases. The experimental results indicated that our algorithm is efficient and outperforms other existing classical methods.

2. Results

We have applied our CFOCM method on the Database of Interacting Proteins (DIP) data and Gavin data. In this section, we will first discuss parameter t affecting the performance of CFOCM. Next, we perform comprehensive comparisons with various existing classical methods and analyse the results in detail. Finally, we explore the functional definition of the complex-core as a whole, contributing to the biological significance of the detected complexes.

2.1. Evaluation Metrics

The neighborhood affinity score $NS(p,r)$ can also be devoted to measure the overall similarity between a predicted complex p and a real complex r, and if $NS(p,r) \geq \omega$, p and r are considered to be matching. On the one hand, the greater setting value of ω means the more stringent matching of between the predicted complex and the real complex in the benchmark, probably resulting in a sharp decline in all the prediction measure values; on the other, the smaller value could not only lead to identify the low-confidence predicted complex as the real complex, which is also not reasonable. In our experiments, we set ω to 0.2 the same as most literatures do [5,7,11,13,15,28], which provides easy and fair comparisons between results of various algorithms.

Let P and R represent the set of predicted complexes and the real complexes in benchmarks, respectively. $N_{cp} = \{p \in P | \exists r \in B, NS(p,r) \geq \omega\}$ denotes the predicted complexes matching at least one real complex, and $N_{cr} = \{r \in R | \exists p \in P, NS(r,p) \geq \omega\}$ denotes the real complexes matching at least one predicted complex. In addition, then the performance of a clustering algorithm can be measured using precision, recall, and F-measure, which can be calculated as follows:

$$\text{Precison} = \frac{|N_{cp}|}{|P|},$$

$$\text{Recall} = \frac{|N_{cr}|}{|R|},$$

$$\text{F-measure} = 2 \times \text{Precison} \times \text{Recall} / (\text{Precision} + \text{Recall}),$$

where Precision means the ratio of predicted protein complexes that are matched with the real complexes, Recall means the rate of real complexes that are successfully detected and F-measure evaluates the overall performance.

2.2. Optimization of the Parameter t

Recall that the process of mining cores from PPI network in Algorithm 1 of CFOCM employs a user-defined parameter t calculated by $NS(mc_i, mc_j)$ to decide whether a certain candidate core mc_j should be merged into the family of the current candidate core mc_i. In general, CFOCM can predict more complexes with the bigger value of t; nevertheless, this may lead to compromise on the quality of the predicted complexes, and thus how to choose a relatively appropriate t to achieve a balance between the predicted complexes' quality and quantity needs to be probed. Here, varying t from 0.2 to 0.6 with the interval 0.01, the F-measure values of each predicted complex set are computed, and help us to intuitively observe that the variation of t affects the performance of our CFOCM method and selects the relatively suitable t as well (see Figure 1).

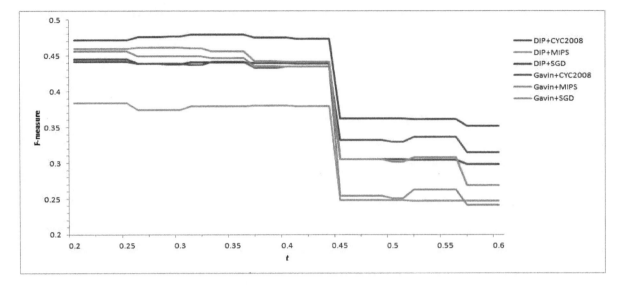

Figure 1. The effect of t, showing how the variation of parameter t affects the performance of our proposed overlapping protein complexes prediction method based on core–attachment structure and function annotations (CFOCM) in terms of F-measure.

In Figure 1, all the curves of different CFOCMs, based on DIP data or Gavin data, validated in benchmark set MIPS or SGD or CYC2008, are comparably smooth and steady when the t varies from 0.2 to 0.44. However, the curves change abruptly near $t = 0.45$, and the causation of this phenomenon can be rationally explained with the NS score of two candidate cores being 4/9 ($\approx$0.44) in which the number of proteins are both three and two of them are the overlapping; that is to say, these two cores can not be put into the same family if t is larger than 4/9, resulting in a rapid increase of low-confidence detected cores with size 3 and a sharp decease of recall value and F-measure score as well. For example, under $t = 0.44$, CFOCM based on DIP and Gavin network generates 751,453 complexes respectively, while under $t = 0.45$ generates 2629, 1703 complexes respectively, conforming to the above analysis and interpretation.

As stated above, t should definitely not be set to larger than 0.44 as increasing abundant low-confidence three-size cores, and actually the performance of CFOCM does not change significantly when $t \in [0.2, 0.44]$. Still, demand for more complexes shows a preference to a larger t; otherwise, if there is demand for a fewer number of complexes, a preference is shown for a smaller t. For example, CFOCM predicts 545 complexes with average matching of 156 real complexes in MIPS when $t = 0.2$, while predicting 751 complexes matching 205 real complexes in MIPS when $t = 0.44$. In the following part, either in DIP data or Gavin data, the t of our CFOCM algorithm is set to 0.4.

2.3. Comparison Experiments on Different Datasets

For performance evaluation, the comparison experiments between CFOCM and six representative algorithms including MCL, MCODE, RNSC, CORE, COACH and ClusterONE are performed on both DIP data , Gavin data and Srihari data. Note that the parameters of these six comparative methods are set to the default values. Figure 2, Table 1, Figure 3, Table 2 ,Figure 4, and Table 3 exhibit the overall comparison results in terms of Precision, Recall and F-measure on DIP data, Gavin data and Srihari data, respectively.

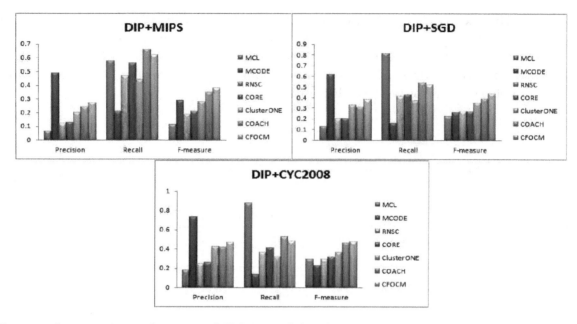

Figure 2. Comparative performance of CFOCM and the other six methods in DIP data using benchmark MIPS, SGD, CYC2008, respectively.

Table 1. Results of various approaches using DIP data.

Algorithms	MCL	MCODE	RNSC	COER	ClusterONE	COACH	CFOCM
# complexes	4838	63	543	592	341	746	748
N_p (MIPS)	305	31	65	78	69	179	205
N_b (MIPS)	117	42	96	113	89	134	126
N_p (SGD)	621	39	106	117	112	231	285
N_b (SGD)	262	53	134	138	121	176	168
N_p (CYC2008)	853	46	134	153	145	311	351
N_b (CYC2008)	358	55	149	168	132	215	196

In Figure 2, no matter whether benchmarks MIPS or SGD or CASP2008 are used, MCODE achieves the highest precision that is far beyond other methods. However, since the number of predicted protein complexes is very limited and also matches with fewer real complexes, resulting in much low recall and F-measure values. In addition, CORE, RNSC, and ClusterONE are observed to attain high recall values, but, nevertheless, the F-measure values of them merely end up with relatively lower F-measure value due to their very low precision values. In fact, CFOCM and COACH demonstrate their distinctive competitive advantages in F-measure as a result of balanced precisions and recalls. Moreover, it is obvious that CFOCM remarkably outperforms COACH in F-measure when using benchmark MIPS and SGD. Meanwhile, both CFOCM and COACH are based on core–attach structure, it may indicate that the protein complex detection method seems more appropriate when taking consideration of the inherent organization of complex. As Table 1 shows, CFOCM detects moderate number of complexes, many of which correctly match with the real complexes and have a high coverage rate of real complexes as well.

In order to evaluate the robustness of algorithm CFOCM, comparison experiments are also carried on Gavin network, which is different from the DIP network for containing much fewer and more densely connected proteins. Figure 3 illustrates the results for Gavin data, CFOCM shows even better performance for Gavin data, which achieves the highest precision values when using benchmark MIPS and CYC2008, and, apparently, CFOCM obtains the best F-measure value for every benchmark. This may suggest that CFOCM indeed works on dense network as well. For each method, the total number of identified complexes, the number of correct predictions N_p matching at least a real complex, and the number of real complexes N_b matching at least a predicted one are listed in Table 2, reaching similar conclusions that are consistent with DIP data.

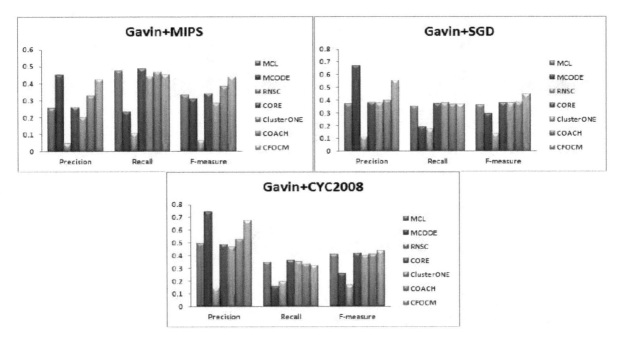

Figure 3. Comparative performance of CFOCM and the other six methods in Gavin data using benchmarks MIPS, SGD, CYC2008, respectively.

Table 2. Results of various approaches using Gavin data.

Algorithms	MCL	MCODE	RNSC	COER	ClusterONE	COACH	CFOCM
# complexes	232	69	476	267	292	326	453
N_p (MIPS)	59	31	22	69	65	106	191
N_b (MIPS)	96	47	21	98	80	94	91
N_p (SGD)	86	46	53	101	109	130	250
N_b (SGD)	114	61	55	120	121	118	119
N_p (CYC2008)	115	51	68	130	136	171	305
N_b (CYC2008)	142	63	79	148	143	135	131

For further evaluation, Srihari data derived from three different repositories are also used for comparison, and the results are showed in Figure 4 and Table 3. Similar conclusions can be reached as in DIP and Gavin data, except that both the Precision value and Recall value of CFOCM are better than COACH, and this may indicate that CFOCM has more potential on composite data.

Table 3. Results of various approaches using Srihari data.

Algorithms	MCL	MCODE	RNSC	COER	ClusterONE	COACH	CFOCM
# complexes	4732	88	552	525	773	726	758
N_p (MIPS)	325	26	78	92	117	219	225
N_b (MIPS)	168	42	102	111	131	150	152
N_p (SGD)	654	36	108	176	224	299	322
N_b (SGD)	292	44	184	189	217	231	240
N_p (CYC2008)	846	46	138	218	275	397	452
N_b (CYC2008)	362	57	154	236	272	281	290

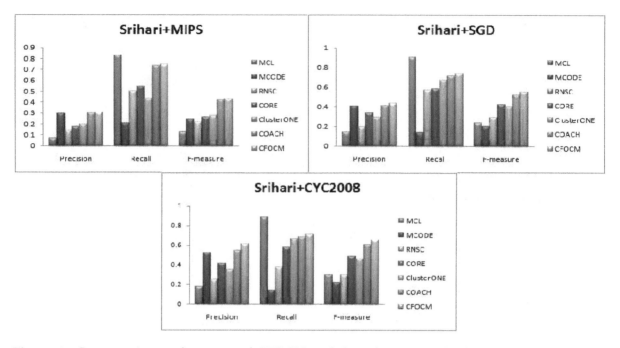

Figure 4. Comparative performance of CFOCM and the other six methods in Srihari data using benchmarks MIPS, SGD, CYC2008, respectively.

In a word, either in relatively sparse DIP networks or in relatively dense Gavin data even using a composite data set, CFOCM is able to identify a suitable number of protein complexes, and, meanwhile, the predicted complexes are also biologically meaningful as a consequence of cooperating the protein function annotations into our model, so it compellingly performs better than other existing methods in term of F-measure. Thus, we can come to the conclusion that CFOCM is efficient and has strong adaptability and robustness to different types of data.

3. Discussion

3.1. The Effectiveness of Functional Annotation

As the assumption of the complex-core described before, the proteins in each CFOCM detected core must be functional related to a certain common GO item, namely either annotated with that GO item or annotated with a GO item that is functionally interdependent with that GO item. To estimate the contribution of this, comparison experiments between CFOCM and CFOCM without use (unCFOCM) are conducted. As the results listed in Table 4 (DIP) and Table 5 (Gavin), unCFOCM in all the tests predicts much more biological meaningless complexes on account of not using GO annotation, leading to lower F-measure values. In other words, owing to the requirement of functional relevance within the discovered cores, CFOCM is capable of filtering abundant low-confidence protein complexes, and the detected protein complexes are supposed to be more biologically significant. Therefore, the cores detected by CFOCM should share some common functions, which is more in conformity with the original definition of the complex core, and it is greatly obliged to help finding more accurate protein complexes.

3.2. Case Studies

This section illustrates two predicted protein complexes, namely the Glycine decarboxylase complex and the RNA polymerase I complex as Figure 5. The Glycine decarboxylase complex is a small-sized complex responsible for the oxidation of glycine by mitochondria, and it consists of four proteins including YDR019C, YMR18W, YAL044C and YFL018C. As showed, CFOCM successfully identified these four proteins, in which YDR019C, YMR18W, and YAL044C are recognized as core proteins and YFL018C is detected as an attachment to the core. In another case, the RNA polymerase I complex is a larger complex comprised of 14 proteins, and CFOCM could also completely identify all the proteins in this complex with 100% precision, in which all proteins except YHR143W-A are detected as members of the core having more dense connections with each other and sharing more functional relevance as well.

Table 4. Results of CFOCM and CFOCM without using Gene Ontology (GO) (unCFOCM) on DIP data.

Algorithms + Benchmark	# Complexes	N_p	N_b	Precision	Recall	F-Measure
CFOCM + MIPS	748	205	126	0.2741	0.6207	0.3802
unCFOCM + MIPS	862	213	130	0.2471	0.6404	0.3566
CFOCM + SGD	748	285	168	0.381	0.5201	0.4398
unCFOCM + SGD	862	297	175	0.3445	0.5418	0.4212
CFOCM + CYC2008	748	351	196	0.4693	0.4804	0.4748
unCFOCM + CYC2008	862	363	201	0.4211	0.4926	0.4541

Table 5. Results of CFOCM and CFOCM without using Gene Ontology (GO) (unCFOCM) on Gavin data.

Algorithms + Benchmark	# Complexes	N_p	N_b	Precision	Recall	F-Measure
CFOCM + MIPS	453	191	91	0.4216	0.4483	0.4345
unCFOCM + MIPS	551	197	92	0.3575	0.4532	0.3997
CFOCM + SGD	453	250	119	0.5519	0.3684	0.4419
unCFOCM + SGD	551	262	124	0.4755	0.3839	0.4248
CFOCM + CYC2008	453	305	131	0.6733	0.3211	0.4348
unCFOCM + CYC2008	551	321	138	0.5826	0.3382	0.4280

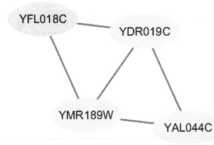

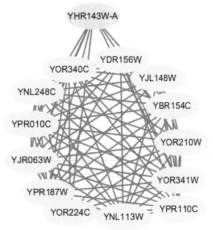

Figure 5. The Glycine decarboxylase complex and the RNA polymerase I complex as detected by CFOCM. The yellow nodes represent proteins within the complex core, while the blue node proteins represent proteins that are attachments.

4. Materials and Methods

4.1. Terminologies

A PPI Network typically can be represented as an undirected graph $G = (V, E)$, where V and $E = \{(u, v) | u, v \in V)$ represent proteins and protein–protein interactions, respectively. A graph $G' = (V', E')$ is regarded as a subgraph of G if $V' \subseteq V$ and $E' \subseteq E$. v's direct interacting neighbors in graph G is denoted as $N_v = \{u|(u,v) \in E, \ u \in V\}$, and $N_v^{G'} = \{u|(u,v) \in E, \ u \in V'\}$ is v's neighbors in subgraph G'. Subgraph G' external boundary nodes are defined as $V_{ob}(G') = \{v| <v,w> \in E(G),$ $v \in V(G) \backslash V(G'), w \in V(G')\}$.

A neighborhood affinity score metric [25], denoted as $NS(G', G'')$, is imported to measure the similarity between two overlapping graphs $G' = (V', E')$ and $G'' = (V'', E'')$,

$$NS(G', G'') = \frac{|V_{G'} \cap V_{G''}|^2}{|V_{G'}| \times |V_{G''}|},$$

where, if $NS(G', G'') >= t$ (t is a predefined threshold), we may declare cluster $G' = (V', E')$ and cluster $G'' = (V'', E'')$ can be further merged as a result of their high topological similarity.

As is well known, GO is composed of three orthogonal ontologies capturing knowledge about biological process, molecular function and cellular component, and each ontology consists of controlled and structured biological terms that can be used to annotate genes and proteins. Some GO item pairs are highly functionally related—for example, sharing a common parent node, or one is just a near ancestor of the other, while other GO item pairs may possess much weaker relationships or even be functionally independent. Therefore, the urgent need is to design a metric to quantify the functional interdependence between two GO items. Fortunately, Ref. [18] has done what we want (see the formula below):

$$fr_{i,j} = \frac{re_{i,j} - ee_{i,j}}{\sqrt{ee_{i,j}(1 - (\sum\limits_{k \in GI} ee_{i,k}/|E|))(1 - (\sum\limits_{k \in GI} ee_{k,j}/|E|))}},$$

where $re_{i,j}$ represents the real number of edges in G connecting one protein annotated with GO item i and the other annotated with item j, $ee_{i,j}$ represents the expected number of edges that one protein is annotated with item i and the other annotated with item j in G, hence it equals (Number of edges in G with one protein annotated with i)*(Number of edges in G with one protein annotated with j to the others)/$|E|$, and GI represents the whole GO items set. Ref. [18] also indicates that item i and j are functionally interdependent if $fr_{i,j} > 1.96$; otherwise, they are considered to be functionally independent.

A protein complex is pervasively modeled as an induced subgraph of PPI network G, the proteins in which have dense intra-connections and are sparely connected to the rest of the network, thus we introduce a new and effective closeness function to quantify the probability that G' is complex based on network topology:

$$cf(G') = density(G') \times \left(\frac{1}{|G'|} * \sum_{v \in G'} \frac{|N_v^{G'}|}{|N_v|}\right),$$

where $density(G') = \frac{2 \times |E'|}{|V'| \times (|V'|-1)}$ is the density of graph G', depicted to quantify the richness of edges in G', and $\frac{|N_v^{G'}|}{|N_v|}$ corresponds to the percentage of v's direct neighbors located within G'. If $\frac{|N_v^{G'}|}{|N_v|}$ equals 1, all the neighbors of v are in G', so there is a high tendency that v should be a member of G'. If equals 0, v has little chance to be a member of G'. Consequently, the expression in the bracket represents the mean possibility of each node being retained in G'. Compared with previous closeness function based on the density of G', cf not only assesses the inner denseness of G', but also takes the ratio of G' inner edges and outer edges into consideration, hence manifesting superiority in appraising the likelihood of G' to be a real complex.

4.2. Description of CFOCM Algorithm

Most of the protein complexes contain core–attachment structure, and the proteins in the core share similar topology and are highly functionally related, while the attach proteins are usually located in the periphery of the core [14]. As the differences between core proteins and attach proteins, therefore, our core–attachment based algorithm CFOCM for protein complexes identification, comprised of two necessary phases, which first detects the protein complexes' cores and then selects attach proteins to the discovered cores.

4.2.1. The Complex Cores Detection

Protein-complex core plays a key role for complex performing biological function, and determines the cellular role and significance of the complex in the context to a large extent [14]. The results of biological analysis also indicate that most protein complex cores own some significant distinguishing features: including a small group of proteins which are densely intra-connected and sparsely to outsides, allowing overlaps between cores, possession of some common functions, showing an altitudinal mRNA co-expression patterns. In this paper, however, only the former three features are used to portray the cores discovered by CFOCM, and our detected cores satisfy the following assumption.

Assumption 1. *A subgraph $G' = (V', E')$ is a protein-complex core unless if satisfying the followed conditions:*

1. *The topology of G' meets: $|G'| >= 3$, G' reaches the local optimum that there does not exist any neighbor node v that satisfies $\text{cf}(G' + \{v\}) > \text{cf}(G')$ or $\text{cf}(G' - \{v\}) > \text{cf}(G')$, and no such G'' exists if $G' \subseteq G''$ and G'' is a complex core.*
2. *If G' has overlaps with G'', then $\text{NS}(G', G'') < t$ must be satisfied; otherwise, G' and G'' could combine together.*
3. *G' needs to be biologically significant: mx is defined as the the maximum common GO item annotating a maximum number of nodes in G', $\forall v \in V'$, v is either annotated by mx or annotated by a GO item gi interdependent with mx, which satisfies $\text{fr}(gi, mx) > 1.96$.*

Different from traditional methods exploring each core protein separately, our above complex-core assumption is more plausible for considering all proteins in the core as a whole. Benefiting from this renovation ensures that each protein in the core owns similar topology and contributes to the enforcement of core's biological functions. Conditions 1, 2, and 3 guarantees the maximizes closeness function value of core, the nearest distance can be retained between different cores, and participation of at least one common biological functions, respectively. Specifically, most traditional literature is mainly focused on the assurance of highly functional similarity between each protein pair in the core, which will result in neglecting that the core as a whole should perform some common functions, while this flaw is certainly renovated by our integrated global view of the core.

Algorithm 1 illustrates that the overall framework to detect protein-complex cores, and, without question, the discovered cores comply with definitions in Assumption 1. We first compute the functional interdependence between each GO items pair by the definition fr in line 1. Then, in line 2, we identify all cliques that are fully connected subgraphs by using a complete enumeration method [29], based on the fact that a k-clique can be obtained by adding a vertex to the clique with k-1 vertices and the 2-cliques can be initialized as the edges in the graph, but only the maximal cliques are reserved at last, and a k-clique is regarded as a maximal k-clique only in the case that it cannot be enlarged by adding any vertex. After that, lines 4–19 mining complex cores by a iteration process on the basis of the two aforementioned pretreatment works. Here, a concept of candidate-core family is presented, containing the core itself and its similar candidate-cores with the neighborhood affinity score NS less than a predefined threshold t. For each certain candidate-core, its family set is obtained in lines 8–13, and a more reasonable combined candidate-core comes into being through algorithm Merge_Similar_Cores in line 14. The details of Merge_Similar_Cores algorithm are described in Algorithm 2. Still, in lines 15–17, if the current generated candidate-core already exists in the generated candidate-core set, we simply discard it; otherwise, we add it to the candidate-core set. After these steps, though, there unavoidably exist some incorrect manipulations, excessive overlapping and biological meaningless candidate cores are substantially removed, and the overwhelming majority of the vertexes in retained cores are densely connected internally, possess similar topology and attend to share at least one common GO annotated function.

Algorithm 1: Complex cores detection algorithm.

  Require: The PPI network $G = (V, E)$;

      Neighborhood affinity score threshold t.

  Ensure: The detected complex cores set CS.

  1: calculate each GO item pair functional interdependence fr;

  2: find all the maximum cliques MC in G;

  3: $CS = MC$;

  4: **repeat**

  5:     $MC = CS$;

  6:     $CS = \{\}$;

  7:     **for** mc_i in MC **do**

  8:         $F_{mc_i} = \{mc_i\}$; $\{F_{mc_i}$ stores the cliques similar with $mc_i\}$

  9:         **for** mc_j in MC **do**

 10:            **if** $\text{NS}(mc_i, mc_j) >= t$ **then**

 11:              $F_{mc_i} = F_{mc_i} \cup \{mc_j\}$;

 12:            **end if**

 13:         **end for**

 14:         $c = Merge\_Similar\_Cores(F_{mc_i})$;

 15:         **if** c is not exists in CS **then**

 16:            $CS = CS \cup \{c\}$;

 17:         **end if**

 18:     **end for**

 19: **until** not exists any two elements c_i and c_j in CS satisfying $\text{NS}(c_i, c_j) >= t$

 20: **return** CS;

A crucial artifice, not described in Algorithm 1, is applied in the process of detecting cores. First, for each maximal cliques set with the same number of vertexes, we generate their corresponding new candidate cores by executing steps in lines 4–19, and then form the final detected cores via the same steps on these different-sized generated cores. Without using this, the smaller cliques may be annexed by the larger similar cliques so that they barely contribute to the generation of the new candidate core. Actually, this artifice is proved to be an effective means of improving the predicting performance.

4.2.2. Similar Complex Cores Merge

Given the family F_{mc} of the candidate core mc, the Merge_Similar_Cores algorithm will filter the proteins that can not help to preserve the topology of the core or are functionally independent with other proteins in the core and return a new candidate core.

Our Merge_Similar_Cores algorithm works as follows. To begin with, we extract the proteins PS from the input family of a candidate-core in line 1, and find the GO item m disappeared in the GO annotations of maximal proteins in line 2. Afterwards, in lines 3–7, we remove proteins that are neither annotated by the common item m nor have a GO item functional interdependent with item m, and this procedure ensures that the returned candidate-core has a high probability of owning at least one common GO function because the proteins in the returned candidate-core either have the common GO item m or a GO item j exists that is functionally interdependent with m. Finally, in lines 8–10, we iteratively delete a protein p from the PS until no such protein p exists, satisfying

$cf(PS - \{p\}) > cf(PS)$, and ensuring that the remaining proteins reach the local optimum, which is relatively richly inner-connected and sparsely connected to the outside.

Algorithm 2: $c = $ Merge_Similar_Cores (F_{mc}).

  1: get all proteins PS contained in F_{mc};
  2: find the GO item m which annotating maximum number of proteins in PS;
  3: **for** each p in PS **do**

  4:        **if** p is not annotated by m and exists no GO item j annotating p satisfying: $fr_{m,j} < 1.96$ **then**

  5:          $PS = PS - \{p\}$;
  6:        **end if**
  7: **end for**
  8: **while** exists $\max\limits_{p \in PS} cf(PS - \{p\}) > cf(PS)$ **do**
  9:      $PS = PS - \{\arg\max\limits_{p \in PS} cf(PS - \{p\})\}$;
10: **end while**
11: **return** $c = PS$;

Each input candidate-core family goes through these steps, and a newer candidate-core has been formed. In addition, Figure 6 also provides an example to illustrate the process of our proposed Merge_Similar_Cores algorithm.

4.2.3. Attach-Proteins Screening

After the foregoing phase of our CFOCM method, the protein-complex cores have already been mined from PPI network $G = (V, E)$. In the second phase, we will form the final predicted complex by appending reliable peripheral proteins to the discovered cores. Given a protein complex core c, for each external boundary protein p of current core c, the following Assumption 2 presents whether p should be an attachment to the core c or not.

Assumption 2. *A external boundary protein p is affirmed as an attachment to the complex core c if satisfying* $cf(c + \{p\}) > cf(c)$.

From the above assumption, the external boundary protein p improves the closeness function cf of the current cluster selected as an attachment. Through appending some attachment proteins to the current core, the topology of core can still be reserved, and thus all the proteins in each final predicted complex are densely connected and sparsely connected to the outside. Algorithm 3 is the pseudo code description.

Algorithm 3: Attach-proteins screening algorithm.

Require: Protein complex cores CS.
Ensure: The predicted complexes $Complexes$.
  1: $Complexes = \{\}$;
  2: **for** each c in CS **do**

  3:        **while** exists $\max\limits_{v \in Neighbors(c)} cf(c \cup \{v\}) > cf(c)$ **do**

  4:          $c = c \cup \{\arg \max\limits_{v \in Neighbors(c)} cf(c \cup \{v\})\}$;
  5:        **end while**
  6:        $Complexes = Complexes \cup \{c\}$;
  7: **end for**
  8: **return** $Complexes$;

ID ▽	GO
A	GO:01, GO:02, GO:04
B	GO:02, GO:04
C	GO:02
D	GO:02
E	GO:03, GO:04
F	GO:05
G	GO:02, GO:06
H	GO:02

fr	GO:01	GO:02	GO:03	GO:04	GO:05	GO:06
GO:01	INF	0.87	1.23	1.28	0.21	0
GO:02	0.87	INF	1.34	3.26	0	0.04
GO:03	1.23	1.34	INF	0.76	0	0.03
GO:04	1.28	3.26	0.76	INF	0	0.02
GO:05	0.21	0	0	0	INF	0.01
GO:06	0.05	0.04	0.03	0.02	0.01	INF

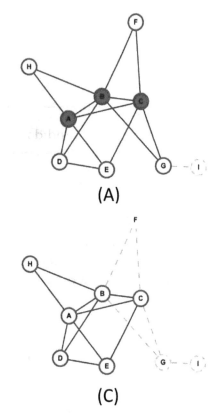

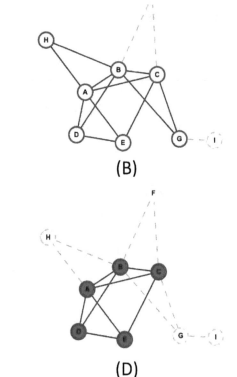

Figure 6. The diagram of Merge_Similar_Cores algorithm. In the example, (**A**) is the family graph of clique {A,B,C}, including cliques {{A,B,C},{A,B,D},{A,B,H},{A,C,E},{B,C,F},{B,C,G}}, and the proteins set is {A,B,C,D,E,F,G,H}. In (**B**), the common Gene Ontology (GO) item is GO:02, and reserve vertex E as $fr_{GO:02,GO:04} > 1.96$, while drop vertex F is $fr_{GO:02,GO:05} < 1.96$. In (**C**), drop vertex G is $\arg\max_{p \in PS} cf(PS - \{p\}) = G$. In (**D**), drop vertex H is $\arg\max_{p \in PS} cf(PS - \{p\}) = H$, and returns the next candidate-core A,B,C,D,E, as no remove operation can improve the cf.

4.3. Data Sources

Three publicly available yeast PPI networks, namely the Database of Interacting Proteins (DIP) data [30], Gavin data [14] and Srihari data collected by Srihari et al. [31], are used to evaluate the performance of our method CFOCM in protein complex prediction. DIP consists of 17,203 PPIs involving 4930 proteins, while Gavin data contains fewer proteins but is more densely connected, which consists of 6531 high-quality interactions among 1430 proteins. Srihiri data contains 20,000 interactions covering 3680 proteins derived from the BioGRID, IntAct, and MINT repositories.

Table 6. Three protein-protein interaction (PPI) networks used in the experiments.

Dataset	#Proteins	#Interactions	Average Node Degree
DIP	4930	17203	6.98
Gavin	1430	6531	9.13
Srihari	3680	20,000	10.87

To validate our predicted complexes, three reference sets of real complexes, denoted as the Munich Information Center for Protein Sequence (MIPS) [32], Saccharomyces Genome Database (SGD) [33], and CYC2008 [34], are selected as benchmarks. MIPS consists of 203 protein complexes manually curated from the literature, SGD contains 323 complexes derived from Gene Ontology-based complex annotations, and CYC2008 consists of 408 hand-curated complexes reliably backed by small-scale experiments.

The yeast GO annotation dataset is downloaded from the SGD database, and the submission data is February 2014.

5. Conclusions

In this paper, we have proposed a novel algorithm CFOCM for protein complex identification from the protein–protein interaction network. According to the fact that there some proteins involved in more than one biological function or cellular processes, CFOCM implements allowing overlaps between detected complexes. Meanwhile, CFOCM also takes into account the inherent core–attachment structure in the protein complexes. Moreover, CFOCM ensures topological similarity and functional interdependence between each pair of proteins within detected cores.

Comparison experiments between CFOCM and the other six state-of-the-art methods are carried out in DIP networks, Gavin networks and Srihari data, and the results of all tests show that CFOCM significantly outperforms the others. Moreover, CFOCM has been demonstrated to be capable of filtering the low-confidence or biological insignificant protein complexes via comparing with unCFOCM without consideration that the proteins in a complex core should occupy some common functions. In a word, CFOCM is efficient, robust, and it is applicable for helping biologists search for new biological meaningful protein complexes.

The follow-up works are ongoing. For instance, since some proteins still have not been functionally annotated, and we intend to find a more suitable strategy to handle this data problem, and design a parallel version of CFOCM to accelerate the operating speed. In addition, how to extend CFOCM to detect protein complexes and functional modules in dynamic PPI networks, which can be constructed by incorporating gene expression data, is also a promising direction.

Acknowledgments: This work is supported by the Program for New Century Excellent Talents in University (Grant NCET-10-0365), the National Nature Science Foundation of China (Grant 60973082, 11171369, 61272395, 61370171, 61300128), the National Nature Science Foundation of Hunan Province (Grant 12JJ2041), the Planned Science and Technology Project of Hunan Province (Grant 2009FJ3195, 2011FJ3123) and supported by the Fundamental Research Funds for the Central Universities, Hunan University.

Author Contributions: Bo Li and Bo Liao conceived and designed the experiments; Bo Li performed the experiments; Bo Li and Bo Liao analyzed the data; Bo Li contributed reagents/materials/analysis tools; Bo Li wrote the paper.

References

1.   Gavin, A.C.; Basche, M.; Krause, R.; Grandi, P.; Marzioch, M.; Bauer, A.; Schultz, J.; Rick, J.M.; Michon, A.M.; Cruciat, C.M.; et al. Functional organization of the yeast proteome by systematic analysis of protein complexes. *Nature* **2002**, *415*, 141–147.

2.   Eichler, E.E.; Flint, J.; Gibson, G.; Kong, A.; Leal, S.M.; Moore, J.H.; Nadeau, J.H. Missing heritability and strategies for finding the underlying causes of complex disease. *Nat. Rev. Genet.* **2010**, *11*, 446–450.

3.   Rigaut, G.; Shevchenko, A.; Rutz, B.; Wilm, M.; Mann, M.; Séraphin, B. A generic protein purification method for protein complex characterization and proteome exploration. *Nat. Biotechnol.* **1999**, *17*, 1030–1032.

4.   Ho, Y.; Gruhler, A.; Heilbut, A.; Bader, G.D.; Moore, L.; Adams, S.L.; Millar, A.; Taylor, P.; Bennett, K.; Boutilier, K.; et al. Systematic identification of protein complexes in Saccharomyces cerevisiae by mass spectrometry. *Nature* **2002**, *415*, 180–183.

5.   King, A.D.; Preulj, N.; Jurisica, I. Protein complex prediction via cost-based clustering. *Bioinformatics* **2004**, *20*, 3013–3020.

6.   Ou-Yang, L.; Dai, D.Q.; Zhang, X.F. Protein complex detection via weighted ensemble clustering based on Bayesian nonnegative matrix factorization. *PLoS ONE* **2013**, *8*, e62158.

7.   Aldecoa, R.; Priulj, I. Jerarca: Efficient analysis of complex networks using hierarchical clustering. *PLoS ONE* **2010**, *5*, e11585.

8.   Wang, J.; Li, M.; Chen, J.; Pan, Y. A fast hierarchical clustering algorithm for functional modules discovery in protein interaction networks. *IEEE/ACM Trans. Comput. Biol. Bioinform.* **2011**, *8*, 607–620.

9.   Pizzuti, C.; Rombo, S.E. A Coclustering Approach for Mining Large Protein-Protein Interaction Networks. *IEEE/ACM Trans. Comput. Biol. Bioinform.* **2012**, *9*, 717–730.

10. Bader, G.D.; Hogue, C.W.V. An automated method for finding molecular complexes in large protein interaction networks. *BMC Bioinform.* **2003**, *4*, 2.

11. Adamcsek, B.; Palla, G.; Farkas, I.J.; Derényi, I.; Vicsek, T. CFinder: Locating cliques and overlapping modules in biological networks. *Bioinformatics* **2006**, *22*, 1021–1023.

12. Palla, G.; Derinyi, I.; Farkas, I.; Vicsek, T. Uncovering the overlapping community structure of complex networks in nature and society. *Nature* **2005**, *435*, 814–818.

13. Nepusz, T.; Yu, H.; Paccanaro, A. Detecting overlapping protein complexes in protein-protein interaction networks. *Nat. Methods* **2012**, *9*, 471–472.

14. Gavin, A.C.; Aloy, P.; Grandi, P.; Krause, R.; Boesche, M.; Marzioch, M.; Rau, C.; Jensen, L.J.; Bastuck, S.; Dümpelfeld, B.; et al. Proteome survey reveals modularity of the yeast cell machinery. *Nature* **2006**, *440*, 631–636.

15. Leung, H.C.M.; Xiang, Q.; Yiu, S.M.; Chin, F.Y. Predicting protein complexes from PPI data: A core-attachment approach. *J. Comput. Biol.* **2009**, *16*, 133–144.

16. Wu, M.; Li, X.; Kwoh, C.K.; Ng, S.K. A core-attachment based method to detect protein complexes in PPI networks. *BMC Bioinform.* **2009**, *10*, 169.

17. Chen, B.; Wu, F.X. Identifying Protein Complexes Based on Multiple Topological Structures in PPI Networks. *IEEE Trans. NanoBiosci.* **2013**, *12*, 165–172.

18. Lam, W.W.M.; Chan, K.C.C. Discovering functional interdependence relationship in PPI networks for protein complex identification. *IEEE Trans. Biomed. Eng.* **2012**, *59*, 899–908.

19. Zhang Y, Lin H, Yang Z, Wang J. Construction of Ontology Augmented Networks for Protein Complex Prediction. *PLoS ONE* **2013**, *8*, e62077.

20. Hu, A.L.; Chan, K.C.C. Utilizing Both Topological and Attribute Information for Protein Coplex Identification in PPI Networks. *IEEE/ACM Trans. Comput. Biol. Bioinform.* **2013**, *10*, 780–792.

21. Li, M.; Wu, X.; Wang, J.; Pan, Y. Towards the identification of protein complexes and functional modules by integrating PPI network and gene expression data. *BMC Bioinform.* **2012**, *13*, 109.

22. Chen, B.; Fan, W.; Liu, J.; Wu, F.X. Identifying protein complexes and functional modules from static PPI networks to dynamic PPI networks. *Brief. Bioinform.* **2014**, *15*, 177–194.

23. Bernigen, D.; Pers, T.H.; Thorrez, L.; Huttenhower, C.; Moreau, Y.; Brunak, S. Concordance of gene expression in human protein complexes reveals tissue specificity and pathology. *Nucleic Acids Res.* **2013**, *41*, e171.

24. Babu, M.; Vlasblom, J.; Pu, S.; Guo, X.; Graham, C.; Bean, B.D.; Burston, H.E.; Vizeacoumar, F.J.; Snider, J.; Phanse, S.; et al. Interaction landscape of membrane-protein complexes in Saccharomyces cerevisiae. *Nature* **2012**, *489*, 585–589.

25. Li, X.; Wu, M.; Kwoh, C.K.; Ng, S.K. Computational approaches for detecting protein complexes from protein interaction networks: A survey. *BMC Genom.* **2010**, *11* (Suppl. 1), S3.

26. Srihari, S.; Leong, H.W. A survey of computational methods for protein complex prediction from protein interaction networks. *J. Bioinform. Comput. Biol.* **2013**, *11*, 1230002.

27. Ji, J.; Zhang, A.; Liu, C.; Quan, X.; Liu, Z. Survey: Functional module detection from protein-protein interaction networks. *IEEE Trans. Knowl. Data Eng.* **2013**, *26*, 261–277.

28. Dongen, S.M.V. Graph Clustering by Flow Simulation. Utrecht University Repository, 2000. Available online: https://dspace.library.uu.nl/bitstream/handle/1874/848/full.pdf?sequence=1&isAllowed=y (accessed on 6 September 2017).

29. Spirin, V.; Mirny, L.A. Protein complexes and functional modules in molecular networks. *Proc. Natl. Acad. Sci. USA* **2003**, *100*, 12123–12128.

30. Salwinski, L.; Miller, C.S.; Smith, A.J.; Pettit, F.K.; Bowie, J.U.; Eisenberg, D. The database of interacting proteins: 2004 update. *Nucleic Acids Res.* **2004**, *32* (Suppl. 1), D449–D451.

31. Srihari, S.; Yong, C.H.; Patil, A.; Wong, L. Methods for protein complex prediction and their contributions towards understanding the organisation, function and dynamics of complexes. *FEBS Lett.* **2015**, *589*, 2590–2602. doi:10.1016/j.febslet.2015.04.026.

32. Mewes, H.W.; Frishman, D.; Gmldener, U.; Mannhaupt, G.; Mayer, K.; Mokrejs, M.; Morgenstern, B.; Münsterkötter, M.; Rudd, S.; Weil, B. MIPS: A database for genomes and protein sequences. *Nucleic Acids Res.* **2002**, *30*, 31–34.

33. Cherry, J.M.; Adler, C.; Ball, C.; Chervitz, S.A.; Dwight, S.S.; Hester, E.T.; Jia, Y.; Juvik, G.; Roe, T.; Schroeder, M.; et al. SGD: Saccharomyces genome database. *Nucleic Acids Res.* **1998**, *26*, 73–79.

34. Pu, S.; Wong, J.; Turner, B.; Cho, E.; Wodak, S.J. Up-to-date catalogues of yeast protein complexes. *Nucleic Acids Res.* **2009**, *37*, 825–831.

PCVMZM: Using the Probabilistic Classification Vector Machines Model Combined with a Zernike Moments Descriptor to Predict Protein–Protein Interactions from Protein Sequences

Yanbin Wang [1,†], Zhuhong You [1,*,†], Xiao Li [1,*], Xing Chen [2], Tonghai Jiang [1] and Jingting Zhang [3]

[1] Xinjiang Technical Institutes of Physics and Chemistry, Chinese Academy of Science, Urumqi 830011, China; wangyanbin15@mails.ucas.ac.cn (Y.W.); jth@ms.xjb.ac.cn (T.J.)

[2] School of Information and Control Engineering, China University of Mining and Technology, Xuzhou 221116, China; xingchen@amss.ac.cn

[3] Department of Mathematics and Statistics, Henan University, Kaifeng 100190, China; zhangjingting15@mails.ucas.ac.cn

* Correspondence: zhuhongyou@ms.xjb.ac.cn (Z.Y.); xiaoli@ms.xjb.ac.cn (X.L.)

† These authors contributed equally to this work.

Academic Editor: Christo Z. Christov

Abstract: Protein–protein interactions (PPIs) are essential for most living organisms' process. Thus, detecting PPIs is extremely important to understand the molecular mechanisms of biological systems. Although many PPIs data have been generated by high-throughput technologies for a variety of organisms, the whole interatom is still far from complete. In addition, the high-throughput technologies for detecting PPIs has some unavoidable defects, including time consumption, high cost, and high error rate. In recent years, with the development of machine learning, computational methods have been broadly used to predict PPIs, and can achieve good prediction rate. In this paper, we present here PCVMZM, a computational method based on a Probabilistic Classification Vector Machines (PCVM) model and Zernike moments (ZM) descriptor for predicting the PPIs from protein amino acids sequences. Specifically, a Zernike moments (ZM) descriptor is used to extract protein evolutionary information from Position-Specific Scoring Matrix (PSSM) generated by Position-Specific Iterated Basic Local Alignment Search Tool (PSI-BLAST). Then, PCVM classifier is used to infer the interactions among protein. When performed on PPIs datasets of *Yeast* and *H. Pylori*, the proposed method can achieve the average prediction accuracy of 94.48% and 91.25%, respectively. In order to further evaluate the performance of the proposed method, the state-of-the-art support vector machines (SVM) classifier is used and compares with the PCVM model. Experimental results on the *Yeast* dataset show that the performance of PCVM classifier is better than that of SVM classifier. The experimental results indicate that our proposed method is robust, powerful and feasible, which can be used as a helpful tool for proteomics research.

Keywords: proteins; position-specific scoring matrix; probabilistic classification vector machines

---

1. Introduction

Recognition of protein–protein interactions (PPIs) is essential for elucidating the function of proteins and further understanding the various biological processes in cells. In the last decade, a variety of biological methods have been used for large-scale PPIs detection, such as tandem affinity purification [1], yeast two-hybrid systems [2,3], and protein chip [4]. For the limit of the experimental technique, these methods have some disadvantages, including high cost and time-intensive, as well as high rates of both false-positive and false-negative. Hence, computational methods for the detection of protein interactions have become hot research topics of proteomics research. So far, a number of computational methods have been presented for the detection of PPIs based on different data types, such as protein domains, protein structure information, genomic information and phylogenetic profiles [5–13]. However, these approaches cannot be achieved unless prior information of the protein is available. Hence, the mentioned methods are not widespread. Compared to the rapid growth of a large number of protein sequences, other data that can be used to predict the PPIs are scarce. Therefore, computational methods using only protein amino acid sequence information for PPIs prediction is especially interesting [14]. Bock and Gough used a support vector machine (SVM) with protein sequence descriptors to predict PPIs [15]. Martin et al. proposed an approach to predict PPIs by using signature product, which is a descriptor that extends from signature descriptors [16]. Najafabadi et al. attempted to solve this problem with Bayesian network [17]. Shen et al. adopted a SVM model to predict PPI network by combining Skernel function of protein pairs with a conjoint triad feature [18]. Yu-An Huang et al. developed a method by combining discrete cosine transform and using weighted sparse representation-based classifier to predict PPIs, and it has achieved very exciting prediction accuracy when applying this method to detecting yeast PPIs [19]. Yan-Zhi Guo et al. also obtained promising prediction results by adopting support vector machine and auto covariance [20]. Loris Nanni et al. developed several matrix-based protein representation methods, including [21–25]. Other feature extraction approaches based on protein sequence have been proposed in [26–34]. In this study, a novel computational approach for predicting PPIs from amino acid sequences based on a probabilistic classification vector machines model (PCVM) and a Zernike moments descriptor (PCVMZM) was proposed. The major improvement is the development of a more accurate protein sequence representation. Specifically, we employed the Zernike moments feature representation on a Position-Specific Scoring Matrix (PSSM) to extract the evolutionary information from protein sequence, and then a probabilistic classification vector machines classifier is used to infer the PPIs. In more detail, a PSSM representation is used to represent each protein. Afterward, for the sake of obtaining more representative information, we apply a Zernike moments descriptor to extract features in each protein PSSM and use Zernike moments of 12-order information and generate a 42-dimensional feature vector. Finally, we adopt the machine learning method called PCVM to accomplish classification. The proposed method was applied to *Yeast* and *H. Pylori* PPIs datasets. The experiments have shown that a PCVM prediction model with a Zernike moments descriptor yields fantastic performance. By further contrast experiment, we found that our proposed method was superior to the state-of-the-art SVM, which clearly shows that the proposed approach is trustworthy in predicting PPIs [35–39].

2. Results and Discussion

2.1. Evaluation Measure

The proposed method is evaluated against the following criteria: The Accuracy (Acc), Sensitivity (Sen), Precision (Pre), and Matthew's correlation coefficient (MCC). All the computational formula is defined as follows:

$$\text{Accuracy} = \frac{TP + TN}{TP + FP + TN + FN} \tag{1}$$

$$\text{Sensitivity} = \frac{TP}{TP + FN} \tag{2}$$

$$\text{Precision} = \frac{TP}{TP + FP} \tag{3}$$

$$\text{MCC} = \frac{(TP \times TN) - (FP \times FN)}{\sqrt{(TP + FN) \times (TN + FP) \times (TP + FP) \times (TN + FN)}} \tag{4}$$

where TP represents the number of true positive, that true samples are predicted correctly, TN represents the number of true negative that true noninteracting pairs are predicted correctly. FP represents the number of false positive that non-interacting pairs are predicted to be interaction. FN represents the number of false negative that interacting pairs are predicted to be non-interacting. In addition, the receiver operating characteristic (ROC) curve [40] is applied to evaluate the performance of our method. The area under an ROC curve (AUC) [41] also is computed.

2.2. Assessment of Prediction

In order to make our method more reliable, five-fold cross-validation was adopted to divide a whole dataset into five parts. Hence, we obtained five models through separate experiments for each data set. The prediction result of PCVM prediction models with a Zernike moments description of protein sequence on *Yeast* and *H. Pylori* datasets are shown in Tables 1 and 2. From Table 1, we can see that our proposed method achieved a good performance on the *Yeast* dataset. Its average accuracy, sensitivity, precision, and MCC are 94.48%, 95.13%, 93.92% and 89.58%, respectively. When using our proposed method on the *H. Pylori* dataset, as shown in Table 2, we also achieved some satisfactory results of average accuracy, sensitivity, precision, and MCC of 91.25%, 92.05%, 90.60% and 84.04%, respectively.

Table 1. Fivefold cross validation results using the proposed method on *Yeast* dataset.

Testing Set	Acc (%)	Sen (%)	Pre (%)	MCC (%)
1	96.38	97.21	95.57	93.02
2	94.05	95.23	92.77	88.81
3	93.07	96.73	90.27	87.06
4	94.46	94.20	94.71	89.53
5	94.42	92.26	96.26	89.46
Average	94.48 ± 1.2	95.13 ± 2.0	93.92 ± 2.4	89.58 ± 2.2

Table 2. Fivefold cross validation results using the proposed method on *H. Pylori* dataset.

Testing Set	Acc (%)	Sen (%)	Pre (%)	MCC (%)
1	89.54	92.11	86.82	81.24
2	92.11	92.68	91.41	85.46
3	91.08	91.16	91.16	83.75
4	91.42	92.25	90.34	84.31
5	92.12	92.04	93.23	85.42
Average	91.25 ± 1.1	92.05 ± 0.6	90.06 ± 2.4	84.04 ± 1.7

From the experimental results, it can be seen that our proposed approach is robust, accurate and practical for predicting PPIs. The outstanding performance for detecting PPIs can be put down to the feature extraction and the classification model of our proposed method. It is effective that Zernike moments are used for feature extraction, and the PCVM model is accurate and robust in dealing with classification problems.

2.3. Comparison with the Support Vector Machine (SVM)-Based Method

In order to further evaluate the prediction performance of the proposed entire model, the SVM model is adopted based on the *Yeast* dataset to predict PPIs using the same Zernike moments to extract feature, and then, we compared the classification result between PCVM and SVM. We employed the SVM through the library for Support Vector Machines (LIBSVM) tool [42]. SVM have two parameters, c and g, respectively. A grid search method is used to optimize parameters c and g. In our experiment, a radial basis function is used as the kernel function and the initial value c and g was set to 0.4 and 0.5.

Table 3 gives the prediction results of five-fold cross-validation over two different classification methods on the *Yeast* dataset. From Table 3, we can see that the classification method of SVM achieved 89.31% average accuracy, 87.54% average sensitivity, 90.81% average precision, 80.91% average MCC. While the classification results of the PCVM method achieved 94.48% average accuracy, 95.13% average sensitivity, 93.92% average precision, 89.58% average MCC. Experimental results show that PCVM classification method is significantly better than the SVM classification method. Comparison of ROC curves performed between RVM and SVM on the *Yeast* dataset from Figures 1 and 2, we have experimental data obtained that the PCVM classifier is more accurate and robust than the SVM classifier for detecting PPIs.

Table 3. Five-fold cross-validation results by using two models on the *Yeast* dataset.

Model	Testing Set	Acc (%)	Sen (%)	Pre (%)	MCC (%)
Probabilistic Classification Vector Machines (PCVM)	1	96.38	97.21	95.57	93.02
	2	94.05	95.23	92.77	88.81
	3	93.07	96.73	90.27	87.06
	4	94.46	94.20	94.71	89.53
	5	94.42	92.26	96.26	89.46
	Average	94.48 ± 1.2	95.13 ± 2.0	93.92 ± 2.4	89.58 ± 2.2
Support Vector Machin (SVM)	1	89.23	87.75	90.27	80.76
	2	90.48	88.73	91.49	82.74
	3	87.62	87.37	88.07	78.30
	4	89.63	88.05	90.97	81.40
	5	89.60	85.79	93.23	81.32
	Average	89.31 ± 1.7	87.54 ± 1.1	90.81 ± 1.9	80.91 ± 1.62

The main improvement is attributed to three points: (1) the main advantage of PCVM is that the truncated Gaussian priors are adopted to generate robust and sparse results—in other words, the number of weight vectors is less than SVM. Hence, the complexity of the model is reduced, besides, the model is more general; (2) The parameter optimization procedure of the PCVM based on EM algorithm and probabilistic inference not only can improve the performance, but also save the effort to do cross-validation; (3) The PCVM model is simpler and easier to be understood, because the number of basic functions does not grow linearly with the number of training points. In general, the PCVM is a sparse model that makes up the shortcoming of SVM without deskilling the generalization performance and provides probabilistic outputs. Here it is, our proposed approach can produce satisfactory results.

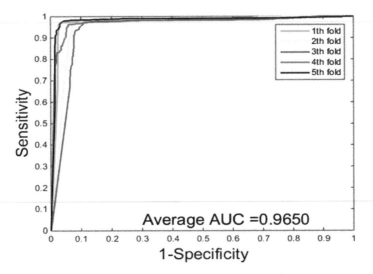

Figure 1. Receiver operating characteristic (ROC) curves performed of a probabilistic classification vector machines model (PCVM) on the *Yeast* dataset.

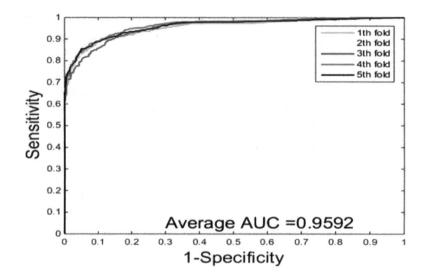

Figure 2. ROC curves performed of support vector machine (SVM) on the *Yeast* dataset.

2.4. Comparison with Other Methods

In recent years, many classification methods have been developed to predict PPIs. To further validate the performance of our proposed method, we compared the predictive performance of our method with other existing several well-known methods. The achieved results of five-fold cross-validation of different methods on the *Yeast* dataset and *H. pylori* dataset are shown in Tables 4 and 5. From Table 4, the prediction accuracy of other previous methods on the *Yeast* dataset varies from 75.08% to 93.92%, while the proposed method achieved higher value of 94.48%. Similarly, the sensitivity and MCC of our method are also higher than those of other methods. We can find similar results on the *H. pylori* dataset in Table 5. Our proposed method achieves 91.25% accuracy, which is higher than the other five methods with the highest prediction accuracy of 87.50%. The same is true for precision, sensitivity and MCC. All prediction results in Tables 4 and 5 indicate that the PCVM classifier is stable and robust and can improve the prediction performance compared with the state-of-the-art methods. The improvement of prediction performance of our method may derive from the novel feature extraction method which extracts the highly discriminative information, and the use of PCVM classifier which ensures accurate and stable prediction.

Table 4. Practical predicting results of different methods on the *Yeast* dataset.

Model	Testing Set	Acc (%)	Sen (%)	Pre (%)	MCC (%)
Guo [20]	Auto Covariance (ACC)	89.33 ± 2.67	89.93 ± 3.68	88.87 ± 6.16	N/A
	auto covariance (AC)	87.36 ± 1.38	87.30 ± 4.68	87.82 ± 4.33	N/A
Yang [23]	Cod1	75.08 ± 1.13	75.81 ± 1.20	74.75 ± 1.23	N/A
	Cod2	80.04 ± 1.06	76.77 ± 0.69	82.17 ± 1.35	N/A
	Cod3	80.41 ± 0.47	78.14 ± 0.90	81.66 ± 0.99	N/A
	Cod4	86.15 ± 1.17	81.03 ± 1.74	90.24 ± 1.34	N/A
You [24]	Principal Component Analysis-Ensemble Extreme Learning Machines (PCA-EELM)	87.00 ± 0.29	86.15 ± 0.43	87.59 ± 0.32	77.36 ± 0.44
Wong [30]	Rotation Forest (RF) + Property Response-Local Phase Quantization (PR-LPQ)	93.92 ± 0.36	91.10 ± 0.31	96.45 ± 0.45	88.56 ± 0.63
Proposed Method	PCVM	94.48 ± 1.20	95.13 ± 2.00	93.92 ± 2.40	89.58 ± 2.20

Table 5. Practical predicting results of different methods on the *H. Pylori* dataset.

Model	Acc (%)	Sen (%)	Pre (%)	MCC (%)
Nanni [23]	83.00	86.00	85.10	N/A
Nanni [32]	84.00	86.00	84.00	N/A
Nanni and Lumini [25]	86.60	86.70	85.00	N/A
Z-H You [29]	87.50	88.95	86.15	78.13
L Nanni [24]	84.00	84.00	84.00	N/A
Proposed Method	91.25	92.05	90.06	84.04

3. Materials and Methodology

3.1. Dataset

Up to now, many databases of PPIs data have been generated, such as Database of Interaction Proteins (DIP) [43], Molecular Interaction Database (MINT) [44], and Biomolecular Interaction Network Database (BIND) [45]. To evaluate our approach, we used two publicly available datasets: *Yeast* and *H. Pylori*, which were extracted from Database of Interaction Proteins (DIP). In order to ensure the reliability of the tests, we extract 5594 positive protein pairs to constitute the positive dataset and 5594 negative protein pairs to constitute the negative protein dataset from the *Yeast* dataset. Analogously, we extract 1458 positive protein pairs to constitute the positive dataset and 1458 negative protein pairs to constitute the negative protein dataset from the *H. Pylori* dataset. Therefore, the *Yeast* dataset consists of 11,188 protein pairs and the *H. Pylori* dataset consists of 2916 protein pairs.

3.2. Position-Specific Scoring Matrix

A Position-Specific Scoring Matrix (PSSM) was usually adopted to find distantly related proteins, protein disulfide, protein quaternary structural attributes and protein folding patterns [46–49]. In this paper, we also adopt PSSM to predict PPIs. Here, each protein was transformed into a PSSM matrix by employing the Position-Specific Iterated Basic Local Alignment Search Tool (PSI-BLAST) [50,51]. A PSSM is represented as

$$PSSM = (N_1, N_2, \ldots, N_i, \ldots, N_{20}) \tag{5}$$

where $N_i = (N_{1i}, N_{2i}, \ldots, N_{Li})^T$, $(i = 1, 2, \ldots, 20)$. A PSSM contains $L \times 20$ elements, where L denotes the length of an amino acid sequence and 20 columns are owing to 20 amino acids. The N_{ij} of the PSSM element is indicated as a score of jth amino acid in the ith position of the given protein sequence and it can be expressed as $N_{ij} = \sum_{k=1}^{20} p(i,k) \times q(j,k)$ where $p(i,k)$ is the appearing frequency value of the k_{th} amino acid at position i of the probe, and $q(j,k)$ represents the value of Dayhoff's mutation matrix [52] between the j_{th} and the k_{th} amino acids. Consequently, the higher the score, the better the conserved position [53–55].

In our study, the experiment datasets were built by using PSI-BLAST to transform each protein into a PSSM for detecting PPIs. To obtain more extensive homologous sequences, the e-value parameter of PSI-BLAST was set to 0.001 and chose three iterations. As a result, the PSSM of a protein sequence can be represented as a $M \times 20$ matrix, where M is the number of residues and each column represents an amino acid [56–59].

3.3. Zernike Moments

Zernike moments have an exciting performance in the field of image recognition for extract image feature, because it is robust against rotation and it can represent information from different angles. In this paper, we first introduced Zernike moments to extract significant information from protein sequences. In this section, Zernike moments and their principal properties are described, and we illustrate how to achieve the rotation invariance. Finally, we describe the process of feature selection.

3.3.1. Invariance of Normalized Zernike Moment

The principle of Zernike moments [60–63] is Zernike polynomials [64–66], that is a set of complete orthogonal polynomials within the unit circle. In two-dimensional space, these polynomials can be expressed as $\{V_{nm}(x,y)\}$ and expression is as follows:

$$V_{nm}(x,y) = V_{nm}(\rho,\theta) = R_{nm}(\rho)e^{jm\theta} \qquad \text{for } \rho \leq 1 \tag{6}$$

where n is a nonnegative integer and m is an integer subject to constraints $n-|m|$ even, $|m| \leq n$. Here, $\{R_{nm}(\rho)\}$ is a radial polynomial in the form of

$$R_{nm}(\rho) = \sum_{s=0}^{(n-|m|/2)}(-1)^s\frac{(n-s)!}{s!\left(\frac{n+|m|}{2}-s\right)!\left(\frac{n+|m|}{2}-s\right)!}\rho^{n-2s} \tag{7}$$

Note that $R_{n,-m}(\rho) = R_{nm}(\rho)$. The set of polynomials are orthogonal, i.e.,

$$\int_0^{2\pi}\int_0^1 V_{nm}^*(\rho,\theta)V_{pq}(\rho,\theta)\rho d\rho d\theta = \frac{\pi}{n+1}\delta_{np}\delta_{mq} \tag{8}$$

With

$$\delta_{ab} = \begin{cases} 1 & a=b \\ 0 & otherwise \end{cases} \tag{9}$$

The two-dimensional Zernike moments for continuous function $f(\rho,\theta)$ are the projection of $f(\rho,\theta)$ onto these orthogonal basis function and denoted by

$$A_{nm} = \frac{n+1}{\pi}\int_0^{2\pi}\int_0^1 f(\rho,\theta)V_{nm}^*(\rho,\theta)\rho d\rho d\theta \tag{10}$$

Correspondingly, for a digital function, the two-dimensional Zernike moments are represented by

$$A_{nm} = \frac{n+1}{\pi}\sum_{(\rho,\theta)\in unit\ circle}\sum f(\rho,\theta)V_{nm}^*(\rho,\theta) \tag{11}$$

To compute the Zernike moments of a PSSM matrix [67–70], the center of the matrix is taken as the origin and coordinates are mapped into a unit circle, i.e., $x^2+y^2 \leq 1$. Those values of matrix falling outside the unit disk are not used in the computation. Note that $A_{nm}^* = A_{n,-m}$.

3.3.2. Introduction of a Zernike Moments Descriptor

When we define $f'(\rho,\theta)$ as the rotated function, the equivalence between original and rotated function is

$$f'(\rho,\theta) = f(\rho,\theta-\alpha) \tag{12}$$

The Zernike moments A'_{nm} of the rotated function $f'(\rho,\theta)$ become

$$A'_{nm} = A_{nm}e^{-jm\alpha} \tag{13}$$

Equation (13) indicates that Zernike moments only need phase shift on rotation. Therefore, the magnitude of the Zernike moment, $|A'_{nm}|$, can be adopted as rotation-invariant feature.

Therefore, after moving the origin of PSSM matrix into the centroid, we can compute the Zernike moments and the magnitudes of the moments are rotation-invariant [71,72].

3.3.3. Feature Selection

According to the foregoing, we have known that the magnitudes of Zernike moments can be used as rotation-invariant features. One problem that must be considered is how big should N be?

The lower-order moments extract gross information and high details information are captured by higher-order moments. In our experiments, N is set to 12. We can obtain 42 features from each protein sequence. The feature vector $\vec{F}$ be represented as:

$$\vec{F} = [|A_{11}|, |A_{22}|, \ldots\ldots, |A_{NM}|]^T \tag{14}$$

where $|A_{nm}|$ represent the Zernike moments magnitude. Here, we do not consider the case of $m = 0$, because they do not include useful information regarding the PPIs and Zernike moments with $m < 0$ have not been considered, because they are inferred through $A_{n,-m} = A_{nm}^*$. Hence, the dimension of the feature vector $\vec{F}$ is 42 [73]. The obtained Zernike moments is shown in Table 6.

Table 6. List of Zernike Moments (ZMs) sorted by n and m in sequence for the case where $(n, m) = (12, 12)$.

N	Moments	No.	N	Moments	No.
1	A_{11}	1	7	$A_{71}, A_{73}, A_{75}, A_{77}$	4
2	A_{22}	1	8	$A_{82}, A_{84}, A_{86}, A_{88}$	4
3	A_{31}, A_{33}	2	9	$A_{91}, A_{93}, A_{95}, A_{97}, A_{99}$	5
4	A_{42}, A_{44}	2	10	$A_{10,2}, A_{10,4}, A_{10,6}, A_{10,8}, A_{10,10}$	5
5	A_{51}, A_{53}, A_{55}	3	11	$A_{11,1}, A_{11,3}, A_{11,5}, A_{11,7}, A_{11,9}, A_{11,11}$	6
6	A_{62}, A_{64}, A_{66}	3	12	$A_{12,2}, A_{12,4}, A_{12,6}, A_{12,8}, A_{12,10}, A_{12,12}$	6

3.4. Related Machine Learning Models

In the field of machine learning, the Support Vector Machines (SVM) [74] are acknowledged as an excellent supervision model in pattern recognition, classification, and regression analysis. However, there are certain apparent disadvantages when using this method: (1) the count of support vectors grows linearly with the scale of the training set; (2) Outputs of the SVMs are not probabilistic; (3) The parameters of kernel function need to be optimized by cross-validation, the procedure wastes a lot of computing resources. Compared with SVM, the Relevance Vector Machines (RVM) [75] based on Bayesian technique can avoid these problems. The RVM method takes advantage of the Bayesian automatic relevance determination (ARD) [76] framework and gives a zero-mean Gaussian prior over every weight w_i to produce a sparse solution. However, for a classification problem, the zero-mean Gaussian prior are given over weights for negative and positive classes, which leads to a problem that some training points belonging to negative classes may be given positive weights and vice-versa. Under this circumstance, it may give rise to produce some unreliable vectors for the decision of RVMs. For the sake of addressing this problem and proposing an appropriate probabilistic model for predicting PPIs, we first adopt the Probabilistic Classification Vector Machine (PCVM) classifier which gives different priors over weights for training points that belong to different classes, i.e., the non-negative, left-truncated Gaussian is used for the positive class and the non-positive, right-truncated Gaussian is used for the negative class. PCVM provides many advantages: (1) PCVM produces the probabilistic outputs for each test point; (2) It is effective that PCVM used expectation maximization (EM) algorithm to optimizing kernel parameters; (3) PCVM introduced a sparser model leading to faster performance in the test stage.

3.5. PCVM Algorithm

PCVM is a classification model that supervised learning. Hence, we need a set of input-target training pairs $\{x_i, y_i\}_{i=1}^N$, where $y_i = \{-1, +1\}$ to train a learning model $f(x; w)$, which is defined by parameters W. The model is a linear combination of N basis functions and is represented as

$$f(x; w) = \sum_{i=1}^N w_i \varnothing_{i,\theta}(x) + b \tag{15}$$

where the $\{\varnothing_{1,\theta}(x),\ldots\ldots\varnothing_{N,\theta}(x)\}$ is basis function, (wherein θ represent the parameter vector of the basis function), the $W = (w_1,\ldots\ldots,w_N)^T$ is the parameter of the PCVM model, the b is the bias.

In this paper, we adopt the radial basis function (RBF) [77] as the basis and adopt the probit link function $\psi(x) = \int_{-\infty}^{x} N(t|0,1)dt$ to obtain the binary outputs. Finally, mapping the f (x; w) into $\psi(x)$, the expression of the PCVM model becomes:

$$\text{L}(X;\ w,\ b) = \psi\left(\sum_{i=1}^{N} w_i\varnothing_{i,\theta}(x) + b\right) = \psi(\Phi_\theta(X)W + b) \tag{16}$$

A truncated Gaussian distribution as a prior is employed over each weight w_i as follow

$$p(W|\alpha) = \prod_{i=1}^{N} p(w_i|\alpha_i) = \prod_{i=1}^{N} N_t(w_i|0,\alpha_i^{-1}) \tag{17}$$

A zero-mean Gaussian distribution as a prior is employed over the bias b:

$$p(b|\beta) = N\left(b|0,\ \beta^{-1}\right) \tag{18}$$

The $N_t(w_i|0,\alpha_i^{-1})$ is a truncated Gaussian function, α_i is the precision of the corresponding parameter w_i, β represents the precision of the normal distribution of b. When $y_i = +1$, the truncated prior is a non-negative, left-truncated Gaussian, and when $y_i = -1$, the prior is a non-positive, right-truncated Gaussian. This can be represented as

$$p(w_i|\alpha_i) = \begin{cases} 2N(w_i|0,\alpha_i^{-1}) & y_iw_i \geq 0 \\ 0 & others \end{cases} \tag{19}$$

The gamma distribution is adopted as the hyper prior of α and β. Using the EM algorithm, assign the parameters of a PCVM model, such as parameters b, W and θ. The EM algorithm is an iterative algorithm, which is used to estimate the maximum likelihood or maximum posterior probability involving latent variables. For more details about the PCVM theory, please refer to [78,79].

3.6. Initial Parameter Selection and Training

The PCVM algorithm has only one parameter, θ, which can be optimized automatically in the training process. However, the EM algorithm is susceptible to initial point and trap in local maxima. Choosing the best initialization point is an effective method to avoid the local maxima. We train a PCVM model with eight initialization points over the five training folds of each data. Hence, we obtain a 5 × 8 matrix of parameters, where the rows represent the folds and the columns represent the initializations. For each row, we select the results of the lowest test error. Hence, we find only five points, and then, we select the medium over those parameters. We have experimental obtained the optimal initial value θ which is seted as 3.6 on the *Yeast* dataset and 1.18 on the *H. pylori* dataset.

4. Conclusions

Considering time, efficiency and economy, the use of computational methods based on protein amino acid sequences to predict PPIs has attracted the attention of researchers. The computational method is playing an important role in proteomics research, because it saves manpower and material resources and is more accurate and efficient. In this paper, we introduce an accurate computational method based on protein sequence. It is established by using a PCVM classifier combined with a Zernike moments descriptor on the PSSM. The experiments showed that the performance of our proposed method achieves a high classification accuracy and is superior to the SVM. The main improvements of the developed approach come from adopting a Zernike moments descriptor as feature extraction approach that can capture multi-angle useful and representative information. More than this, the use of a PCVM classifier ensures more reliable and accurate recognition, because the use of the truncated Gaussian priors can lead to obtaining robust and sparse results—the number of support vectors is less than SVM, and the probabilistic outputs produced by PCVM can assess the uncertainty of prediction on the skewed dataset. In addition, the parameter optimization procedure of the PCVM not only can improve the performance, but also save effort to do cross-validation. Due to the outstanding performance of the Zernike moments descriptor and PCVM, our method can improve the PPIs accuracy rate. All in all, our proposed method is highly efficient and stable and can be a useful tool for predicting PPIs.

Acknowledgments: This work is supported in part by the National Science Foundation of China, under Grants 61373086, 11301517 and 61572506. The authors would like to thank all the editors and anonymous reviewers for their constructive advices.

Author Contributions: Yanbin Wang and Zhuhong You conceived the algorithm, carried out analyses, prepared the data sets, carried out experiments, and wrote the manuscript. Xiao Li, Xing Chen, Tonghai Jiang and Jingting Zhang designed, performed and analyzed experiments. All authors read and approved the final manuscript.

References

1.	Puig, O.; Caspary, F.; Rigaut, G.; Rutz, B.; Bouveret, E.; Bragado-Nilsson, E.; Wilm, M.; Seraphin, B. The tandem affinity purification (TAP) method: A general procedure of protein complex purification. *Methods* **2001**, *24*, 218–229. [CrossRef] [PubMed]

2.	Staudinger, J.; Zhou, J.; Burgess, R.; Elledge, S.J.; Olson, E.N. PICK1: A perinuclear binding protein and substrate for protein kinase C isolated by the yeast two-hybrid system. *J. Cell Biol.* **1995**, *128*, 263–271. [CrossRef] [PubMed]

3.	Koegl, M.; Uetz, P. Improving yeast two-hybrid screening systems. *Brief. Funct. Genom.* **2007**, *6*, 302–312. [CrossRef] [PubMed]

4.	Zhu, H.; Snyder, M. Protein chip technology. *Curr. Opin. Chem. Biol.* **2003**, *7*, 55–63. [CrossRef]

5.	Pazos, F.; Valencia, A. Similarity of phylogenetic trees as indicator of protein-protein interaction. *Protein Eng. Des. Sel.* **2001**, *14*, 609–614. [CrossRef]

6.	Wang, B.; Chen, P.; Huang, D.S.; Li, J.J.; Lok, T.M.; Lyu, M.R. Predicting protein interaction sites from residue spatial sequence profile and evolution rate. *FEBS Lett.* **2006**, *580*, 380–384. [CrossRef] [PubMed]

7.	Maleki, M.; Hall, M.; Rueda, L. Using structural domains to predict obligate and non-obligate protein-protein interactions. *CIBCB* **2012**, 252–261. [CrossRef]

8.	Huang, C.; Morcos, F.; Kanaan, S.P.; Wuchty, S.; Chen, D.Z.; Izaguirre, J.A. Predicting protein–protein interactions from protein domains using a set cover approach. *IEEE/ACM Trans. Comput. Biol. Bioinform.* **2007**, *4*, 78–87. [CrossRef] [PubMed]

9.	Jansen, R.; Yu, H.; Greenbaum, D.; Kluger, Y.; Krogan, N.J.; Chung, S.; Emili, A.; Snyder, M.; Greenblatt, J.F.; Gerstein, M. A Bayesian networks approach for predicting protein-protein interactions from genomic data. *Science* **2003**, *302*, 449–453. [CrossRef] [PubMed]

10.	Qin, S.; Cai, L. Predicting protein–protein interaction based on protein secondary structure information using Bayesian classifier. *J. Inn. Mongolia Univ. Sci. Technol.* **2010**, *1*, 021. (In Chinese).

11.	Cai, L.; Pei, Z.; Qin, S.; Zhao, X. Prediction of protein–protein interactions in *Saccharomyces cerevisiae* Based on Protein Secondary Structure. *iCBEB* **2012**, 413–416. [CrossRef]

12.	You, Z.H.; Yu, J.Z.; Zhu, L.; Li, S.; Wen, Z.K. A MapReduce based parallel SVM for large-scale predicting protein–protein interactions. *Neurocomputing* **2014**, *145*, 37–43. [CrossRef]

13.	You, Z.H.; Zheng, Y.; Han, K.; Huang, D.S.; Zhou, X. A semi-supervised learning approach to predict synthetic genetic interactions by combining functional and topological properties of functional gene network. *BMC Bioinform.* **2010**, *11*, 1–13. [CrossRef] [PubMed]

14.	Zou, Q.; Hu, Q.; Guo, M.; Wang, G. HAlign: Fast multiple similar DNA/RNA sequence alignment based on the centre star strategy. *Bioinformatics* **2015**, *31*, 2475. [CrossRef] [PubMed]

15.	Bock, J.R.; Gough, D.A. Whole-proteome interaction mining. *Bioinformatics* **2003**, *19*, 125–134. [CrossRef] [PubMed]

16. Martin, S.; Roe, D.; Faulon, J.L. Predicting protein–protein interactions using signature products. *Bioinformatics* **2005**, *21*, 218–226. [CrossRef] [PubMed]

17. Najafabadi, H.S. Sequence-based prediction of protein–protein interactions by means of codon usage. *Genome Biol.* **2008**, *9*, 1–9. [CrossRef] [PubMed]

18. Shen, J.; Zhang, J.; Luo, X.; Zhu, W.; Yu, K.; Chen, K.; Li, Y.; Jiang, H. Predicting protein–protein interactions based only on sequences information. *Proc. Natl. Acad. Sci. USA* **2007**, *104*, 4337–4341. [CrossRef] [PubMed]

19. Huang, Y.A.; You, Z.H.; Xin, G.; Leon, W.; Wang, L. Using Weighted Sparse Representation Model Combined with Discrete Cosine Transformation to Predict Protein-Protein Interactions from Protein Sequence. *BioMed Res. Int.* **2015**, *2015*, 1–10. [CrossRef] [PubMed]

20. Guo, Y.; Yu, L.; Wen, Z.; Li, M. Using support vector machine combined with auto covariance to predict protein-protein interactions from protein sequences. *Nucleic Acids Res.* **2008**, *36*, 3025–3030. [CrossRef] [PubMed]

21. Nanni, L.; Lumini, A. An ensemble of support vector machines for predicting the membrane protein type directly from the amino acid sequence. *Amino Acids* **2008**, *35*, 573–580. [CrossRef] [PubMed]

22. Nanni, L.; Lumini, A. An ensemble of K-local hyperplanes for predicting protein-protein interactions. *Bioinformatics* **2006**, *22*, 1207–1210. [CrossRef] [PubMed]

23. Nanni, L. Fusion of classifiers for predicting protein–protein interactions. *Neurocomputing* **2005**, *68*, 289–296. [CrossRef]

24. Nanni, L.; Brahnam, S.; Lumini, A. High performance set of PseAAC and sequence based descriptors for protein classification. *J. Theor. Biol.* **2010**, *266*, 1–10. [CrossRef] [PubMed]

25. Nanni, L.; Lumini, A. A genetic approach for building different alphabets for peptide and protein classification. *BMC Bioinform.* **2008**, *9*, 45. [CrossRef] [PubMed]

26. You, Z.H.; Li, J.; Gao, X.; He, Z.; Zhu, L.; Lei, Y.K.; Ji, Z. Detecting protein-protein interactions with a novel matrix-based protein sequence representation and support vector machines. *BioMed Res. Int.* **2015**, *2015*, 1–9. [CrossRef] [PubMed]

27. You, Z.H.; Chan, K.C.C.; Hu, P. Predicting protein–protein interactions from primary protein sequences using a novel multi-scale local feature representation scheme and the random forest. *PLoS ONE* **2015**, *10*, e0125811. [CrossRef] [PubMed]

28. Wang, L.; You, Z.H.; Chen, X.; Li, J.Q.; Yan, X.; Zhang, W.; Huang, Y.A. An ensemble approach for large-scale identification of protein- protein interactions using the alignments of multiple sequences. *Oncotarget* **2016**, *8*, 5149–5159. [CrossRef] [PubMed]

29. You, Z.; Le, Y.; Zh, L.; Xi, J.; Wang, B. Prediction of protein-protein interactions from amino acid sequences with ensemble extreme learning machines and principal component analysis. *BMC Bioinform.* **2013**, *14*, S10. [CrossRef] [PubMed]

30. Wong, L.; You, Z.H.; Ming, Z.; Li, J.; Chen, X.; Huang, Y.A. Detection of Interactions between Proteins through Rotation Forest and Local Phase Quantization Descriptors. *Int. J. Mol. Sci.* **2016**, *17*, 21. [CrossRef] [PubMed]

31. Lei, Y.K.; You, Z.H.; Ji, Z.; Zhu, L.; Huang, D.S. Assessing and predicting protein interactions by combining manifold embedding with multiple information integration. *BMC Bioinform.* **2012**, *13*, S3. [CrossRef] [PubMed]

32. Nanni, L. Letters: Hyperplanes for predicting protein-protein interactions. *Neurocomputing* **2005**, *69*, 257–263. [CrossRef]

33. You, Z.H.; Li, S.; Gao, X.; Luo, X.; Ji, Z. Large-scale protein-protein interactions detection by integrating big biosensing data with computational model. *BioMed Res. Int.* **2014**, *2014*, 598129. [CrossRef] [PubMed]

34. Huang, Y.A.; You, Z.H.; Li, X.; Chen, X.; Hu, P.; Li, S.; Luo, X. Construction of Reliable Protein–Protein Interaction Networks Using Weighted Sparse Representation Based Classifier with Pseudo Substitution Matrix Representation Features. *Neurocomputing* **2016**, *218*, 131–138. [CrossRef]

35. An, J.Y.; You, Z.H.; Chen, X.; Huang, D.S.; Yan, G.Y. Robust and accurate prediction of protein self-interactions from amino acids sequence using evolutionary information. *Mol. BioSyst.* **2016**, *12*, 3702–3710. [CrossRef] [PubMed]

36. Pan, J.B.; Hu, S.C.; Wang, H.; Zou, Q.; Ji, Z.L. PaGeFinder: Quantitative identification of spatiotemporal pattern genes. *Bioinformatics* **2012**, *28*, 1544–1545. [CrossRef] [PubMed]

37. Zou, Q.; Li, X.B.; Jiang, W.R.; Lin, Z.Y.; Li, G.L.; Chen, K. Survey of MapReduce frame operation in bioinformatics. *Brief. Bioinform.* **2014**, *15*, 637. [CrossRef] [PubMed]
38. Zeng, X.; Zhang, X.; Zou, Q. Integrative approaches for predicting microRNA function and prioritizing disease-related microRNA using biological interaction networks. *Brief. Bioinform.* **2016**, *17*, 193–203. [CrossRef] [PubMed]
39. Li, P.; Guo, M.; Wang, C.; Liu, X.; Zou, Q. An overview of SNP interactions in genome-wide association studies. *Brief. Funct. Genom.* **2015**, *14*, 143–155. [CrossRef] [PubMed]
40. Fawcett, T. An introduction to ROC analysis. *Pattern Recognit. Lett.* **2006**, *27*, 861–874. [CrossRef]
41. Huang, J.; Ling, C.X. Using AUC and accuracy in evaluating learning algorithms. *Knowl. Data Eng. Trans.* **2005**, *17*, 299–310.
42. Chang, C.C.; Lin, C.J. LIBSVM: A library for support vector machines. *ACM Trans. Intell. Syst. Technol.* **2007**, *2*, 389–396. [CrossRef]
43. Quan, Z.; Li, J.; Li, S.; Zeng, X.; Wang, G. Similarity computation strategies in the microRNA-disease network: A survey. *Brief. Funct. Genom.* **2016**, *15*, 55.
44. Licata, L.; Briganti, L.; Peluso, D.; Perfetto, L.; Iannuccelli, M.; Galeota, E.; Sacco, F.; Palma, A.; Nardozza, A.P.; Santonico, E. MINT, the molecular interaction database: 2012 update. *Nucleic Acids Res.* **2012**, *40*, D857–D861. [CrossRef] [PubMed]
45. Bader, G.D.; Donaldson, I.; Wolting, C.; Ouellette, B.F.F.; Pawson, T.; Hogue, C.W.V. BIND—The Biomolecular Interaction Network Database. *Nucleic Acids Res.* **2001**, *29*, 242–245. [CrossRef] [PubMed]
46. Jones, D.T. Protein secondary structure prediction based on position-specific scoring matrices. *J. Mol. Biol.* **1999**, *292*, 195–202. [CrossRef] [PubMed]
47. Maurer-Stroh, S.; Debulpaep, M.; Kuemmerer, N.; de la Paz, M.L.; Martins, I.C.; Reumers, J.; Morris, K.L.; Copland, A.; Serpell, L.; Serrano, L. Exploring the sequence determinants of amyloid structure using position-specific scoring matrices. *Nat. Methods* **2010**, *7*, 237–242. [CrossRef] [PubMed]
48. Henikoff, J.G.; Henikoff, S. Using substitution probabilities to improve position-specific scoring matrices. *Bioinformatics* **1996**, *12*, 135–143. [CrossRef]
49. Paliwal, K.K.; Sharma, A.; Lyons, J.; Dehzangi, A. A Tri-Gram Based Feature Extraction Technique Using Linear Probabilities of Position Specific Scoring Matrix for Protein Fold Recognition. *J. Theor. Biol.* **2014**, *13*, 44–50. [CrossRef] [PubMed]
50. Altschul, S.F.; Madden, T.L.; Schäffer, A.A.; Zhang, J.; Zhang, Z.; Miller, W.; Lipman, D.J. Gapped BLAST and PSI-BLAST: A new generation of protein database search programs. *Nucleic Acids Res.* **1997**, *25*, 3389–3402. [CrossRef] [PubMed]
51. Huang, Q.Y.; You, Z.H.; Zhang, X.F.; Yong, Z. Prediction of Protein-Protein Interactions with Clustered Amino Acids and Weighted Sparse Representation. *Int. J. Mol. Sci.* **2015**, *16*, 10855–10869. [CrossRef] [PubMed]
52. Dayhoff, M. A model of evolutionary change in proteins. *Atlas Protein Seq. Struct.* **1977**, *5*, 345–352.
53. Bhagwat, M.; Aravind, L. PSI-BLAST tutorial. *Methods Mol. Biol.* **2007**, *395*, 177–186. [PubMed]
54. Xiao, R.Q.; Guo, Y.Z.; Zeng, Y.H.; Tan, H.F.; Tan, H.F.; Pu, X.M.; Li, M.L. Using position specific scoring matrix and auto covariance to predict protein subnuclear localization. *J. Biomed. Sci. Eng.* **2009**, *2*, 51–56. [CrossRef]
55. An, J.Y.; Meng, F.R.; You, Z.H.; Fang, Y.H.; Zhao, Y.J.; Ming, Z. Using the Relevance Vector Machine Model Combined with Local Phase Quantization to Predict Protein-Protein Interactions from Protein Sequences. *BioMed Res. Int.* **2016**, *2016*, 1–9. [CrossRef] [PubMed]
56. Kim, W.Y.; Kim, Y.S. A region-based shape descriptor using Zernike moments. *Signal Process. Image Commun.* **2000**, *16*, 95–102. [CrossRef]
57. Liao, S.X.; Pawlak, M. On the accuracy of Zernike moments for image analysis. *IEEE Trans. Pattern Anal. Mach. Intell.* **1998**, *20*, 1358–1364. [CrossRef]
58. Li, S.; Lee, M.C.; Pun, C.M. Complex Zernike moments features for shape-based image retrieval. *IEEE Trans. Syst. Man Cybern. Part A Syst. Hum.* **2009**, *39*, 227–237. [CrossRef]
59. Georgiou, D.N.; Karakasidis, T.E.; Megaritis, A.C. A short survey on genetic sequences, chou's pseudo amino acid composition and its combination with fuzzy set theory. *Open Bioinform. J.* **2013**, *7*, 41–48. [CrossRef]
60. Liu, T.; Qin, Y.; Wang, Y.; Wang, C. Prediction of Protein Structural Class Based on Gapped-Dipeptides and a Recursive Feature Selection Approach. *Int. J. Mol. Sci.* **2015**, *17*, 15. [CrossRef] [PubMed]

61. Wang, S.; Liu, S. Protein Sub-Nuclear Localization Based on Effective Fusion Representations and Dimension Reduction Algorithm LDA. *Int. J. Mol. Sci.* **2015**, *16*, 30343–30361. [CrossRef] [PubMed]

62. Georgiou, D.N.; Karakasidis, T.E.; Nieto, J.J.; Torres, A. A study of entropy/clarity of genetic sequences using metric spaces and fuzzy sets. *J. Theor. Biol.* **2010**, *267*, 95. [CrossRef] [PubMed]

63. Hse, H.; Newton, A.R. Sketched symbol recognition using Zernike moments. In Proceedings of the 17th International Conference on Pattern Recognition, Cambridge, UK, 23–26 August 2004; Volume 1, pp. 367–370.

64. Noll, R.J. Zernike polynomials and atmospheric turbulence. *JOsA* **1976**, *66*, 207–211. [CrossRef]

65. Wang, J.Y.; Silva, D.E. Wave-front interpretation with Zernike polynomials. *Appl. Opt.* **1980**, *19*, 1510–1518. [CrossRef] [PubMed]

66. Schwiegerling, J.; Greivenkamp, J.E.; Miller, J.M. Representation of videokeratoscopic height data with Zernike polynomials. *JOsA* **1995**, *12*, 2105–2113. [CrossRef]

67. Chong, C.W.; Raveendran, P.; Mukundan, R. A comparative analysis of algorithms for fast computation of Zernike moments. *Pattern Recognit.* **2003**, *36*, 731–742. [CrossRef]

68. Singh, C.; Walia, E.; Upneja, R. Accurate calculation of Zernike moments. *Inf. Sci.* **2013**, *233*, 255–275. [CrossRef]

69. Hwang, S.K.; Billinghurst, M.; Kim, W.Y. Local Descriptor by Zernike Moments for Real-Time Keypoint Matching. *Image Signal Process.* **2008**, *2*, 781–785.

70. Liao, S.X.; Pawlak, M. A study of Zernike moment computing. *Asian Conf. Comput. Vis.* **2006**, *98*, 394–401.

71. Khotanzad, A.; Hong, Y.H. Invariant Image Recognition by Zernike Moments. *IEEE Trans. Pattern Anal. Mach. Intell.* **1990**, *12*, 489–497. [CrossRef]

72. Kim, H.S.; Lee, H.K. Invariant image watermark using Zernike moments. *IEEE Trans.Circuits Syst. Video Technol.* **2003**, *13*, 766–775.

73. Zou, Q.; Zeng, J.C.; Cao, L.J.; Ji, R.R. A Novel Features Ranking Metric with Application to Scalable Visual and Bioinformatics Data Classification. *Neurocomputing* **2016**, *173*, 346–354. [CrossRef]

74. Burges, C.J.C. A Tutorial on Support Vector Machines for Pattern Recognition. *Data Min. Knowl. Discov.* **1998**, *2*, 121–167. [CrossRef]

75. Bishop, C.M.; Tipping, M.E.; Nh, C.C. Variational Relevance Vector Machines. *Adv. Neural Inf. Process. Syst.* **2000**, *12*, 299–334.

76. Li, Y.; Campbell, C.; Tipping, M. Bayesian automatic relevance determination algorithms for classifying gene expression data. *Bioinformatics* **2002**, *18*, 1332–1339.

77. Wei, L.Y.; Tang, J.J.; Zou, Q. Local-DPP: An Improved DNA-binding Protein Prediction Method by Exploring Local Evolutionary Information. *Inf. Sci.* **2017**, *384*, 135–144. [CrossRef]

78. Chen, H.; Tino, P.; Yao, X. Probabilistic classification vector machines. *IEEE Trans. Neural Netw.* **2009**, *20*, 901–914. [CrossRef] [PubMed]

79. Chen, H.; Tino, P.; Xin, Y. Efficient Probabilistic Classification Vector Machine With Incremental Basis Function Selection. *IEEE Trans. Neural Netw. Learn. Syst.* **2014**, *25*, 356–369. [CrossRef] [PubMed]

15

An Ameliorated Prediction of Drug–Target Interactions Based on Multi-Scale Discrete Wavelet Transform and Network Features

Cong Shen [1,2], **Yijie Ding** [1,2], **Jijun Tang** [1,2,3,*], **Xinying Xu** [4] **and Fei Guo** [1,2,*]

1   School of Computer Science and Technology, Tianjin University, Tianjin 300350, China; congshen@tju.edu.cn (C.S.); wuxi_dyj@tju.edu.cn (Y.D.)

2   Tianjin University Institute of Computational Biology, Tianjin University, Tianjin 300350, China

3   Department of Computer Science and Engineering, University of South Carolina, Columbia, SC 29208, USA

4   College of Information Engineering, Taiyuan University of Technology, Taiyuan 030024, Shanxi, China; xuxinying@tyut.edu.cn

*   Correspondence: tangjijun@tju.edu.cn (J.T.); fguo@tju.edu.cn (F.G.)

Abstract: The prediction of drug–target interactions (DTIs) via computational technology plays a crucial role in reducing the experimental cost. A variety of state-of-the-art methods have been proposed to improve the accuracy of DTI predictions. In this paper, we propose a kind of drug–target interactions predictor adopting multi-scale discrete wavelet transform and network features (named as DAWN) in order to solve the DTIs prediction problem. We encode the drug molecule by a substructure fingerprint with a dictionary of substructure patterns. Simultaneously, we apply the discrete wavelet transform (DWT) to extract features from target sequences. Then, we concatenate and normalize the target, drug, and network features to construct feature vectors. The prediction model is obtained by feeding these feature vectors into the support vector machine (SVM) classifier. Extensive experimental results show that the prediction ability of DAWN has a compatibility among other DTI prediction schemes. The prediction areas under the precision–recall curves (AUPRs) of four datasets are 0.895 (Enzyme), 0.921 (Ion Channel), 0.786 (guanosine-binding protein coupled receptor, GPCR), and 0.603 (Nuclear Receptor), respectively.

Keywords: drug–target interactions; discrete wavelet transform; network property; support vector machine

1. Introduction

Although the PubChem database [1] has stored millions of chemical compounds, the number of compounds having target protein information are limited. Drug discovery (finding new drug–target interactions, DTIs) requires much more cost and time via biochemical experiments. Hence, some efficient computational methods for predicting potential DTIs are used to cover the shortage of traditional experimental methods. There are three categories of the DTIs prediction approaches: molecular docking, matrix-based, and feature vector-based methods. Cheng et al. [2] and Rarey et al. [3] developed molecular docking methods, which were based on the crystal structure of the target binding site (3D structures). Docking simulations quantitatively estimate the maximal affinity achievable by a drug-like molecule, and these calculated values correlate with drug discovery outcomes. However, docking simulations depend on the spatial structure of targets and are usually time-consuming because of the screening technique. In contrast to docking methods, the other two kinds of computational methods (matrix-based and feature vector-based methods) can achieve the large-scale prediction of DTIs.

Compared with molecular docking, matrix-based methods of chemical structure similarity are more popular. Many matrix-based approaches are becoming popular in the area of DTI predicition. The bipartite graph learning (BGL) [4] model was firstly proposed by Yamanishi et al. They developed a new supervised method to infer unknown DTIs by integrating chemical space and genomic space into a unified space. Bleakley and Yamanishi et al. [5] raised the bipartite local model (BLM) to solve the DTI prediction problem in chemical and genomic spaces, and applied the bipartite model to transform prediction into a binary classification [5]. Mei et al. [6] improved the BLM with neighbor-based interaction-profile inferring (BLM-NII). The NII strategy inferred label information or training data from neighbors when there was no training data readily available from the query compound/protein itself. Laarhoven et al. designed kernel regularized least squares (RLS), in which they defined Gaussian interaction profile (GIP) kernels on the profiles of drugs and targets to predict DTIs [7]. Xia et al. raised Laplacian regularized least square based on interaction network (NetLapRLS) [8] to improve the prediction performance of RLS. Zheng et al. built a DTI predictor with collaborative matrix factorization (CMF) [9], which can incorporate multiple types of similarities from drugs and those from targets at once. Laarhoven et al. [10] also proposed weighted nearest neighbor with Gaussian interaction profile kernels (WNN-GIP) to predict DTIs. The WNN constructed an interaction score profile for a new drug compound using chemical and interaction information about known compounds in the dataset. Another matrix factorization-based method—kernelized Bayesian matrix factorization with twin kernels (KBMF2K) [11]—was proposed by Gönen, M. The novelty of KBMF2K came from the joint Bayesian formulation of projecting drug compounds and target proteins into a unified subspace using the similarities and estimating the interaction network in that subspace. Neighborhood regularized logistic matrix factorization (NRLMF) was raised by Liu et al. [12]. NRLMF focused on modeling the probability that a drug would interact with a target by logistic matrix factorization, where the properties of drugs and targets were represented by drug-specific and target-specific latent vectors, respectively. Nevertheless, the drawback of pairwise kernel method is the high computational complexity on the occasion of a large numbers of samples. In addition, matrix-based methods did not consider the physical and chemical properties of the target protein. These properties reflect some particular relationship between targets and the molecular structure of drugs.

To handle the above problem, other machine learning approaches of feature vector-based method was raised. Cao et al. firstly proposed several works to predict DTIs via drug (molecular fingerprint), target (sequence descriptors), and network information [13,14]. They used composition (C), transition (T), and distribution (D) and Molecular ACCess System (MACCS) fingerprint to describe target sequence and drug molecule, respectively. The above features were fed into random forest (RF) to detect DTIs.

In this article, we propose a new DTI predictor based on signal compression technology. The target sequence can be regarded as biomolecule signal of a cell. To further extract effective features from the target sequence, we utilize discrete wavelet transform (DWT) as a spectral analysis tool to compress the signal of the target sequence. According to Heisenberg's uncertainty principle, the velocity and location of moving quanta cannot be determined at the same time. Similarly, in a time–frequency coordinate system, the frequency and location of a signal cannot be determined at the same time. Wavelet transform can be based on the scale of the transformation and offset in different frequency bands, given different resolution. This is an effective scenario in practice. We also use MACCS fingerprint to describe the drug. Further more, network feature provides the relationship between drug–target pairs. Many models (e.g., BLM, BLM-NII, NetLapRLS, CMF, KBMF2K, NRLMF, and Cao's work [14]) were built with network information. Therefore, our feature contains sequence (DWT feature), drug (MACCS feature), and network (net feature). Moreover, we combine the above three types of features with support vector machine (SVM) and feature selection (FS) to develop a predictor of DTIs. We evaluate our method on four benchmark datasets including Enzyme, Ion Channel, guanosine-binding protein coupled receptor (GPCR), and Nuclear receptor. The result shows that our method achieves better prediction performance than outstanding approaches.

2. Results

We evaluated our method (DAWN) on balanced DTI datasets, described by Cao's work [14]. We analyzed the performance of features (including MACCS, DWT, and net feature). Then, we compared DAWN with other outstanding methods, including BLM [5], RLS [7], BGL [4], NetLapRLS [8], and Cao's work [14]. In addition, we also tested DAWN on imbalanced DTI datasets, compared with NetLapRLS [8], BLM-NII [6], CMF [9], WNN-GIP [10], KBMF2K [11], and NRLMF [12]. We found that DAWN achieved better values of AUCs.

2.1. Dataset

To evaluate the performance and scalability of our method, we adopted enzyme, ion channels, GPCR, and nuclear receptors used by Yamanishi et al. [4] as the gold standard datasets. These datasets come from the Kyoto Encyclopedia of Genes and Genomes (KEGG) database [15]. The information of drug–target interactions comes from KEGG BRITE [15], BRENDA [16], Super Target [17], and DrugBank databases [18]. Table 1 presents some quantitative descriptors about the golden datasets, including the number of drugs (n), number of targets (m), number of interactions, and ratio of n to m.

Table 1. Statistics of DTI datasets [4].

	Drugs (n)	Targets (m)	Interactions	Ratio (n/m)
Enzyme	445	664	2926	0.67
IC	210	204	1476	1.03
GPCR	223	95	635	2.35
Nuclear receptors	54	26	90	2.08

IC: ion channel; GPCR: guanosine-binding protein coupled receptor.

2.1.1. Balanced Dataset

In Cao's study [14], all real drug–target interaction pairs were used as the positive samples. For negative examples, they selected random, unknown interacting pairs from these drug and protein molecules. DAWN was tested on Cao's four balanced benchmark datasets (including Enzyme, Ion channels, GPCRs, and Nuclear receptors).

2.1.2. Imbalanced Dataset

The gold standard datasets only contain positive examples (interaction pairs). Hence, non-interaction drug–target pairs are considered as negative examples. Because the number of non-interaction pairs is larger than interaction pairs, the ratio between majority and minority examples is much greater than 1.

2.2. Evaluation Measurements

Three parameters were adopted as criteria: overall prediction accuracy (ACC), sensitivity (SN), and specificity (Spec).

- Accuracy:

$$ACC = \frac{TP + TN}{TP + FP + TN + FN} \tag{1}$$

- Sensitivity or Recall:

$$SN = \frac{TP}{TP + FN} \tag{2}$$

- Specificity:

$$Spec = \frac{TN}{TN + FP} \tag{3}$$

TP represents the number of positive samples predicted correctly. Similarly, we have TN, FP and FN, which represent the number of negative samples predicted correctly, the number of negative samples predicted as positive, and the positive samples predicted as negative, respectively.

In signal detection theory, a receiver operating characteristic (ROC), or simply ROC curve, is a graphical plot illustrating the performance of a binary classifier system as its varied discrimination threshold. A ROC curve can be used to illustrate the relation between sensitivity and specificity.

Area under the precision–recall curve (PRC) (AUPR) is an average of the precision weighted by a given threshold probability. We employed both ROC and the area under the precision–recall curve (PRC), because the representation of PRC is more effective than ROC on highly imbalanced or skewed datasets. Area under the ROC curve (AUC) and AUPR can quantitatively describe sensitivity against specificity and precision against recall, respectively.

2.3. Experimental Results on Balanced Datasets

2.3.1. Performance Analysis of Feature

In order to analyze the performance of MACCS, DWT, and net features, we tested these features on four balanced datasets (each set contains 10 balanced subsets) through five-fold cross-validation. Results of DWT + MACCS, DWT + MACCS (with FS), DWT + NET + MACCS, and DWT + NET + MACCS (with FS) are shown in Table 2. Because the datasets are balanced, the evaluation of ACC or AUC can measure overall performance. DWT + NET + MACCS (with FS) had the best performance of ACC on Enzyme (0.938), IC (0.943), GPCR (0.890), and Nuclear receptor (0.860), respectively. The performance (AUC) of DWT + NET + MACCS (Enzyme: 0.977, IC: 0.978, GPCR: 0.934, Nuclear receptor: 0.866) was better than DWT + MACCS (Enzyme: 0.925, IC: 0.929, GPCR: 0.872, Nuclear receptor: 0.816). The feature DWT + NET + MACCS indeed improved the prediction performance by adding network information. In addition, the performance (AUC) of DWT + NET + MACCS (with FS) (Enzyme: 0.980, IC: 0.983, GPCR: 0.950, Nuclear receptor: 0.931) was better than DWT + NET + MACCS (without FS) (Enzyme: 0.977, IC: 0.978, GPCR: 0.934, Nuclear receptor: 0.866).

Table 2. Comparison of the prediction performance between different features on balanced datasets.

Dataset	Feature	ACC	Sn	SP	AUC
Enzyme	DWT + MACCS	0.867 ± 0.002	0.861 ± 0.004	0.873 ± 0.003	0.925 ± 0.003
	DWT + MACCS (FS)	0.895 ± 0.001	0.901 ± 0.003	0.889 ± 0.003	0.949 ± 0.001
	DWT + NET + MACCS	0.932 ± 0.003	0.933 ± 0.002	0.933 ± 0.002	0.977 ± 0.002
	DWT + NET + MACCS (FS)	0.938 ± 0.002	0.938 ± 0.002	0.939 ± 0.004	0.980 ± 0.001
IC	DWT + MACCS	0.864 ± 0.003	0.868 ± 0.004	0.861 ± 0.005	0.929 ± 0.004
	DWT + MACCS (FS)	0.879 ± 0.004	0.891 ± 0.004	0.866 ± 0.007	0.935 ± 0.003
	DWT + NET + MACCS	0.940 ± 0.004	0.932 ± 0.005	0.943 ± 0.006	0.978 ± 0.003
	DWT + NET + MACCS (FS)	0.943 ± 0.002	0.938 ± 0.003	0.949 ± 0.003	0.983 ± 0.001
GPCR	DWT + MACCS	0.826 ± 0.005	0.831 ± 0.003	0.822 ± 0.007	0.872 ± 0.004
	DWT + MACCS (FS)	0.836 ± 0.006	0.846 ± 0.007	0.827 ± 0.009	0.892 ± 0.005
	DWT + NET + MACCS	0.872 ± 0.004	0.872 ± 0.005	0.872 ± 0.003	0.934 ± 0.005
	DWT + NET + MACCS (FS)	0.890 ± 0.005	0.888 ± 0.009	0.891 ± 0.011	0.950 ± 0.002
Nuclear receptor	DWT + MACCS	0.750 ± 0.011	0.619 ± 0.013	0.879 ± 0.021	0.816 ± 0.015
	DWT + MACCS (FS)	0.791 ± 0.017	0.790 ± 0.018	0.793 ± 0.036	0.850 ± 0.016
	DWT + NET + MACCS	0.805 ± 0.021	0.767 ± 0.017	0.837 ± 0.013	0.866 ± 0.011
	DWT + NET + MACCS (FS)	0.860 ± 0.009	0.855 ± 0.013	0.867 ± 0.024	0.931 ± 0.009

DWT: discrete wavelet transform; FS: feature selection; NET: network features; MACCS: drug features of molecular access system.

It is clear that FS plays a key role in elevating the prediction of our method. The FS can enhance generalization by reducing the overfitting. Obviously, the performance of DWT + NET + MACCS (with FS) can be seen from Figures 1 and 2. Network topology can be a useful supplement to improve prediction effect.

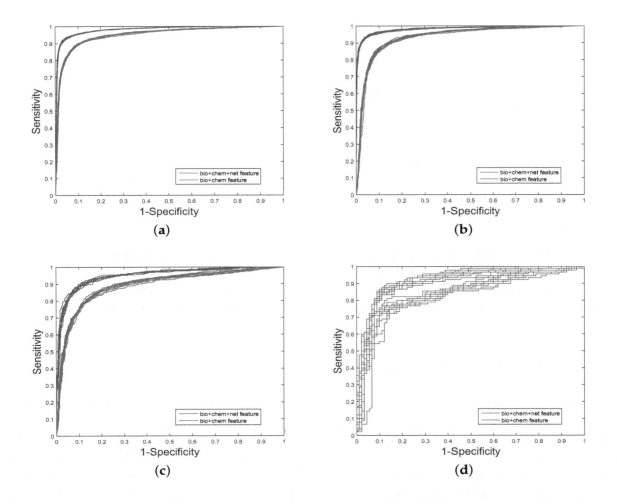

Figure 1. The area under the Receiver Operating characteristic Curve (ROC) values obtained on balanced datasets (with FS). The blue curve is the combined feature of MACCS (chem), DWT (bio), and net. The red curve is the combined feature of MACCS (chem) and DWT (bio); (**a**) Enzyme's ROC curve with network feature; (**b**) IC 's ROC curve with network feature; (**c**) GPCR's ROC curve with network feature; (**d**) Nuclear receptor's ROC curve with network feature.

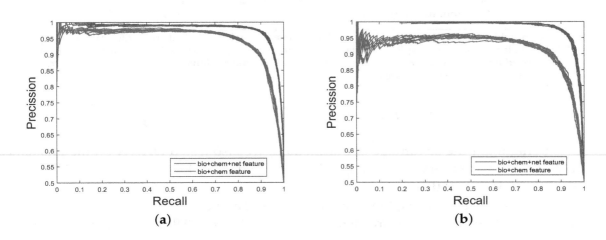

Figure 2. *Cont.*

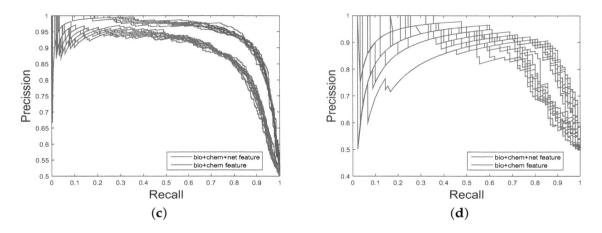

Figure 2. The area under the precision–recall (PR) curve (AUPR) values obtained on balanced datasets (with FS). The blue curve is the combined feature of MACCS (chem), DWT (bio), and net. The red curve is the combined feature of MACCS (chem) and DWT (bio); (**a**) Enzyme's PR curve with network feature; (**b**) IC's PR curve with network feature; (**c**) GPCR's PR curve with network feature; (**d**) Nuclear receptor's PR curve with network feature.

2.3.2. Comparing with Existing Methods

On the balanced datasets [14], we compare DAWN with other common methods by five-fold cross validation. These methods contain BLM [5], RLS [7], BGL [4], NetLapRLS [8] and Cao's work [14]. The detailed results are listed in Table 3. DAWN achieved the best values of AUCs on Enzyme (0.980) and Nuclear receptor (0.931), respectively. Although the AUC value of DAWN on Ion channel and GPCR datasets were not higher than Cao's work [14] and BLM, we still have a competitive prediction rate. Recapitulating about the aforementioned description, DAWN has a competitive ability among these works.

Table 3. The mean AUC values of five methods on balanced datasets.

Methods	Enzyme	IC	GPCR	Nuclear Receptor
Cao's work [14]	0.979	**0.987**	0.951	0.924
BGL	0.904	0.851	0.899	0.843
BLM	0.976	0.973	**0.955**	0.881
NetLapRLS	0.956	0.947	0.931	0.856
RLS	0.978	0.984	0.954	0.922
DAWN (our method)	**0.980**	0.983	0.950	**0.931**

Results excerpted from [14]. The best results in each column are in bold faces. BGL: bipartite graph learning; BLM: bipartite local model; NetLapRLS: Laplacian regularized least square based on interaction network; RLS: regularized least square. DAWN: prediction of Drug–tArget interactions based on multi-scale discrete Wavelet transform and Network features.

2.4. Experimental Results on Imbalanced Datasets

In order to highlight the advantage of our method, we also tested DAWN on the imbalanced datasets of DTIs by 10-fold cross validation. DAWN was compared with NetLapRLS [8], BLM-NII [6], CMF [9], WNN-GIP [10], KBMF2K [11], and NRLMF [12]. The detailed results are listed in Table 4. Because the datasets are imbalanced, the evaluation of AUC and AUPR were both used to measure overall performance. DAWN achieved average AUCs of 0.981, 0.990, 0.952, and 0.906, and the AUPR values of DAWN were 0.895, 0.921, 0.786, and 0.603 on Enzyme, Ion channel, GPCR, and Nuclear receptor, respectively. The AUC value of DAWN on the Enzyme dataset was 0.981 and AUPR was 0.895, and only the NRLMF (AUC: 0.987, AUPR: 0.892) method was comparable. On Ion channel and GPCR datasets, we also had best or second-best results. For AUPR value on Nuclear receptor, NRLMF was higher than DAWN. The Nuclear receptor dataset is smaller than the other three datasets. The size of the dataset might be a reason for DAWN's performance. Therefore, the DAWN method that adopted the mean of DWT was not as effective as larger datasets. However, among methods in Table 4, none could give markedly higher prediction performance on all four datasets in both AUC and AUPR. Therefore, it is fair to claim that our strategy has comparable performance. Further, Figures 3 and 4 show the curves of AUC and AUPR on imbalanced datasets through 10-fold cross validation. Related datasets, codes, and figures of our algorithm are available at https://github.com/6gbluewind/DTI_DWT.

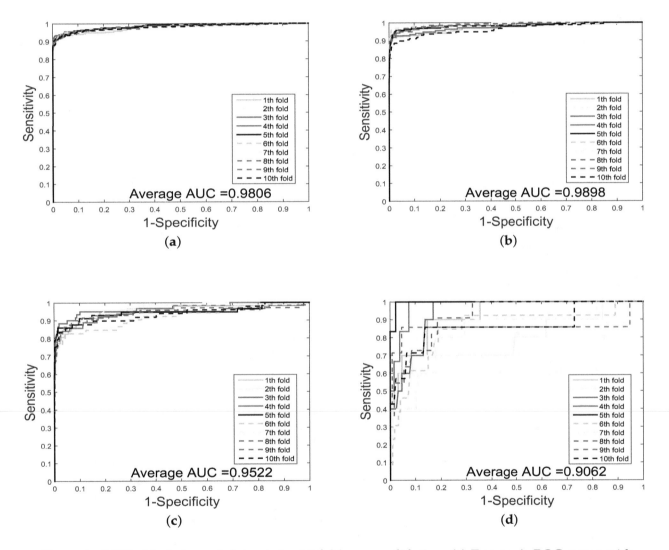

Figure 3. ROC of imbalanced datasets by 10-fold cross-validation; (**a**) Enzyme's ROC curve with network feature; (**b**) IC's ROC curve with network feature; (**c**) GPCR's ROC curve with network feature; (**d**) Nuclear receptor's ROC curve with network feature.

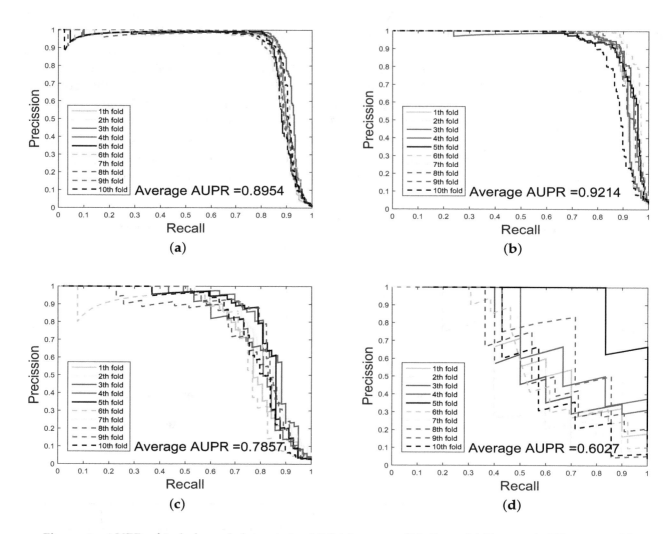

Figure 4. AUPR of imbalanced datasets by 10-fold cross-validation. (**a**) Enzyme's PR curve with network feature. (**b**) IC's PR curve with network feature. (**c**) GPCR's PR curve with network feature. (**d**) Nuclear receptor's PR curve with network feature.

Table 4. Overall AUC and AUPR values of different methods on imbalanced dataset for four species.

Evaluation	Method	Enzyme	Ion Channel	GPCR	Nuclear Receptor
AUC	NetLapRLS	0.972 ± 0.002	0.969 ± 0.003	0.915 ± 0.006	0.850 ± 0.021
	BLM-NII	0.978 ± 0.002	0.981 ± 0.002	0.950 ± 0.006	0.905 ± 0.023
	WNN-GIP	0.964 ± 0.003	0.959 ± 0.003	0.944 ± 0.005	0.901 ± 0.017
	KBMF2K	0.905 ± 0.003	0.961 ± 0.003	0.926 ± 0.006	0.877 ± 0.023
	CMF	0.969 ± 0.002	0.981 ± 0.002	0.940 ± 0.007	0.864 ± 0.026
	NRLMF	$\mathbf{0.987} \pm 0.001$	$\underline{0.989} \pm 0.001$	$\mathbf{0.969} \pm 0.004$	$\mathbf{0.950} \pm 0.011$
	DAWN	$\underline{0.981} \pm 0.004$	$\mathbf{0.990} \pm 0.014$	$\underline{0.952} \pm 0.009$	$\underline{0.906} \pm 0.067$
AUPR	NetLapRLS	0.789 ± 0.005	0.837 ± 0.009	0.616 ± 0.015	0.465 ± 0.044
	BLM-NII	0.752 ± 0.011	0.821 ± 0.012	0.524 ± 0.024	$\underline{0.659} \pm 0.039$
	WNN-GIP	0.706 ± 0.017	0.717 ± 0.020	0.520 ± 0.021	0.589 ± 0.034
	KBMF2K	0.654 ± 0.008	0.771 ± 0.009	0.578 ± 0.018	0.534 ± 0.050
	CMF	0.877 ± 0.005	$\mathbf{0.923} \pm 0.006$	0.745 ± 0.013	0.584 ± 0.042
	NRLMF	$\underline{0.892} \pm 0.006$	0.906 ± 0.008	$\underline{0.749} \pm 0.015$	$\mathbf{0.728} \pm 0.041$
	DAWN	$\mathbf{0.895} \pm 0.011$	$\underline{0.921} \pm 0.036$	$\mathbf{0.786} \pm 0.023$	0.603 ± 0.087

Results excerpted from [12]. The best results in each column are in bold faces and the second best results are underlined. BLM-NII: improved BLM with neighbor-based interaction-profile inferring; CMF: collaborative matrix factorization; KBMF2K: kernelized Bayesian matrix factorization with twin kernels; NRLMF: neighborhood regularized logistic matrix factorization; WNN-GIP: weighted nearest neighbor with Gaussian interaction profile kernels.

2.5. Predicting New DTIs

In this experiment, the balanced DTIs were set as training data sets. We ranked the remaining non-interacting pairs and selected the top five non-interacting pairs as predicted interactions. We utilized four well-known biological databases (including ChEMBL (C) [19], DrugBank (D) [18], KEGG (K) [15] and Matador (M) [20]) as references to verify whether or not the predicted new DTIs are true. The predicted novel interactions by DAWN can be ranked based on the interaction probabilities, which are shown in Table 5. The potential DTIs may be present in one or several databases. For example, the secondly ranked DTI of GPCR (D00563: hsa3269) belongs to DrugBank and Matador databases. In addition, the DTI databases (the above four databases) are still being updated, and the accuracy of identifying new DTIs by DAWN may be increased.

Table 5. Top five new DTIs predicted by DAWN on four data sets.

Dataset	Rank	Drug	Target	Databases
Enzyme	1	D00545	hsa1571	
	2	D03365	hsa1571	
	3	D00437	hsa1559	M
	4	D00546	hsa1571	
	5	D00184	hsa5478	D
Ion channel	1	D00542	hsa6262	
	2	D00542	hsa6263	M
	3	D00349	hsa6263	
	4	D00477	hsa6336	C
	5	D01448	hsa3782	
GPCR	1	D01051	hsa3269	
	2	D00563	hsa3269	D, M
	3	D00563	hsa1812	D
	4	D00715	hsa1129	D, K
	5	D00563	hsa1129	
Nuclear receptor	1	D01689	hsa5241	
	2	D01115	hsa5241	
	3	D00443	hsa5241	D
	4	D00443	hsa367	D
	5	D00187	hsa2099	

C: ChEMBL; D: DrugBank; K: KEGG; M: Matador.

3. Discussion

In this paper, we proposed a new DTIs predictor based on signal compression technology. We encoded the drug molecule by a substructure fingerprint with a dictionary of substructure patterns. Moreover, we applied the DWT to extract features from target sequences. At last, we concatenated the target, drug, and network features to construct predictive model of DTIs.

To evaluate the performance of our method, the DTIs model was compared to other state-of-the-art DTIs prediction methods on four benchmark datasets. DAWN achieved average AUCs of 0.981, 0.990, 0.952, and 0.906, and the AUPR values of DAWN were 0.895, 0.921, 0.786, and 0.603 on Enzyme, Ion channel, GPCR, and Nuclear receptor, respectively. Although our result using feature selection could be a kind of ameliorated prediction, the imbalanced problem of DTIs prediction is not solved very well. SVM is poor on imbalanced data. The AUPR value of DAWN is low on the Nuclear receptor dataset.

4. Materials and Methods

To predict DTIs by machine learning methods, one challenge is to extract effective features from the target protein, drug, and the relationship between drug–target pairs. Considering that DTIs depend on the molecular properties of the drug and the physicochemical properties of target, we use MACCS fingerprints (Open Babel 2.4.0 Released, OpenEye Scientific Software, Inc., Santa Fe, New Mexico, United States) to represent the drug, and extract biological features from the target via DWT.

In addition, the net feature describes the topology information of the DTIs network. We utilize the above features to train the SVM predictor (LIBSVM Version 3.22, National Taiwan University, Taiwan, China) for detecting DTIs.

4.1. Molecular Substructure Fingerprint of Drug

To encode the chemical structure of the drug, we utilize MACCS fingerprints with 166 common chemical substructures. These substructures are defined in the Molecular Design Limited (MDL) system, which can be found from OpenBabel (http://openbabel.org). The MACCS feature is encoded by a binary bits vector, which shows the presence (1) or absence (0) of some specific substructures in a molecule. Please refer to the relevant literature [13,14] for details.

4.2. Biological Feature of Target

4.2.1. Six Physicochemical Properties of Amino Acids

The target sequence can be denoted by $seq = \{r_1, r_2, \cdots, r_i, \cdots, r_L\}$, where $1 \leq i \leq L$. r_i is the i-th residue of sequence seq, and L is the length of sequence seq. In addition, for ease of calculation about feature representation, we select six kinds of physicochemical properties for 20 amino acid types as original target features [21–24]. More specifically, they are hydrophobicity (H), volumes of side chains of amino acids (VSC), polarity (P1), polarizability (P2), solvent-accessible surface area (SASA) and net charge index of side chains (NCISC), respectively. Values of all kinds of amino acid are shown in Table 6.

Table 6. Six physicochemical properties of 20 amino acid types.

Amino Acid	H	VSC	P1	P2	SASA	NCISC
A	0.62	27.5	8.1	0.046	1.181	0.007187
C	0.29	44.6	5.5	0.128	1.461	−0.03661
D	−0.9	40	13	0.105	1.587	−0.02382
E	−0.74	62	12.3	0.151	1.862	0.006802
F	1.19	115.5	5.2	0.29	2.228	0.037552
G	0.48	0	9	0	0.881	0.179052
H	−0.4	79	10.4	0.23	2.025	−0.01069
I	1.38	93.5	5.2	0.186	1.81	0.021631
K	−1.5	100	11.3	0.219	2.258	0.017708
L	1.06	93.5	4.9	0.186	1.931	0.051672
M	0.64	94.1	5.7	0.221	2.034	0.002683
N	−0.78	58.7	11.6	0.134	1.655	0.005392
P	0.12	41.9	8	0.131	1.468	0.239531
Q	−0.85	80.7	10.5	0.18	1.932	0.049211
R	−2.53	105	10.5	0.291	2.56	0.043587
S	−0.18	29.3	9.2	0.062	1.298	0.004627
T	−0.05	51.3	8.6	0.108	1.525	0.003352
V	1.08	71.5	5.9	0.14	1.645	0.057004
W	0.81	145.5	5.4	0.409	2.663	0.037977
Y	0.26	117.3	6.2	0.298	2.368	0.023599

H: hydrophobicity; VSC: volumes of side chains of amino acids; P1: polarity; P2: polarizability; SASA: solvent-accessible surface area; NCISC: net charge index of side chains.

For the sake of facilitating the dealing with the datasets, the amino acid residues are translated and normalized according to Equation (4).

$$P'_{ij} = \frac{P_{ij} - P_j}{S_j} (j = 1, 2, \ldots, 6; \ i = 1, 2, \ldots, 20) \tag{4}$$

where $P_{i,j}$ and P_j indicate the value of the j-th descriptor of amino acid type i and the mean of 20 amino acid types of descriptor value j, respectively, standard deviation (SD) corresponding to S_j.

Each target sequence can be translated into six vectors with each amino acid represented by normalized values of six descriptors. Thus, the seq can be represented as physicochemical matrix $X = [x_1, \ldots, x_{ch}, \ldots, x_6], X \in R^{L \times 6}, x_{ch} \in R^{L \times 1}, ch = 1, 2, \ldots, 6$.

4.2.2. Discrete Wavelet Transform

Discrete wavelet transform (DWT) with its inversion formula was established by physical intuition and practical experience of signal processing [25].

If a signal or a function can be represented as Equation (5), then the signal or function has a linear decomposition. If the formula of expansion is unique, then the set of expansion can be said as a group of basis. If this group of basis is orthogonal or represented as Equation (6), then the coefficient can be computed by inner product as Equation (7).

$$f(t) = \sum_{\ell} a_{\ell} \psi_{\ell}(t), \tag{5}$$

$$(\psi_k(t), \psi_{\ell}(t)) = \int \psi_k(t) \psi_{\ell}(t) dt = 0, k \neq \ell, \tag{6}$$

$$a_k = (f(t), \psi_k(t)) = \int f(t) \psi_k(t) dt, \tag{7}$$

where ℓ and k are the finite or infinite integer indexes, a_{ℓ} and a_k are the real coefficients of the expansion, and $\psi_{\ell}(t)$ and $\psi_k(t)$ are the set of real functions.

For wavelet expansion, we can construct a system with two parameters, then the formula can be transferred as Equation (8):

$$f(t) = \sum_k \sum_j a_{j,k} \psi_{j,k}(t), \tag{8}$$

where j and k are integer index, and $\psi_{j,k}(t)$ is wavelet function, which generally forms a group of orthogonal basis.

The expansion coefficient set $a_{j,k}$ is known as the discrete wavelet transform (DWT) of $f(t)$. Nanni et al. proposed an efficient algorithm to perform DWT by assuming that the discrete signal $f(t)$ is $x_{ch}(n)$.

$$y_{l,high,ch}(n) = \sum_{k=1}^{L} [x_{ch}(k) \cdot h(2n - k)] \tag{9a}$$

$$y_{l,low,ch}(n) = \sum_{k=1}^{L} [x_{ch}(k) \cdot g(2n - k)] \tag{9b}$$

where h and g refer to high-pass filter and low-pass filter, L is the length of discrete signal, $y_{l,low,ch}(n)$ is the approximate coefficient (low-frequency components) of the signal, $l(l = 1,2,3,4)$ is the decomposition level of DWT, $ch(ch = 1,2,3,4,5,6)$ is the physicochemical index, and $y_{l,low,ch}(n)$ is the detailed coefficient (high-frequency components).

DWT can decompose discrete sequences into high- and low-frequency coefficients. Nanni et al. [26] substituted each amino acid of the protein sequence with a physicochemical property. Then, the protein sequence was encoded as a numerical sequence. DWT compresses discrete sequence and removes noise from the origin sequence. Different decomposition scales with discrete wavelet have different results for representing the sequence of the target protein. They used 4-level DWT and calculated the maximum, minimum, mean, and standard deviation values of different scales (four levels of both low- and high-frequency coefficients). In addition, high-frequency components are more noisy while low-frequency components are more critical. Therefore, they extracted the beginning of the first five Discrete Cosine Transform (DCT) coefficients from the approximation coefficients. We utilize Nanni's method to describe the sequence of the target protein. The schematic diagram of a 4-level DWT is shown in Figure 5.

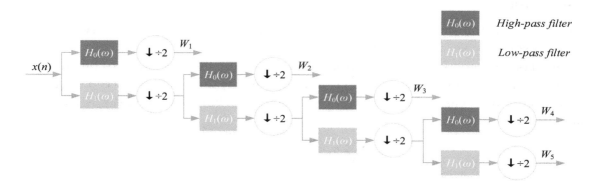

Figure 5. Wavelet decomposition tree.

4.3. Drug–Target Associations from Network

State-of-the-art works such as BLM [5], BLM-NII [6], NetLapRLS [8], CMF [9], KBMF2K [11], NRLMF [12], and Cao's work [14] used DTI network topology information to improve the prediction performance. Therefore, we also consider utilizing net feature to build a DTI predictor.

The DTI network can be conveniently regarded as a bipartite graph. In the network, each drug is associated with n_t targets, and each target is associated with n_d drugs. Excluding target T_j itself, we make a binary vector of all other known targets of D_i in the bipartite network, as well as a separate list of targets not known to be targeted by D_i. Known and unknown targets are labeled by 1 and 0, respectively. For drug D_i, we get $(n_t - 1)$-dimensional binary vector. Similarly, we also get $(n_d - 1)$-dimensional binary vector of target T_j. Thus, we can get a $[(n_d - 1) + (n_t - 1)]$-dimensional vector for describing net feature.

4.4. Feature Selection and Training SVM Model

Not all features are useful for DTIs prediction. Therefore, we apply support vector machine recursive feature elimination and correlation bias reduction (SVM-RFE+CBR) [27,28] to select the important features of DTIs. The SVM-RFE+CBR can estimate the score of importance for each dimensional feature. We rank these features (including MACCS feature, DWT feature, and net feature) by the scores in descending order. Then, we select an optimal feature subset in top k ranked manner to predict DTIs.

Support vector machine (SVM) was originally developed by Vapnik [29] and coworkers, and has shown a promising capability to solve a number of chemical or biological classification problems. SVM and other machine learning algorithms (e.g., random forest, RF, k-nearest neighbor, kNN, etc.) are widely used in computational biology [30–33]. SVM performs classification tasks by constructing a hyperplane in a multidimensional space to differentiate two classes with a maximum margin. The input data of SVM is defined as $\{x_i, y_i\}, i = 1, 2, ..., N$, feature vector $x_i \in R^n$ and labels $y_i \in \{+1, -1\}$.

The classification decision function implemented by SVM is shown as Equation (10).

$$f(\mathbf{x}) = sgn\{\sum_{i=1}^{N} y_i \alpha_i \cdot K(\mathbf{x}, \mathbf{x_i}) + b\} \tag{10}$$

where the coefficient α_i is obtained by solving a convex quadratic programming problem, and $K(\mathbf{x}, \mathbf{x_i})$ is called a kernel function.

Here, we focus on choosing a radial basis function (RBF) kernel [34], because it not only has better boundary response but can also make most high-dimensional data approximate a Gaussian-like distribution. The architecture of our proposed method is shown in Figures 6 and 7.

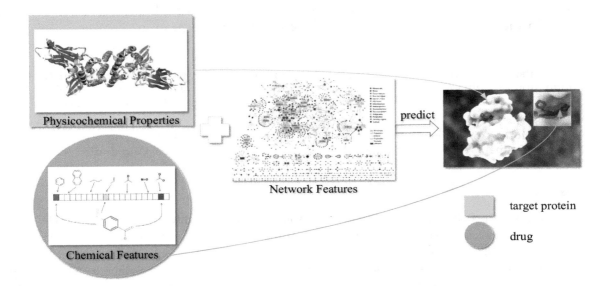

Figure 6. Overview of the drug–target interaction (DTI) prediction.

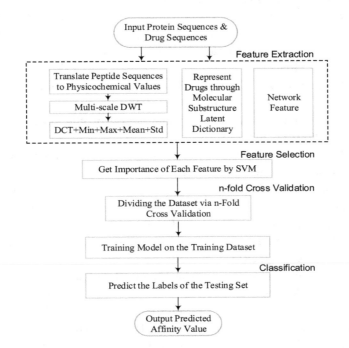

Figure 7. Flow chart. DWT: discrete wavelet transform; DCT: discrete cosine transform; Std: standard deviation; SVM: support vector machine.

5. Conclusions

In this paper, we present a DTI prediction method by using multi-scale discrete wavelet transform and network features. We employ a DWT algorithm to extract target features, and combine them with drug fingerprint and network feature. Our method can achieve satisfactory prediction performances, and our prediction can be a kind of ameliorated prediction by comparing with other existing methods after feature selection. However, the imbalanced problem of DTIs prediction is not solved very well. SVM is poor on imbalanced data. The AUPR value of DAWN is low on the Nuclear receptor dataset.

The prediction accuracy may be further enhanced with the further expansion of more refined representation of the structural and physicochemical properties or a better machine learning model (such as sparse representation and gradient boosting decision tree) for predicting drug–target interactions. In the future, we will build the classification by the strategy of bootstrap sampling and weighting sub-classifiers.

Acknowledgments: This research and this article's publication costs are supported by a grant from the National Science Foundation of China (NSFC 61,772,362, 61,402,326), Peiyang Scholar Program of Tianjin University (no. 2016XRG-0009), and the Tianjin Research Program of Application Foundation and Advanced Technology (16JCQNJC00200).

Author Contributions: Cong Shen, Yijie Ding and Fei Guo conceived the study. Cong Shen and Yijie Ding performed the experiments and analyzed the data. Cong Shen, Yijie Ding and Fei Guo drafted the manuscript. All authors read and approved the manuscript.

References

1. Sayers, E.W.; Barrett, T.; Benson, D.A.; Bryant, S.H.; Canese, K.; Chetvernin, V.; Church, D.M.; DiCuccio, M.; Edgar, R.; Federhen, S.; et al. Database resources of the national center for biotechnology information. *Nucleic Acids Res.* **2009**, *37*, 5–15.
2. Cheng, A.C.; Coleman, R.G.; Smyth, K.T.; Cao, Q.; Soulard, P.; Caffrey, D.R.; Salzberg, A.C.; Huang, E.S. Structure-based maximal affinity model predicts small-molecule druggability. *Nat. Biotechnol.* **2007**, *25*, 71–75.
3. Rarey, M.; Kramer, B.; Lengauer, T.; Klebe, G. A fast flexible docking method using an incremental construction algorithm. *J. Mol. Biol.* **1996**, *261*, 470–489.
4. Yamanishi, Y.; Araki, M.; Gutteridge, A.; Honda, W.; Kanehisa, M. Prediction of drug–target interaction networks from the integration of chemical and genomic spaces. *Bioinformatics* **2008**, *24*, i232–i240.
5. Bleakley, K.; Yamanishi, Y. Supervised prediction of drug–target interactions using bipartite local models. *Bioinformatics* **2009**, *25*, 2397–2403.
6. Mei, J.P.; Kwoh, C.K.; Yang, P.; Li, X.L.; Zheng, J. Drug-target interaction prediction by learning from local information and neighbors. *Bioinformatics* **2013**, *29*, 238–245.
7. Van, L.T.; Nabuurs, S.B.; Marchiori, E. Gaussian interaction profile kernels for predicting drug–target interaction. *Bioinformatics* **2011**, *27*, 3036–3043.
8. Xia, Z.; Wu, L.Y.; Zhou, X.; Wong, S.T. Semi-supervised drug-protein interaction prediction from heterogeneous biological spaces. *BMC Syst. Biol.* **2010**, *4*, 6–17.
9. Zheng, X.; Ding, H.; Mamitsuka, H.; Zhu, S. Collaborative matrix factorization with multiple similarities for predicting drug–target interactions. In Proceedings of the 19th ACM SIGKDD International Conference on Knowledge Discovery and Data Mining, Chicago, IL, USA , 11–14 August 2013; pp. 1025–1033.
10. Van, L.T.; Marchiori, E. Predicting Drug-Target Interactions for New Drug Compounds Using a Weighted Nearest Neighbor Profile. *PLoS ONE* **2013**, *8*, e66952.
11. Gönen, M. Predicting drug–target interactions from chemical and genomic kernels using Bayesian matrix factorization. *Bioinformatics* **2012**, *28*, 2304–2310.
12. Liu, Y.; Wu, M.; Miao, C.; Zhao, P.; Li, X.L. Neighborhood Regularized Logistic Matrix Factorization for Drug-Target Interaction Prediction. *PLoS Comput. Biol.* **2016**, *12*, e1004760.
13. Cao, D.S.; Liu, S.; Xu, Q.S.; Lu, H.M.; Huang, J.H.; Hu, Q.N.; Liang, Y.Z. Large-scale prediction of drug–target interactions using protein sequences and drug topological structures. *Anal. Chim. Acta* **2012**, *752*, 1–10.
14. Cao, D.S.; Zhang, L.X.; Tan, G.S.; Xiang, Z.; Zeng, W.B.; Xu, Q.S.; Chen, A.F. Computational Prediction of DrugTarget Interactions Using Chemical, Biological, and Network Features. *Mol. Inform.* **2014**, *33*, 669–681.
15. Kanehisa, M.; Goto, S.; Hattori, M.; Aoki-Kinoshita, K.F.; Itoh, M.; Kawashima, S.; Katayama, T.; Araki, M.; Hirakawa, M. From genomics to chemical genomics: New developments in KEGG. *Nucleic Acids Res.* **2006**, *34*, 354–357.
16. Schomburg, I.; Chang, A.; Placzek, S.; Söhngen, C.; Rother, M.; Lang, M.; Munaretto, C.; Ulas, S.; Stelzer, M.; Grote, A.; et al. BRENDA in 2013: Integrated reactions, kinetic data, enzyme function data, improved disease classification: New options and contents in BRENDA. *Nucleic Acids Res.* **2013**, *41*, 764–772.
17. Hecker, N.; Ahmed, J.; Eichborn, J.V.; Dunkel, M.; Macha, K.; Eckert, A.; Gilson, M.K.; Bourne, P.E.; Preissner, R. SuperTarget goes quantitative: Update on drug–target interactions. *Nucleic Acids Res.* **2012**, *40*, 1113–1117.

18. Law, V.; Knox, C.; Djoumbou, Y.; Jewison, T.; Guo, A.C.; Liu, Y.; Maciejewski, A.; Arndt, D.; Wilson, M.; Neveu, V.; et al. DrugBank 4.0: Shedding new light on drug metabolism. *Nucleic Acids Res.* **2014**, *42*, 1091–1097.

19. Gaulton, A.; Bellis, L.J.; Bento, A.P.; Chambers, J.; Davies, M.; Hersey, A.; Light, Y.; McGlinchey, S.; Michalovich, D.; Al-Lazikani, B.; et al. Chembl: A large-scale bioactivity database for drug discovery. *Nucleic Acids Res.* **2012**, *40*, D1100–D1107.

20. Günther, S.; Kuhn, M.; Dunkel, M.; Campillos, M.; Senger, C.; Petsalaki, E.; Ahmed, J.; Urdiales, E.G.; Gewiess, A.; Jensen, L.J.; et al. Supertarget and matador: Resources for exploring drug–target relationships. *Nucleic Acids Res.* **2008**, *36*, 919–922.

21. Ding, Y.; Tang, J.; Guo, F. Predicting protein-protein interactions via multivariate mutual information of protein sequences. *BMC Bioinform.* **2016**, *17*, 389–410.

22. Ding, Y.; Tang, J.; Guo, F. Identification of Protein–Protein Interactions via a Novel Matrix-Based Sequence Representation Model with Amino Acid Contact Information. *Int. J. Mol. Sci.* **2016**, *17*, 1623.

23. Li, Z.; Tang, J.; Guo, F. Learning from real imbalanced data of 14-3-3 proteins binding specificity. *Neurocomputing* **2016**, *217*, 83–91.

24. You, Z.H.; Lei, Y K.; Zhu, L.; Xia, J.F.; Wang, B. Prediction of protein-protein interactions from amino acid sequences with ensemble extreme learning machines and principal component analysis. *BMC Bioinform.* **2013**, *14*, doi:10.1186/1471-2105-14-S8-S10.

25. Mallat, S.G. A Theory for Multiresolution Signal Decomposition: The Wavelet Representation. *IEEE Trans. Pattern Anal. Mach. Intell.* **1989**, *11*, 674–693.

26. Nanni, L.; Brahnam, S.; Lumini, A. Wavelet images and Chou's pseudo amino acid composition for protein classification. *Amino Acids* **2012**, *43*, 657–665.

27. Guyon, I.; Weston, J.; Barnhill, S.; Vapnik, V. Gene selection for cancer classification using support vector machines. *Mach. Learn.* **2002**, *46*, 389–422.

28. Yan, K.; Zhang, D. Feature selection and analysis on correlated gas sensor data with recursive feature elimination. *Sens. Actuators B Chem.* **2015**, *212*, 353–363.

29. Cortes, C.; Vapnik, V. Support-vector networks. *Mach. Learn.* **1995**, *20*, 273–297.

30. Zou, Q.; Zeng, J.C.; Cao, L.J.; Ji, R.R. A Novel Features Ranking Metric with Application to Scalable Visual and Bioinformatics Data Classification. *Neurocomputing* **2016**, *173*, 346–354.

31. Zou, Q.; Wan, S.X.; Ju, Y.; Tang, J.J; Zeng, X.X. Pretata: Predicting TATA binding proteins with novel features and dimensionality reduction strategy. *BMC Syst. Biol.* **2016**, *10* (Suppl. 4), 114.

32. Wei, L.Y.; Tang, J.J.; Zou, Q. Local-DPP: An Improved DNA-binding Protein Prediction Method by Exploring Local Evolutionary Information. *Inf. Sci.* **2017**, *384*, 135–144.

33. Zou, Q.; Li, J.J.; Hong, Q.Q.; Lin, Z.Y.; Wu, Y.; Shi, H.; Ju, Y. Prediction of microRNA-disease associations based on social network analysis methods. *BioMed Res. Int.* **2015**, *2015*, 810514.

34. Chang, C.C.; Lin, C.J. LIBSVM: A Library for support vector machines. *ACM Trans. Intell. Syst. Technol.* **2011**, *2*, 389–396.

16

Identification of Direct Activator of Adenosine Monophosphate-Activated Protein Kinase (AMPK) by Structure-Based Virtual Screening and Molecular Docking Approach

Tonghui Huang *, Jie Sun, Shanshan Zhou, Jian Gao and Yi Liu *

Jiangsu Key Laboratory of New Drug Research and Clinical Pharmacy, School of Pharmacy,
Xuzhou Medical University, Xuzhou 221004, China; jxbp0812@163.com (J.S.);
ZSS1991530@163.com (S.Z.); gaojian@xzhmu.edu.cn (J.G.)
* Correspondence: tonghhuang@xzhmu.edu.cn (T.H.); cbpeliuyinew@163.com (Y.L.)

Abstract: Adenosine monophosphate-activated protein kinase (AMPK) plays a critical role in the regulation of energy metabolism and has been targeted for drug development of therapeutic intervention in Type II diabetes and related diseases. Recently, there has been renewed interest in the development of direct β1-selective AMPK activators to treat patients with diabetic nephropathy. To investigate the details of AMPK domain structure, sequence alignment and structural comparison were used to identify the key amino acids involved in the interaction with activators and the structure difference between β1 and β2 subunits. Additionally, a series of potential β1-selective AMPK activators were identified by virtual screening using molecular docking. The retrieved hits were filtered on the basis of Lipinski's rule of five and drug-likeness. Finally, 12 novel compounds with diverse scaffolds were obtained as potential starting points for the design of direct β1-selective AMPK activators.

Keywords: Adenosine 5′-monophosphate-activated protein kinase; virtual screening; molecular docking; selective activator

1. Introduction

Kidney disease associated with diabetes is the leading cause of chronic kidney disease (CKD) and end-stage kidney disease worldwide and nearly one-third of patients with diabetes develop nephropathy [1]. As the incidence of both types 1 and 2 diabetes rises worldwide, diabetic nephropathy (DN) is likely become a significant health and economic burden for society [2]. Current therapy for diabetic nephropathy includes glycemic optimization using antidiabetics and blood pressure control with blockade of the renin-angiotensin system [3]. However, these strategies are slow but cannot reverse or at least stop the disease progression [4]. Although several clinical trials are currently in progress, there are still no drugs approved for the treatment of DN. Among these ongoing phase 3 clinical trials, atrasentan is still in progress, while bardoxolone methyl and paricalcitol failed to meet the primary endpoint or was terminated on safety concerns [4,5]. Recently, there has been renewed interest in the development of direct β1-selective Adenosine monophosphate-activated protein kinase (AMPK) activators that have the potential to treat diabetic nephropathy [6].

AMPK is master sensor of cellular energy and plays a critical role in the regulation of metabolic homeostasis [7]. AMPK is a heterotrimeric kinase comprised of a highly conserved catalytic α subunit and two regulatory subunits (β and γ) [8]. The α subunit possess a N-terminal serine/threonine catalytic kinase domain (KD) that is followed by an autoinhibitory domain (AID) and a C-terminal β subunit-binding domain [9]. The β subunit serves as a scaffold to bridge α and γ subunits that contains a glycogen binding domain (GBD) and a C-terminal domain [10]. The γ subunit is composed of a β subunit-binding region and two Bateman domains [11]. These seven subunits (α1, α2, β1, β2, γ1, γ2, and γ3) are encoded by separate genes, resulting in 12 different αβγ AMPK heterotrimers [12]. The distinct physiological functions of each AMPK isoforms are not fully understood, but derive from differential expression patterns among different tissues [13]. For instance, the α1 subunit appears to be relatively evenly expressed in kidney, rat heart, liver, brain, lung and skeletal muscle tissues, while the α2 subunit is mainly expressed in skeletal muscle, heart, and liver tissues [14]. Among the two known β subunits, β1 subunit is highly abundant in kidney as suggested by mRNA levels [6].

More recently, a direct AMPK activator PF-06409577 was reported to activate α1β1γ1 and α2β1γ1 AMPK isoforms with EC_{50} of 7.0 nM and 6.8 nM but was much less active against α1β2γ1/α2β2γ1/α2β2γ3 AMPK isoforms with EC50 greater 4000 nM [6]. Besides, compound PF-06409577 exhibited efficacy in a preclinical model of diabetic nephropathy. Compounds A-769662 and 991 possessed similar potency toward AMPK heterotrimers containing a β1 subunit as PF-96409577 [15]. On the other hand, an allosteric site of AMPK has been named allosteric drug and metabolite site (ADaM site) [16], which was constructed by the catalytic kinase domain (KD) of α subunit and the regulatory carbohydrate-binding module (CBM) of β subunit [13,17]. The three known direct AMPK activators (PF-06409577 [6], A-769662 [18], and 991 [19], Figure 1) all bound to the allosteric site and showed better potency for isoforms that contain the β1 subunit. This implies that the allosteric site can be used to design the selective activators of AMPK containing the β1 subunit.

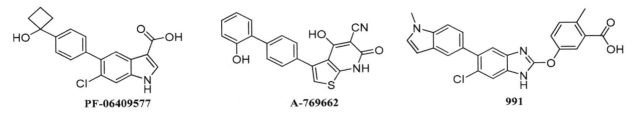

Figure 1. Structures of reported direct AMPK activators.

The present study aims to investigate details of the domain structure and identify new potential β1-selective AMPK activators. Hence, sequence alignment and structural comparison were used to identify the key amino acids that are involved in the interaction with activators and structure difference between different subunits. Furthermore, molecular docking was performed for virtual screening to discover direct β1-selective AMPK activators. The screened retrieved hits were then subjected to several filters such as estimated activity and quantitative estimation of drug-likeness (QED) [20,21]. Finally, 12 compounds with diverse scaffolds were selected as potential hit compounds for the design of novel β1-selective AMPK activators. These findings provided a useful molecular basis for the design and development of novel β1-selective AMPK activators.

2. Results and Discussion

2.1. Sequence Alignment and Structural Comparison

To reveal the possible molecular mechanism for the selective potency of activators against the β1-isoform of AMPK, sequence and secondary structure elements comparison between carbohydrate-binding module of β1 and β2 subunits were investigated. As shown in Figure 2, the sequences that were boxed blue were located within the range of 5 Å of active site. Sequence alignment reveals that β1 and β2 subunits shares 77.1% sequence identity. As shown in Figure 3, superposition with the two subunits reveals a deflexion of sheet1 in β2 subunit as compared with β1 subunit. The Phe-82 of β1 subunit corresponded to Ile-81 in β2 subunit, as well as the Thr-85 to Ser-84,

Gly-86 to Glu-85, which may account for the deflexion of sheet1 in β2 subunit. The large aromatic Phe residues and small Thr and Gly presented a binding surface more capable of accommodating ligand.

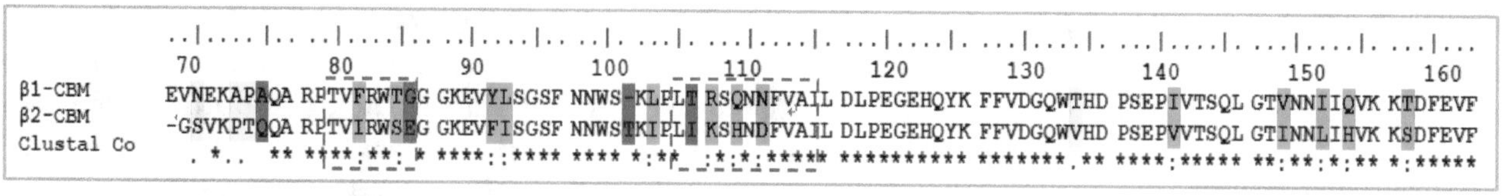

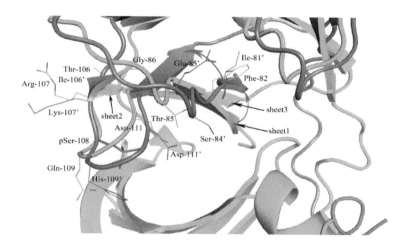

Figure 2. Sequence alignment of carbohydrate-binding module from the β1 and β2 subunits. Asterisks indicate positions that have a single, fully conserved residue. Colon (green) indicates conservation between groups of strongly similar properties. Period (yellow) indicates conservation between groups of weakly similar properties. Blank character (red) indicates conservation between groups of strongly different properties.

Figure 3. Structural comparison of the scope within 5 Å of the active site from β1 and β2 subunits. The α subunit was shown in cartoon and colored by the cyan. The β1 and β2 subunits were shown in cartoon and colored by green and blue, respectively. The sites with different amino acids were shown in line. The Ser-108 was shown in stick and colored by red.

The sheet 2 torsion may attribute to the amino acid sequence differences of the sites of 106 and 107. The most notable is supposed to the Ser108 (red and stick), an autophosphorylation site, phosphorylated serine (pSer108) formed hydrogen bonds with Thr-21, Lys-29, Lys-31, His-109′, and Asn-111′ enhancing the ADaM site stabilization [22], and the phosphate group contributed to the binding of activators [23]. The Gln-109′ and Asn-111′ were mutated to His-109′ and Asp-111′, which abolished original hydrogen bonds and generated a large conformational change. We speculated that the above differences between β1 and β2 may affect the binding of activators to AMPk isoforms.

2.2. Parameter Setting for Molecular Docking

Docking parameters, which exert an important influence on molecular docking-based virtual screening, were optimized in advance. The crystal structure of PF-06409577 bound to the α1β1γ1 AMPK isoform (PDB ID: 5KQ5) and A-769662 bound to the α2β1γ1 AMPK isoform (PDB ID: 4CFF) were chosen as the reference, the docking parameters were adjusted until the docked poses were as close as possible to the original crystallized structures. The ring flexibility was mainly considered in final optimized docking parameters according to the default settings. The overlay of the original ligand from X-ray crystal (stick and magenta) and the conformation from Surflex-Dock results (stick and green) were shown in Figure 4, in which the indole moiety of PF-06409577 and terminal benzene ring of A-769662 generated a little deflection and there was no effect on the interaction between compounds

Identification of Direct Activator of Adenosine Monophosphate-Activated Protein Kinase (AMPK)...

239

and the active site. The hydrogen bond interactions appeared consistent with the original ligands and the root mean square deviation (RMSD) between these two conformations are 0.53 and 0.56 Å, respectively. The molecular docking results indicated that the Surflex-Dock was reliable and could be used for the further virtual screening.

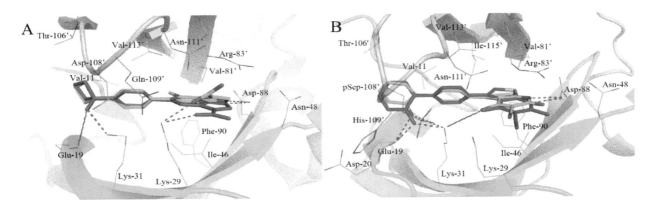

Figure 4. Conformation comparison of the original ligand from X-ray crystal (magenta and stick) and the conformation from Surflex-Dock result (green and stick). (**A**): PF-06409577; (**B**): A-769662. The indole moiety of PF-06409577 and terminal benzene ring of A-769662 generated a little deflection in compared with the original conformation. The hydrogen bond was labeled by red dashed lines.

2.3. High-Throughput Virtual Screening Procedure

To identify new potent activators of AMPK, virtual screening was performed on the active sites as mentioned previously. A chemical library containing with 1,500,000 commercially available compounds (ChemDiv database) was docked to the molecular models of $\alpha1\beta1\gamma1$ and $\alpha2\beta1\gamma1$ AMPK isoforms in silico, respectively. Prior to docking, the ChemDiv database was split into eight subsets for molecular docking. About 600 top ranked compounds with high total-scores were screened and subsequently checked for their binding modes and interactions with the active site, especially the hydrogen bonds formed with the residuals of Asp-88, Lys-29, Lys-31, and Gly-19. Then the potential hit compounds were evaluated for their drug-likeness model scores using Lipinski's rule of five (Table 1). Finally, six potential hits with new scaffolds could serve as activators for $\alpha1\beta1\gamma1$ AMPK isoform and six for $\alpha2\beta1\gamma1$ AMPK isoform were visually chosen from the top potential hits.

Table 1. The docking scores and drug-likeness model scores of selected activators for AMPK ($\alpha1\beta1\gamma1$ and $\alpha2\beta1\gamma1$).

Isoforms	Compound No.	Total-Score	Crash	Polar	Similarity	Number of HBA/HBD	MolLog P	Drug-Likeness Model Score
AMPK ($\alpha1\beta1\gamma1$) activators	F064-1335	10.50	−2.35	3.54	0.44	6/1	3.79	−0.16
	M5653-1884	10.24	−1.43	2.90	0.50	5/1	6.07	0.22
	D454-0135	10.20	−1.58	3.41	0.44	6/2	4.03	−0.31
	M8006-4303	10.07	−1.83	4.29	0.58	6/1	1.01	1.02
	F264-3019	9.93	−1.39	3.19	0.46	6/1	5.44	1.00
	F377-1213	10.03	−1.43	1.32	0.44	6/1	4.12	0.16
	PF-06409577	7.29	−0.07	3.08	0.93	3/3	3.80	0.71
AMPK ($\alpha2\beta1\gamma1$) activators	L267-1138	10.96	−2.46	2.78	0.52	4/1	6.05	−0.08
	F684-0053	10.60	−2.77	4.31	0.54	7/3	2.04	0.54
	C804-0412	10.15	−3.27	3.39	0.54	5/2	2.53	1.00
	M5976-1661	9.46	−0.92	1.32	0.46	6/0	4.84	0.62
	M039-0295	9.35	−1.61	1.48	0.50	6/1	2.66	−0.20
	M5050-0116	9.27	−1.35	2.82	0.49	7/2	5.13	0.38
	A-769662	7.44	−1.46	1.26	0.93	5/3	3.46	0.30
	991	8.38	−0.96	4.08	0.73	4/2	5.38	0.41

2.4. Analysis of Binding Mode of Activators for α1β1γ1 AMPK Isoform

The structures of retrieved hits as activators of α1β1γ1 AMPK isoform are shown in Figure 5. Although these compounds possess different chemical scaffolds, they exhibit similar binding modes at the active site. Among these compounds, compounds F064-1335 and M5653-1884 possess higher docking scores, compounds M8006-4303 and F264-3019 have perfect drug-likeness model scores.

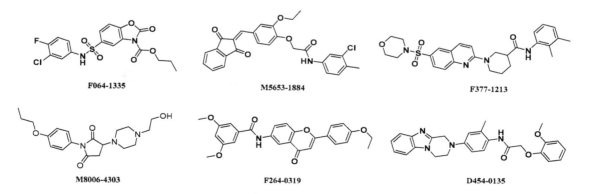

Figure 5. Structures of retrieved hits targeting α1β1γ1 AMPK isoform from ChemDiv database.

As shown in Figure 6A, the compound F064-1335 with the highest docking score (10.50) formed several hydrogen bonds with active site residues. The two oxygen atoms of sulfonamide established a hydrogen bond network with the side chain of Lys-31, Lys-29, and Asn-111'. The carbonyl oxygen atom of the ester group formed two hydrogen bonds with the main chain of Lys-29, the anther oxygen atom of the ester group was bound to the main chain of Asn-48 by a hydrogen bond, which made the alkoxy group trend into a hydrophobic pocket formed by the Lys-51, Ile-52, Val-62, and Leu-47. In addition, the carbonyl oxygen of benzoxazolone ring formed a hydrogen bond with the side chain of Arg-83'.

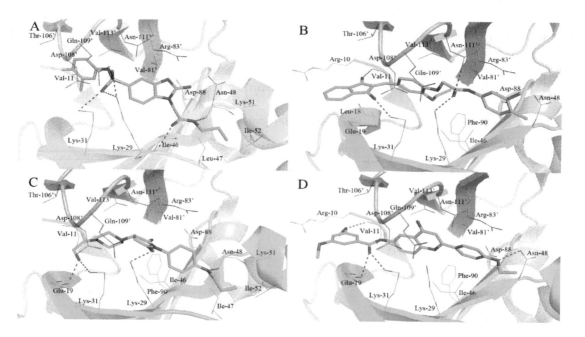

Figure 6. The binding modes of typical hit compounds for α1β1γ1 AMPK isoform. (**A**): F064-1335; (**B**): M5653-1884; (**C**): M8006-4303; (**D**): F264-0391. The α subunit was shown in cartoon and colored by cyan and the β subunit was shown in cartoon and colored by pink. The hydrogen bonds were labeled with red dashed lines.

The compound M5653-1884 with a considerable docking score (10.24) and the bind mode is shown in Figure 6B. Four hydrogen bonds were observed between the compound and the active site residues. One carbonyl oxygen atom of 1,3-indandione formed hydrogen bond with the side chain of Lys-31, another carbonyl oxygen atom formed a hydrogen bond with the main chain of Val-11. The carbonyl oxygen atom of the amide group showed hydrogen bond interactions with the side chain of Lys-29 and Asn-111. In addition, there was a hydrophobic effect with the side chain of Ile-46, Asn-48, Asp-88, and Phe-88.

As shown in Figure 6C, the compound of M8006-4303 exhibited similar binding mode as PF-06409577. The ethanol group attached to the piperazine group participated in two hydrogen bond interactions with the side chain of Gly-19 and Lys-31. The carbonyl oxygen atoms of pyrrolidine-2,5-dione formed a hydrogen bond interaction with the side chain of Lys-29. In addition, the oxygen atom of oxygen butyl associated with the benzene ring accepted a hydrogen bond from the main chain of Asn-48. Within the cavity of the active site, Ile-47, Asn-48, Lys-51, and Ile-52 probably generated a hydrophobic effect.

The binding mode of compound F264-3091 with a prefect drug-likeness score (1.00) was shown in Figure 6D. The oxgen atom of an oxyethyl group on the benzene ring participated in a hydrogen bond with the main chain of Val-11. The carbonyl oxygen atom of the amide group showed two hydrogen bonds with the main chain of Gln-19 and side chain of Lys-31. In addition, one hydrogen bond was formed between the side chain of Asn-48 and the oxyethyl group connected with flavone B-ring while the B-ring showed a stacked cation-π interaction with the side chain of Val-83'.

2.5. Analysis of Binding Mode of Activators for α2β1γ1 AMPK Isoform

The chemical structures of six compounds as activators of α2β1γ1 AMPK isoform are shown in Figure 7. The molecular docking results indicated that all the compounds possess higher docking scores than A-769662 and 991. The binding modes of the representative compound M2958-7438 and M5050-0116 in the active site of α2β1γ1 AMPK isoform are shown in Figure 8.

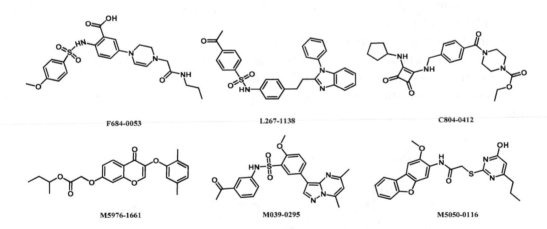

Figure 7. Structures of retrieved hits targeting α2β1γ1 AMPK isoform from ChemDiv database.

As shown in Figure 8A, six hydrogen bonds were formed between the compound M2958-7438 and active site residues, in which the barbituric acid ring formed three hydrogen bonds with the side chain of Asp-88, making prominent contributions to the high docking score (10.04). The oxygen atom of the anisole associated with the barbituric acid ring accepted a hydrogen bond from the side chain of Lys-29, and two oxygen atoms in the linker participated in two hydrogen bonds with the side chain of Lys-31. In addition, the barbituric acid ring generated a stacked cation-π interaction with the side chain of Arg-83'.

The compound M5050-0116 with a docking score of 9.27 and formed four hydrogen bonds with Val-11, Leu-18, Lys-29, and Asn-111'. As shown in Figure 8B, the Lys-29 of α subunit and

Asn-111' of β subunit, simultaneously coordinated the oxygen atom of dibenzofuran with hydrogen bonds. The hydroxyl group attached on pyrimidine anchored in a suitable geometry and formed two hydrogen bonds with the main chain of Glu-19 and the side chain of Val-11. Additionally, the compound M5050-0116 exhibited hydrophobic interactions with several residues, which formed a hydrophobic pocket including Ile-46, Leu-47, Asn-48, Asp-88, and Phe-90.

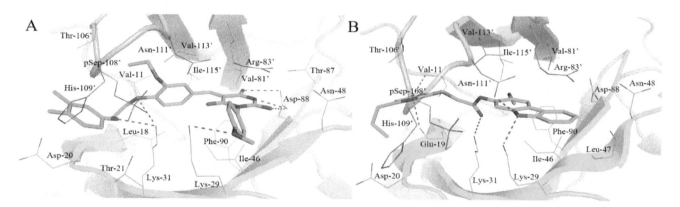

Figure 8. The binding modes of typical activators for α2β1γ1 AMPK isoform. (**A**): M2958-7438; (**B**): M5050-0116. The α subunit was shown in the illustration and colored cyan and the β subunit was shown in the illustration and colored pink. The hydrogen bonds were labeled with red dashed lines.

2.6. Biological Activities

The six screened compounds based on α1β1γ1 AMPK isoform were evaluated for activities against AMPK (α1β1γ1 isoform) at a dosages of 2 μM. A-769662, a known β1-selective AMPK activator, was used as a control. The preliminary in vitro assay (Figure 9) indicated that most of the selected α1β1γ1 AMPK activators displayed promising activation potency against α1β1γ1 AMPK isoform. Compounds D454-0135 and F264-3019 displayed comparable activation activity against AMPK in the comparison with the known β1-selective AMPK activator A-769662. Morever, compounds M8006-4303 and F264-3019 showed stronger activation activities against α1β1γ1 AMPK than A-769662. Compound M563-1884 with a higher logP value and showed a relatively low activation activity among the assayed compounds. This implies that the lipophilicity may play an important role in the bioavailability for the compound.

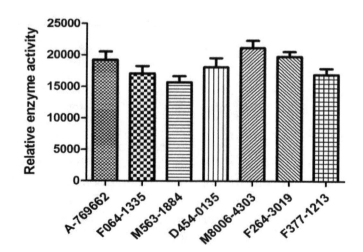

Figure 9. Activation of AMPK (α1β1γ1 isoform) by the screened compounds was measured using Elisa Kit.

3. Materials and Methods

3.1. Sequence Alignment and Structural Comparison

Sequence alignment is an essential method for similarity/dissimilarity analysis of protein, DNA, or RNA sequences [24]. The software used for sequence alignment tasks include HAlign, BioEdit, EMBL-EBI, T-Coffee, and CLUSTAL [25,26]. The crystal structure data of AMPK ($\alpha1\beta1\gamma1$: 5KQ5, 4QFG; $\alpha1\beta2\gamma1$: 4REW and $\alpha2\beta1\gamma1$: 4CFF) were obtained from RCSB Protein Data Bank [6,8,13,19], as well as the amino acid sequences of carbohydrate-binding module. The amino acid sequences of carbohydrate-binding module (CBM) on β subunits were used to study the differences. The sequence alignment between $\alpha1\beta1\gamma1$ isoform (PDB: 4QFG) and $\alpha1\beta2\gamma1$ (PDB:4REW) isoform was performed and edited using BioEdit software (version 7.1.8) [27], which is a user-friendly biological sequence alignment editor and analysis program. The crystallographic structures of AMPK for molecular docking studying were added to the hydrogen atoms and the charge was given to the Gasteiger-Huckel. The crystal structures comparison was conducted by Sybyl X 2.1 (Tripos Associates Inc., S.H. R.: St. Louis, MO, USA.) [28] and the binding modes were generated by PyMOL V0.99 (Schrödinger, New York, NY, USA.) [29]. The polar hydrogen atoms were added to the crystal structures of the AMPK via the biopolymer module and the Gasteiger-Huckel charges were loaded on the atoms of proteins. The protein peptide backbones were shown in cartoon and colored by different colors, the side chains of the nonconservative amino acids were shown in line and colored by chain.

3.2. Molecular Docking

The virtual screening and molecular docking studies were performed using Surflex docking module in Sybyl X 2.1. There were still some deficiencies due to the fact that the receptor was regarded as a rigid structure. Therefore, it was essential to optimize the docking parameters, the co-crystallized ligand was extracted and re-docked into the active site of the AMPK with the varied parameters, and then the conformation of the original ligand and the re-docking ligand were compared. The binding site was defined as a sphere containing the residues that stay within 5 Å from the co-ligand. The maximum conformations per fragment and maximum number of rotatable bonds per molecule were 20 and 100, respectively. Furthermore, the options for pre-dock minimization and post-dock minimization of molecules were omitted, while other parameters were set as default options. The top 20 conformational poses were selected according to the docking score. Dock scores were evaluated by Consensus Score (CScore), which integrates the strengths of individual scoring functions combine to rank the affinity of ligands bound to the active site of a receptor.

3.3. High-Throughput Virtual Screening

High-throughput virtual screening was regarded as an important tool to identify novel lead compounds suitable for specific protein targets [30], and the screened compounds can be easily obtained from commercial sources for biological evaluation as well [31]. The ChemDiv database was supplied by Topscience Co. (Shanghai, China), which includes 1,500,000 compounds was employed for virtual screening through Surflex docking module in Sybyl X 2.1. To accelerate virtual screening, the maximum quantity of conformations was reduced from 20 to 10, the maximum quantity of rotatable bonds was decreased from 100 to 50, and the top six conformations were collected. The same as the molecular docking studies, the default optimization of molecules before and after the docking was canceled. Other parameters were kept as default values. Compounds PF-06409577 and A-769662 were severed as reference molecules, respectively. The compounds with the docking score (≥ 8.0) were extracted for further analyzing the interactions between ligand and active site, to this end, 100 compounds were collected to calculate the drug-likeness model score. Drug-likeness model scores were computed for hit compounds using the MolSoft software (MolSoft, San Diego, CA, USA) [32].

3.4. The In Vitro Activation Assay

The in vitro preliminary kinase assays human α1β1γ1 AMPK were carried out according to the previous experimental method [33]. The screened compounds and α1β1γ1 AMPK isoform were provided by Topscience Co. (Shanghai, China) and Huawei Pharmaceutical Co. Ltd. (Shanghai, China), respectively. Generally, each of the evaluated compounds was dissolved in 10% Dimethyl sulphoxide at 10 μM and diluted to a required concentration with buffer solution. Then, 5 μL of the dilution was added to a 30 μL kinase assay buffer and 5 μL AMPK isoform per well. The solution was mixed at 0 °C for 30 min. Next, 5 μL of AMARA petide and 5 μL Adenosine triphosphate (ATP) were added to the well. The enzymatic reactions were conducted at 30 °C for 30 min. The AMPK activity was determined by quantifying the amount of ATP remaining in assay solution with Kinase-Glo Plus luminescent kinase assay kit (Promega, Madison, WI, USA). The luminescent signal is correlated with the amount of ATP present, while inversely correlated with the kinase activity. The mean values from three independent experiments were used for the expression of relative activities. A-769662, a β1-selective AMPK activator reported by Abbott laboratories, was used as a control.

4. Conclusions

In summary, the sequence alignment and structural comparison was performed to identify the AMPK domain structure detail, which provides a molecular basis of selective AMPK activators on β1-containing isoforms. The key amino acid residues (Phe/Ile82, Thr/Ser85, Gly/Glu86, Thr/Ile106, Arg/Lys107, Gln/His109, Asn/Asp111) may contribute to the selectivity and provide a foundation for structure-based design of new direct β1-selective AMPK activators. Furthermore, the structure-based virtual screening workflow for the identification of selective activators of AMPK (α1β1γ1 and α2β1γ1) was established and six potential hit compounds for α1β1γ1 isoform and α2β1γ1 isoform were obtained, respectively. The preliminary assay indicated that most of the selected α1β1γ1 AMPK activators displayed promising activation potency. Overall, these findings revealed extensive interactions of activators and AMPK for rational design of novel selective AMPK activators. Further in vitro testing of retrieved hits is still in progress in our laboratory.

Acknowledgments: This work was supported by the Postdoctoral Science Foundation funded project (2017M611916), Natural Science Foundation of Jiangsu Province (Grants No. BK20140225), Science and Technology Plan Projects of Xuzhou (Grants No. KC16SG249), Scientific Research Foundation for Talented Scholars of Xuzhou Medical College (No. D2014008), Xuzhou Medical University School of Pharmacy Graduate Student Scientific Research Innovation Projects (No. 2015YKYCX014).

Author Contributions: Tonghui Huang and Jie Sun performed the sequence alignment and structural comparison; Shanshan Zhou and Jian Gao performed virtual screening study; Yi Liu and Tonghui Huang analyzed the data; Tonghui Huang wrote the paper.

References

1.   Gallagher, H.; Suckling, R.J. Diabetic nephropathy: Where are we on the journey from pathophysiology to treatment? *Diabetes Obes. Metab.* **2016**, *18*, 641–647. [CrossRef] [PubMed]

2.   International Diabetes Federation. *IDF Diabetes Atlas—7th Edition*; IDF: Brussels, Belgium, 2015. Available online: http://www.diabetesatlas.org (accessed on 9 June 2016).

3.   Quiroga, B.; Arroyo, D.; de Arriba, G. Present and future in the treatment of diabetic kidney disease. *J. Diabetes Res.* **2015**, *2015*, 1–13. [CrossRef] [PubMed]

4.   Chan, G.C.; Tang, S.C. Diabetic nephropathy: Landmark clinical trials and tribulations. *Nephrol. Dial. Transpl.* **2016**, *31*, 359–368. [CrossRef] [PubMed]

5.   Perezgomez, M.V.; Sanchez-Niño, M.D.; Sanz, A.B.; Martin-Cleary, C.; Ruiz-Ortega, M.; Egido, J.; Navarro-González, J.F.; Ortiz, A.; Fernandez-Fernandez, B. Horizon 2020 in diabetic kidney disease: The clinical trial pipeline for add-on therapies on top of renin angiotensin system blockade. *J. Clin. Med.* **2015**, *4*, 1325–1347. [CrossRef] [PubMed]

6. Cameron, K.O.; Kung, D.W.; Kalgutkar, A.S.; Kurumbail, R.G.; Miller, R.; Salatto, C.T.; Ward, J.; Withka, J.M.; Bhattacharya, S.K.; Boehm, M.; et al. Discovery and preclinical characterization of 6-chloro-5-[4-(1-hydroxycyclobutyl)phenyl]-1H-indole-3-carboxylic Acid (PF-06409577), a direct activator of adenosine monophosphate-activated protein kinase (AMPK), for the potential treatment of diabetic nephropathy. *J. Med. Chem.* **2016**, *59*, 8068–8081. [PubMed]

7. Hardie, D.G.; Ross, F.A.; Hawley, S.A. AMPK: A nutrient and energy sensor that maintains energy homeostasis. *Nat. Rev. Mol. Cell Biol.* **2012**, *13*, 251–262. [CrossRef] [PubMed]

8. Li, X.D.; Wang, L.L.; Zhou, X.E.; Ke, J.Y.; Waal, P.W.; Gu, X.; Tan, M.H.; Wang, D.; Wu, D.; Xu, H.E.; et al. Structural basis of AMPK regulation by adenine nucleotides and glycogen. *Cell Res.* **2015**, *25*, 50–66. [CrossRef] [PubMed]

9. Xiao, B.; Sanders, M.J.; Underwood, E.; Heath, R.; Mayer, F.V.; Carmena, D.; Jing, C.; Walker, P.A.; Eccleston, J.F.; Haire, L.F.; et al. Structure of mammalian AMPK and its regulation by ADP. *Nature* **2011**, *472*, 230–233. [CrossRef] [PubMed]

10. Rana, S.; Blowers, E.C.; Natarajan, A. Small molecule adenosine 5'-monophosphate activated protein kinase (AMPK) modulators and human diseases. *J. Med. Chem.* **2014**, *58*, 2–29. [CrossRef] [PubMed]

11. Miglianico, M.; Nicolaes, G.A.F.; Neumann, D. Pharmacological targeting of AMP-activated protein kinase and opportunities for computer-aided drug design: Miniperspective. *J. Med. Chem.* **2016**, *59*, 2879–2893. [CrossRef] [PubMed]

12. Ross, F.A.; MacKintosh, C.; Hardie, D.G. AMP-activated protein kinase: A cellular energy sensor that comes in 12 flavours. *FEBS J.* **2016**, *283*, 2987–3001. [CrossRef] [PubMed]

13. Calabrese, M.F.; Rajamohan, F.; Harris, M.S.; Caspers, N.L.; Magyar, R.; Withka, J.M.; Wang, H.; Borzilleri, K.A.; Sahasrabudhe, P.V.; Hoth, L.R.; et al. Structural basis for AMPK activation: Natural and synthetic ligands regulate kinase activity from opposite poles by different molecular mechanisms. *Structure* **2014**, *22*, 1161–1172. [CrossRef] [PubMed]

14. Steinberg, G.R.; Kemp, B.E. AMPK in health and disease. *Physiol. Rev.* **2009**, *89*, 1025–1078. [CrossRef] [PubMed]

15. Cameron, K.O.; Kurumbail, R.G. Recent progress in the identification of adenosine monophosphate-activated protein kinase (AMPK) activators. *Bioorg. Med. Chem. Lett.* **2016**, *26*, 5139–5148. [CrossRef] [PubMed]

16. Langendorf, C.G.; Kemp, B.E. Choreography of AMPK activation. *Cell Res.* **2015**, *25*, 5–6. [CrossRef] [PubMed]

17. Giordanetto, F.; Karis, D. Direct AMP-activated protein kinase activators: A review of evidence from the patent literature. *Expert. Opin. Ther. Pat.* **2012**, *22*, 1467–1477. [CrossRef] [PubMed]

18. Cool, B.; Zinker, B.; Chiou, W.; Kifle, L.; Cao, N.; Perham, M.; Dickinson, R.; Adler, A.; Gagne, G.; Iyengar, R.; et al. Identification and characterization of a small molecule AMPK activator that treats key components of type 2 diabetes and the metabolic syndrome. *Cell Metab.* **2006**, *3*, 403–416. [CrossRef] [PubMed]

19. Xiao, B.; Sanders, M.J.; Carmena, D.; Bright, N.J.; Haire, L.F.; Underwood, E.; Patel, B.R.; Heath, R.B.; Wlaker, P.A.; Hallen, S.; et al. Structural basis of AMPK regulation by small molecule activators. *Nat. Commun.* **2013**, *4*, 3017. [CrossRef] [PubMed]

20. Daina, A.; Michielin, O.; Zoete, V. SwissADME: A free web tool to evaluate pharmacokinetics, drug-likeness and medicinal chemistry friendliness of small molecules. *Sci. Rep.* **2017**, *7*, 42717. [CrossRef] [PubMed]

21. Zuegg, J.; Cooper, M.A. Drug-Likeness and Increased Hydrophobicity of Commercially Available Compound Libraries for Drug Screening. *Curr. Top. Med. Chem.* **2012**, *12*, 1500–1513. [CrossRef] [PubMed]

22. Scott, J.W.; Ling, N.M.; Issa, S.M.A.; Dite, T.A.; O'Brien, M.T.; Chen, Z.P.; Galic, S.; Langendorf, C.G.; Steinberg, G.R.; Kemp, B.E.; et al. Small molecule drug A-769662 and AMP synergistically activate naive AMPK independent of upstream kinase signaling. *Chem. Biol.* **2014**, *21*, 619–627. [CrossRef] [PubMed]

23. Sanders, M.J.; Ali, Z.S.; Hegarty, B.D.; Heath, R.; Snowden, M.A.; Carling, D. Defining the mechanism of activation of AMP-activated protein kinase by the small molecule A-769662, a member of the thienopyridone family. *J. Biol. Chem.* **2007**, *282*, 32539–32548. [CrossRef] [PubMed]

24. Kaya, M.; Sarhan, A.; Alhajj, R. Multiple sequence alignment with affine gap by using multi-objective genetic algorithm. *Comput. Methods Progr. Biomed.* **2014**, *114*, 38–49. [CrossRef] [PubMed]

25. Zou, Q.; Hu, Q.H.; Guo, M.Z.; Wang, G.H. HAlign: Fast multiple similar DNA/RNA sequence alignment based on the centre star strategy. *Bioinformatics* **2015**, *31*, 2475–2481. [CrossRef] [PubMed]

26. Zou, Q.; Li, X.B.; Jiang, W.R.; Lin, Z.Y.; Li, G.L.; Chen, K. Survey of MapReduce frame operation in bioinformatics. *Brief. Bioinform.* **2014**, *15*, 637–647. [CrossRef] [PubMed]

27. Gladue, D.P.; Baker-Bransetter, R.; Holinka, L.G.; Fernandez-Sainz, I.J.; O'Donnell, V.; Fletcher, P.; Lu, Z.Q.; Borca, M.V. Interaction of CSFV E2 protein with swine host factors as detected by yeast two-hybrid system. *PLoS ONE* **2014**, *9*, e85324. [CrossRef] [PubMed]

28. Seeliger, D.; de Groot, B.L. Ligand docking and binding site analysis with PyMOL and Autodock/Vina. *J. Comput. Aided Mol. Des.* **2010**, *24*, 417–422. [CrossRef] [PubMed]

29. Asokan, R.; Nagesha, S.N.; Manamohan, M.; Krishnakumar, N.K.; Mahadevaswamy, H.M.; Rebijith, K.B.; Prakash, M.N.; Sharath Chandra, G. Molecular diversity of *Helicoverpa armigera* Hubner (Noctuidae: Lepidoptera) in India. *Orient. Insects* **2012**, *46*, 130–143. [CrossRef]

30. Dammganamet, K.L.; Bembenek, S.D.; Venable, J.W.; Castro, G.G.; Mangelschots, L.; Peeterst, D.C.G.; Mcallister, H.M.; Edwards, J.P.; Disepio, D.; Mirzadegan, T. A prospective virtual screening study: Enriching hit rates and designing focus libraries to find inhibitors of PI3Kδ and PI3Kγ. *J. Med. Chem.* **2016**, *59*, 4302–4313. [CrossRef] [PubMed]

31. Lionta, E.; Spyrou, G.; Vassilatis, D.K.; Cournia, Z. Structure-Based Virtual Screening for Drug Discovery: Principles, Applications and Recent Advances. *Curr. Top. Med. Chem.* **2014**, *14*, 1923–1938. [CrossRef] [PubMed]

32. Drug-Likeness and Molecular Property Prediction. Available online: http://www.molsoft.com/mprop/ (accessed on 9 June 2017).

33. Kashem, M.A.; Nelson, R.M.; Yingling, J.D.; Pullen, S.S.; Prokopowicz, A.S., III; Jones, J.W.; Wolak, J.P.; Rogers, G.R.; Morelock, M.M.; Snow, R.J.; et al. Three Mechanistically Distinct Kinase Assays Compared: Measurement of Intrinsic ATPase Activity Identified the Most Comprehensive Set of ITK Inhibitors. *J. Biomol. Screen.* **2007**, *12*, 70–83. [CrossRef] [PubMed]

Permissions

The contributors of this book come from diverse backgrounds, making this book a truly international effort. This book will bring forth new frontiers with its revolutionizing research information and detailed analysis of the nascent developments around the world.

We would like to thank all the contributing authors for lending their expertise to make the book truly unique. They have played a crucial role in the development of this book. Without their invaluable contributions this book wouldn't have been possible. They have made vital efforts to compile up to date information on the varied aspects of this subject to make this book a valuable addition to the collection of many professionals and students.

This book was conceptualized with the vision of imparting up-to-date information and advanced data in this field. To ensure the same, a matchless editorial board was set up. Every individual on the board went through rigorous rounds of assessment to prove their worth. After which they invested a large part of their time researching and compiling the most relevant data for our readers.

The editorial board has been involved in producing this book since its inception. They have spent rigorous hours researching and exploring the diverse topics which have resulted in the successful publishing of this book. They have passed on their knowledge of decades through this book. To expedite this challenging task, the publisher supported the team at every step. A small team of assistant editors was also appointed to further simplify the editing procedure and attain best results for the readers.

Apart from the editorial board, the designing team has also invested a significant amount of their time in understanding the subject and creating the most relevant covers. They scrutinized every image to scout for the most suitable representation of the subject and create an appropriate cover for the book.

The publishing team has been an ardent support to the editorial, designing and production team. Their endless efforts to recruit the best for this project, has resulted in the accomplishment of this book. They are a veteran in the field of academics and their pool of knowledge is as vast as their experience in printing. Their expertise and guidance has proved useful at every step. Their uncompromising quality standards have made this book an exceptional effort. Their encouragement from time to time has been an inspiration for everyone.

The publisher and the editorial board hope that this book will prove to be a valuable piece of knowledge for researchers, students, practitioners and scholars across the globe.

List of Contributors

Subina Mehta, Caleb W. Easterly, Ray Sajulga, Praveen Kumar, Timothy J. Griffin and Pratik D. Jagtap
Department of Biochemistry, Molecular Biology and Biophysics, University of Minnesota, Minneapolis, MN 55455, USA

Robert J. Millikin, Michael R. Shortreed and Lloyd M. Smith
Department of Chemistry, University of Wisconsin, Madison, WI 53706, USA

Andrea Argentini, Ignacio Eguinoa and Lennart Martens
VIB-UGent Center for Medical Biotechnology, VIB, Ghent University, 9000 Ghent, Belgium

Thomas McGowan and James E. Johnson
Minnesota Supercomputing Institute, University of Minnesota, Minneapolis, MN 55455, USA

Clemens Blank and Bjoern Gruening
Bioinformatics Group, Department of Computer Science, University of Freiburg, 79110 Freiburg im Breisgau, Germany

Caleb Easterly, Subina Mehta, Zachary Brown and Timothy J. Griffin
Department of Biochemistry, Molecular Biology and Biophysics, University of Minnesota, Minneapolis, MN 55455, USA

James Johnson and Thomas McGowan
Minnesota Supercomputing Institute, University of Minnesota, Minneapolis, MN 55455, USA

Carolin A. Kolmeder
Institute of Biotechnology, University of Helsinki, 00014 Helsinki, Finland

Damon May
Department of Genome Sciences, University of Washington, Seattle, WA 98195, USA

Bart Mesuere
Computational Biology Group, Ghent University, Krijgslaan 281, B-9000 Ghent, Belgium

Joshua E. Elias
Department of Chemical & Systems Biology, Stanford University, Stanford, CA 94305, USA

W. Judson Hervey
Center for Bio/Molecular Science & Engineering, Naval Research Laboratory, Washington, DC 20375, USA

Thilo Muth
Bioinformatics Unit (MF1), Department for Methods Development and Research Infrastructure, Robert Koch Institute, 13353 Berlin, Germany

Joel Rudney
Department of Diagnostic and Biological Sciences, University of Minnesota, Minneapolis, MN 55455, USA

Alessandro Tanca
Porto Conte Ricerche Science and Technology Park of Sardinia, 07041 Alghero, Italy

Michael Riffle
Department of Genome Sciences, University of Washington, Seattle, WA 98195, USA
Department of Biochemistry, University of Washington, Seattle, WA 98195, USA

Damon H. May, Emma Timmins-Schiffman, William Stafford Noble and Brook L. Nunn
Department of Genome Sciences, University of Washington, Seattle, WA 98195, USA

Molly P. Mikan
Department of Ocean, Earth, and Atmospheric Sciences, Old Dominion University, Norfolk, VA 23529, USA

Daniel Jaschob
Department of Biochemistry, University of Washington, Seattle, WA 98195, USA

Haitao Ding, Yong Yu and Bo Chen
Key Laboratory for Polar Science of State Oceanic Administration, Polar Research Institute of China, Shanghai 200136, China

Fen Gao
East China Sea Fisheries Research Institute, Shanghai 200090, China

Bożena Futoma-Kołoch, Bartłomiej Dudek, MartynaWańczyk, Kamila Korzekwa and Gabriela Bugla-Płoskońska
Department of Microbiology, Institute of Genetics and Microbiology, University of Wrocław, 51-148 Wrocław, Poland

Katarzyna Kapczyńska, Eva Krzyżewska and Jacek Rybka
Department of Immunology of Infectious Diseases, Hirszfeld Institute of Immunology and Experimental Therapy, Polish Academy of Sciences, 53-114 Wrocław, Poland

Elżbieta Klausa
Regional Centre of Transfusion Medicine and Blood Bank, 50-345 Wrocław, Poland

Jun Zhang and Bin Liu
School of Computer Science and Technology, Harbin Institute of Technology Shenzhen Graduate School, Shenzhen 518055, China

Shunfang Wang, Bing Nie, Kun Yue, Wenjia Li and Dongshu Xu
Department of Computer Science and Engineering, School of Information Science and Engineering, Yunnan University, Kunming 650504, China

Yu Fei
School of Statistics and Mathematics, Yunnan University of Finance and Economics, Kunming 650221, China

Shiheng Lu, Yan Yan, Zhen Li and Lin Liu
Department of Ophthalmology, Ren Ji Hospital, School of Medicine, Shanghai Jiao Tong University, Shanghai 200127, China

Lei Chen
College of Information Engineering, Shanghai Maritime University, Shanghai 201306, China

Jing Yang and Shaopeng Wang
School of Life Sciences, Shanghai University, Shanghai 200444, China

Yuhang Zhang
Institute of Health Sciences, Shanghai Institutes for Biological Sciences, Chinese Academy of Sciences, Shanghai 200031, China

Jinjian Jiang
School of Electronics and Information Engineering, Anhui University, Hefei 230601, China
School of Computer and Information, Anqing Normal University, Anqing 246133, China

Nian Wang
School of Electronics and Information Engineering, Anhui University, Hefei 230601, China

Peng Chen
Institute of Health Sciences, Anhui University, Hefei 230601, China

Chunhou Zheng
School of Electronic Engineering & Automation, Anhui University, Hefei 230601, China

Bing Wang
School of Electrical and Information Engineering, Anhui University of Technology, Ma'anshan 243032, China

Min Li, Yu Tang and Jianxin Wang
School of Information Science and Engineering, Central South University, Changsha 410083, China

Dongyan Li
School of software, Central South University, Changsha 410083, China

Fangxiang Wu
School of Information Science and Engineering, Central South University, Changsha 410083, China
Department of Mechanical Engineering and Division of Biomedical Engineering, University of Saskatchewan, Saskatoon, SK S7N 5A9, Canada

Pu-Feng Du, Wei Zhao and Le-Yi Wei
School of Computer Science and Technology, Tianjin University, Tianjin 300350, China

Yang-Yang Miao
School of Computer Science and Technology, Tianjin University, Tianjin 300350, China
School of Chemical Engineering, Tianjin University, Tianjin 300350, China

Likun Wang
Institute of Systems Biomedicine, Beijing Key Laboratory of Tumor Systems Biology, Department of Pathology, School of Basic Medical Sciences, Peking University Health Science Center, Beijing 100191, China

Qiu-Li Hou, Jin-Xiang Luo, Bing-Chuan Zhang, Gao-Fei Jiang, Wei Ding and Yong-Qiang Zhang
Laboratory of Natural Products Pesticides, College of Plant Protection, Southwest University, Chongqing 400715, China

Bo Li and Bo Liao
College of Computer Science and Electronic Engineering, Hunan University, Changsha 410082, China

Yanbin Wang, Zhuhong You, Xiao Li and Tonghai Jiang
Xinjiang Technical Institutes of Physics and Chemistry, Chinese Academy of Science, Urumqi 830011, China

Xing Chen
School of Information and Control Engineering, China University of Mining and Technology, Xuzhou 221116, China

Jingting Zhang
Department of Mathematics and Statistics, Henan University, Kaifeng 100190, China

Cong Shen, Yijie Ding and Fei Guo
School of Computer Science and Technology, Tianjin University, Tianjin 300350, China
Tianjin University Institute of Computational Biology, Tianjin University, Tianjin 300350, China

Jijun Tang
School of Computer Science and Technology, Tianjin University, Tianjin 300350, China

Tianjin University Institute of Computational Biology, Tianjin University, Tianjin 300350, China
Department of Computer Science and Engineering, University of South Carolina, Columbia, SC 29208, USA

Xinying Xu
College of Information Engineering, Taiyuan University of Technology, Taiyuan 030024, Shanxi, China

Tonghui Huang, Jie Sun, Shanshan Zhou, Jian Gao and Yi Liu
Jiangsu Key Laboratory of New Drug Research and Clinical Pharmacy, School of Pharmacy, Xuzhou Medical University, Xuzhou 221004, China

Index